THE GREAT
ENERGY
TRANSFORMATION
IN CHINA

Other titles in the China Update Book Series include:

The titles are available online at press.anu.edu.au/publications/series/china-update

THE GREAT ENERGY TRANSFORMATION IN CHINA

EDITED BY LIGANG SONG AND YIXIAO ZHOU

Australian
National
University

ANU PRESS

社会科学文献出版社
SOCIAL SCIENCES ACADEMIC PRESS (CHINA)

Australian
National
University

ANU PRESS

Published by ANU Press
The Australian National University
Acton ACT 2601, Australia
Email: anupress@anu.edu.au

Available to download for free at press.anu.edu.au

A catalogue record for this book is available from the National Library of Australia

ISBN (print): 9781760467210
ISBN (online): 9781760467227

WorldCat (print): 1550174711
WorldCat (online): 1550426093

DOI: 10.22459/GETC.2025

Cover design and layout by ANU Press

Contents

List of figures

List of tables

Abbreviations

ANU	The Australian National University
BEV	battery electric vehicle
BRI	Belt and Road Initiative
btce	billion tonnes of standard coal equivalent
CBIRC	China Banking and Insurance Regulatory Commission
CBRC	China Banking Regulatory Commission
CCER	China Certified Emission Reduction program
CGBP	China Green Bond Principles
CGE	computable general equilibrium
CGT	Common Ground Taxonomy
CPC	Communist Party of China
CSRC	China Securities Regulatory Commission
EC	European Commission
ESG	environmental, social and governance
ESO	energy system optimisation
ETS	Emissions Trading System
EU	European Union
EV	electric vehicle
FDI	foreign direct investment
GAINS	greenhouse gas and air pollution interaction and synergies
GDP	gross domestic product
GFC	Global Financial Crisis
GHG	greenhouse gas
Gt	gigatonne
GVC	global value chain
GW	gigawatt
HEL	public health benefits evaluation

IEA	International Energy Agency
IFAD	International Fund for Agricultural Development
IMF	International Monetary Fund
IPCC	Intergovernmental Panel on Climate Change
IPSF	International Platform on Sustainable Finance
kWh	kilowatt hours
LGB	local government bond
LGFV	local government financing vehicle
MEE	Ministry of Ecology and Environment
MoF	Ministry of Finance
MW	megawatt
M-CGT	Multi-Jurisdiction Common Ground Taxonomy
NDC	Nationally Determined Contribution
NDRC	National Development and Reform Commission
NEA	National Energy Administration of China
NFRA	National Financial Regulatory Administration
NPC	National People's Congress
OECD	Organisation for Economic Co-operation and Development
PBOC	People's Bank of China
PHEV	plug-in hybrid electric vehicle
PM	particulate matter
PPP	public–private partnership
PV	photovoltaic
PV–ES–CS	photovoltaic–energy storage–charging station
R&D	research and development
RMB	renminbi
SOE	state-owned enterprise
TPEC	total primary energy consumption
TWh	terawatt hours
UK	United Kingdom
UN	United Nations
US	United States
USMCA	United States–Mexico–Canada Agreement
WHO	World Health Organization
WTO	World Trade Organization

Acknowledgements

The China Economy Program (CEP) at the Crawford School of Public Policy, The Australian National University (ANU), acknowledges the financial support provided by BHP for the China Update 2024. We thank CEP Project Manager, Tunye Qiu, from the Crawford School, for his program support and assistance in communicating with the contributors to the event and the book. We sincerely thank our chapter authors for their valuable contributions. We would also like to thank colleagues from ANU Press—notably, Nathan Hollier, Elouise Ball, Gabrielė Gaižutytė and Teresa Prowse—for their expeditious publication of the 2024 book. Our copyeditor, Jan Borrie, has consistently lent her professionalism and expertise to the China Update book series throughout the years, and her meticulous work is truly appreciated by the series' contributors. Thanks also go to our Crawford School colleagues, including Professor Janine O'Flynn, Director of the Crawford School; Professor Paul Burke, Deputy Director of the Crawford School; and Professor Jane Golley, head of the Arndt–Corden Department of Economics, for their support in participating in the 2024 update event. We thank Adriana Paulovcinova, Senior Communication and Engagement Officer in the Crawford School, for her great help with organising the update event. We thank Yun Wei and Guanghan Wu from the Social Sciences Academic Press of the Chinese Academy of Social Sciences, in Beijing, for their long-term support in translating and publishing the Chinese version of the book each year—making this important research available to a wider readership.

Contributors

Xianneng Ai
Peking University

Moyu Chen
Peking University

Yarui Deng
China University of
Petroleum

Yinghao Kong
China National Offshore
Oil Corporation

Weifeng Larry Liu
The Australian National
University

Boshu Li
China National Offshore
Oil Corporation

Shanjun Li
Cornell University

Boyu Liu
Shanghai Jiao Tong
University

Christoph Nedopil
Griffith University

Ha Pham
Cornell University

Yu Sheng
The Australian National
University

Ligang Song
The Australian National
University

David I. Stern
The Australian National
University

Chuyu Sun
China University of
Petroleum

Aizhao Wang
Peking University

Feng Wang
Shanghai Jiao Tong
University

Ran Wang
University of International
Business and Economics

Yuerong Wang
Hong Kong University of
Science and Technology

Zhen Wang
China National Offshore
Oil Corporation

Christine Wong
National University of
Singapore

Jinjun Xue
Nagoya University

Lin Yang
Hong Kong University of
Science and Technology

Yang Yao
Peking University

Haitao Yin
Shanghai Jiao Tong
University

Ling Yu
Peking University

Mengdi Yue
Green Finance &
Development Center

Hongjun Zhang
China University of
Petroleum

ZhongXiang Zhang
Tianjin University

Xiaoli Zhao
China University of
Petroleum

Yang Zhou
University of
International Business
and Economics

Yixiao Zhou
The Australian National
University

1

China's energy transformation: Experiences and outlook

Ligang Song and Yixiao Zhou

The Chinese economy experienced several major transformations following rural reforms in the late 1970s. They include the rise of township and village enterprises in the 1980s, the expansion of the private sector in the decades that followed and, importantly, China's energy transformation: a broad move towards a cleaner, low-carbon and innovation-driven economy, underpinned by significant energy sector reform and investment. This has become a central pillar of China's next stage of development and a key driver of its growing global influence.

Historically, China's rapid industrialisation and urbanisation were supported by a heavy reliance on fossil fuels, particularly coal. In 2010, coal accounted for more than 70 per cent of China's total energy consumption, making the country the largest producer and consumer of coal as well as the world's top carbon emitter (IEA 2022). As the world's largest energy consumer and emitter, China has responded to international calls for climate action by adopting an ambitious national strategy. The commitment to reach carbon peak before 2030 and carbon neutrality before 2060 has catalysed a sweeping overhaul of its energy system. Yet, this transformation is not limited to decarbonisation. It embodies a broader structural reorientation from fossil fuel reliance to diversified energy supply, from scale expansion to quality upgrading and from addressing energy scarcity to enhancing energy efficiency and security. At the same time, China's energy transition is tightly intertwined with its digital transformation, as the rollout of smart grids, intelligent transport systems and low-carbon industrial platforms reinforces the complementarities between green and digital technologies.

Since 2010, China has made remarkable progress in its clean energy transition. It has built the world's largest green electricity system and launched the largest national carbon emissions trading market. The share of non-fossil sources in the country's installed power capacity continues to grow steadily, supported by the rapid deployment of solar photovoltaics, wind power, electric vehicles (EVs) and energy storage systems.

This has taken place at a pace and scale unmatched globally. For example, the share of Chinese-made EVs in new EV sales globally rose from less than 10 per cent in 2010 to more than 60 per cent in 2024. Likewise, China's share of global renewable electricity capacity increased from less than 20 per cent in 2010 to more than 40 per cent by 2024, highlighting its expanding leadership in global clean energy infrastructure (Figure 1.1).

This energy transformation has profound implications for the rest of the world as well. Chinese firms now play a central role in the global production and export of green technologies. They are leading participants in key segments such as solar panels, wind turbines, lithium-ion batteries and EV components, serving as critical suppliers within the global clean energy value chain. By 2021, China's share in all the manufacturing stages of solar panels (such as polysilicon, ingots, wafers, cells and modules) exceeded 80 per cent (IEA 2022). In the meantime, China has expanded its green infrastructure footprint abroad, particularly in developing and emerging economies, through initiatives such as the Belt and Road Initiative (BRI). These developments have contributed to significant reductions in the global cost of renewable energy technologies and have provided crucial support for countries seeking to improve energy access and accelerate their own transitions to low-carbon energy systems.

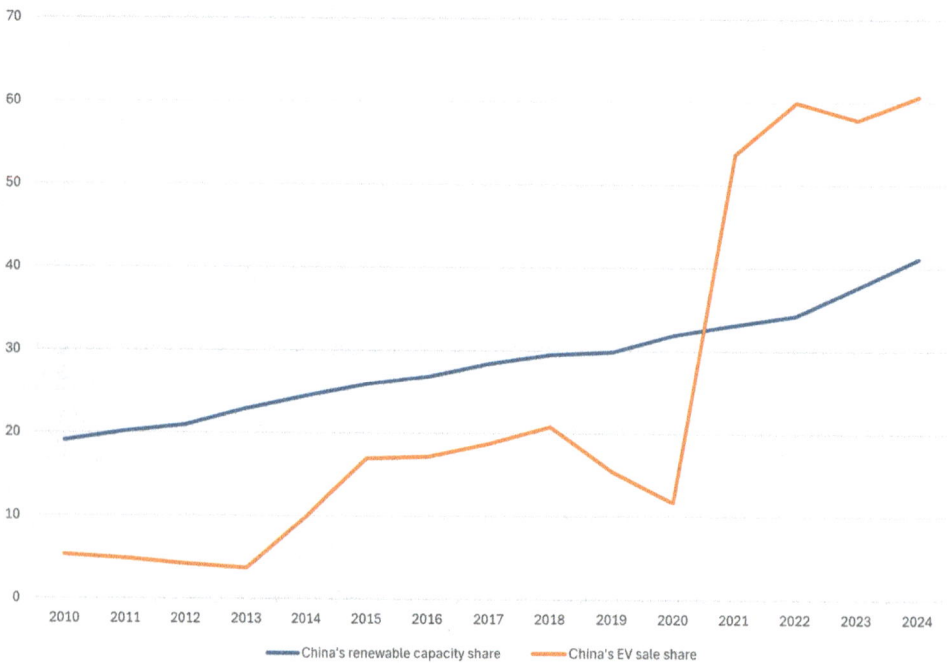

Figure 1.1: China's share in global renewable energy capacity and EV sales, 2010–24 (per cent)

Notes: 'EV' includes both battery electric vehicles and plug-in hybrid electric vehicles. Renewable capacity share refers to China's share of total global installed renewable electricity generation capacity.

Sources: Authors' construction from data in CEIC Database (www.ceicdata.com/en) and International Energy Agency (www.iea.org/data-and-statistics).

China's growing role in green energy is unfolding amid an increasingly complex global environment marked by geopolitical tensions and shifts in trade policy. What was once viewed primarily as a technical and economic transition is now increasingly shaped by broader strategic considerations. Several advanced economies are pursuing policies aimed at strengthening domestic clean energy industries, including efforts to re-establish manufacturing bases, implement export controls and introduce tighter regulations on foreign investment. These trends raise concerns about the potential fragmentation of global supply chains, limitations on technological exchange and a weakening of multilateral cooperation at a time when collective action is critical to advance the global energy transition.

Despite these challenges, China's energy transformation provides a valuable example of how coordinated policy efforts, targeted innovation and sustained industrial development can support progress towards decarbonisation. This experience also highlights the important contribution that developing countries can make to the evolution of global energy governance, provided that international cooperation frameworks remain accessible, balanced and sufficiently supported by financial and technical resources.

In this overview, we will explore China's energy transformation by examining its domestic achievements, the industrial and technological foundations of the transition and the country's expanding role in the global green economy. It also considers the challenges arising from shifting geopolitical dynamics and outlines strategic directions for deepening international cooperation in the years ahead.

Achievements and structural shifts

China's energy transformation has delivered major structural changes in both energy production and energy consumption. Through large-scale investments, long-term planning and ambitious policy targets, China is shifting away from coal dominance towards a more diverse and cleaner energy mix.

By the end of 2024, China's installed capacity for non-fossil energy sources—including wind, solar, hydro, nuclear and biomass—had surpassed 1.8 billion kilowatts (1.8 terawatts), making up more than 56 per cent of the country's total installed power generation capacity (State Council 2025). This shift is also visible in power consumption patterns: non-fossil sources accounted for about 40 per cent of electricity consumed in 2023. At the same time, more than 80 per cent of public buses now run on new energy, reflecting growing electrification in the transport sector. The EV market has expanded rapidly, with the share of EVs in total new car sales rising from less than 5 per cent in 2020 to more than 45 per cent in 2023, supported by industrial policy and strong domestic demand.

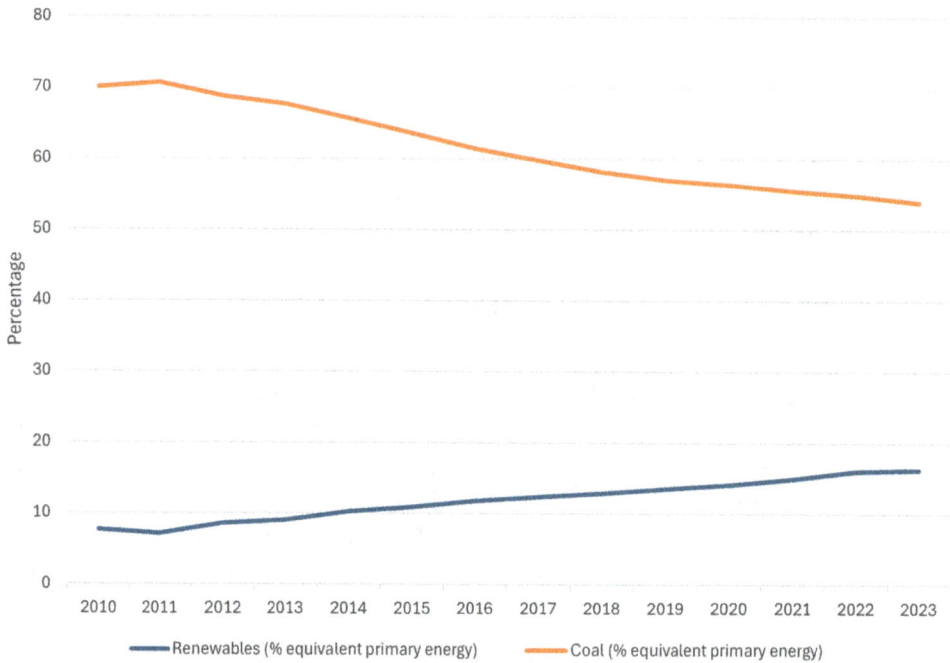

Figure 1.2: Trends in coal and renewable energy shares of China's primary energy consumption, 2010–23 (per cent)

Source: Authors' construction using 2024 data from the Energy Institute (www.energyinst.org/ statistical-review/resources-and-data-downloads).

China's primary energy structure is also becoming cleaner. The share of coal in total primary energy consumption dropped to 55.3 per cent in 2023—more than 12 percentage points lower than a decade earlier. In contrast, clean energy sources made up 16.2 per cent of primary energy consumption—up from 7.6 per cent in 2010 (Figure 1.2). Energy efficiency has improved significantly as well. Between 2006 and 2023, energy consumption per unit of GDP decreased by 43.8 per cent, indicating that economic growth has become less dependent on energy use.

China now operates the largest green electricity system in the world and has launched a national carbon Emissions Trading System, which is currently the largest by volume of emissions covered. This market-based approach is expected to play a growing role in promoting emission reductions. Together, these outcomes show that China's energy system is moving towards being cleaner, more efficient and increasingly electrified.

Drivers of the energy transition

China's energy transformation is closely tied to its domestic technological and industrial capabilities. Over the past decade, an extensive ecosystem of low-carbon technologies has developed in the economy. Sustained public investment and growing manufacturing scale have enabled rapid deployment of renewable energy across the economy and laid the foundation for structural decarbonisation.

At the centre of this transformation are three strategic sectors often referred to as the 'new trio' or 'new three': solar photovoltaic manufacturing, lithium-ion battery production and EVs. These industries have benefited from early public policy support and increasing technological sophistication. Additionally, the rapid cost declines in wind and solar energy reflect China's industrial capability. Between 2013 and 2023, the average cost of producing wind turbines and solar panels fell by 60 and 80 per cent, respectively, largely due to improvements in manufacturing efficiency, supply chain coordination and localised innovation. This cost efficiency is also reflected in China's consistently lower installed cost of onshore wind power compared with the global average over the past two decades (Figure 1.3).

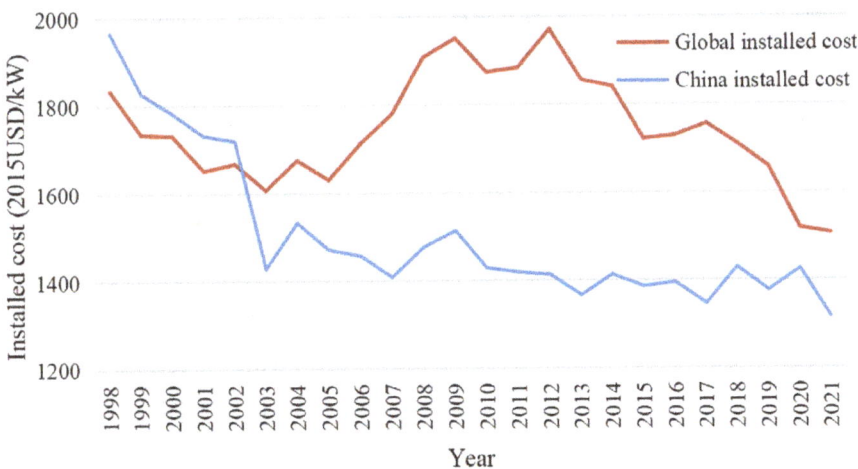

Figure 1.3: Trends in global and Chinese onshore wind installation costs
Source: Zhang et al. (2024).

Policy frameworks have played an instrumental role in scaling up this technological progress. China has made clear commitments to renewable energy and carbon neutrality, setting ambitious targets that accelerate demand for advanced energy storage solutions. It aims to achieve the peak of carbon dioxide emissions by 2030 and carbon neutrality by 2060. To achieve these goals, the Chinese Government has introduced a comprehensive set of policies, including national renewable energy targets, mandatory integration requirements for energy storage in new projects, subsidies for clean energy technologies and preferential taxes and credits for renewable energy firms (Afshan et al. 2024). In addition, regulatory support such as grid priority access for renewables, renewable portfolio standards and capacity auctions has helped to guide investment and coordinate industrial upgrading. These measures are complemented by innovation-driven initiatives embedded in successive five-year plans, which promote sector-wide alignment across electricity, transport and digital infrastructure systems.

China in the global green economy

China's energy transformation is not only reshaping its domestic landscape but also has far-reaching influence on the global green economy. Through large-scale implementation of renewable technologies, leadership in clean energy manufacturing and active international engagement, China has become a central player in global efforts to accelerate the transition to low-carbon energy systems.

Leadership in green manufacturing and export capacity

China has emerged as the world's largest producer and exporter of core green technologies. As of 2023, it accounted for more than 80 per cent of global photovoltaic (PV) module production and 70 per cent of wind power equipment, with exports reaching more than 200 countries and regions (State Council 2024). This is reflected in Figure 1.4, which shows that China's solar PV export volume tripled between 2017 and 2023. At the same time, the average export price of PV modules has steadily declined, falling from more than US$0.35 per watt in 2017 to below US$0.15 per watt by early 2024.

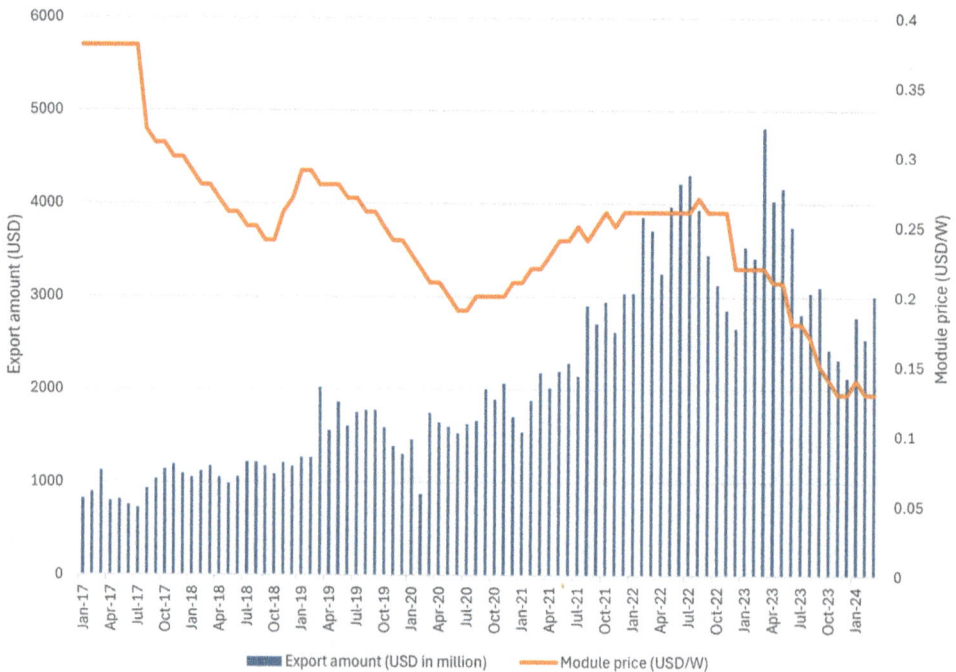

Figure 1.4: Solar PV export volumes
Source: Authors' construction using data from EMBER Energy (ember-energy.org/data/).

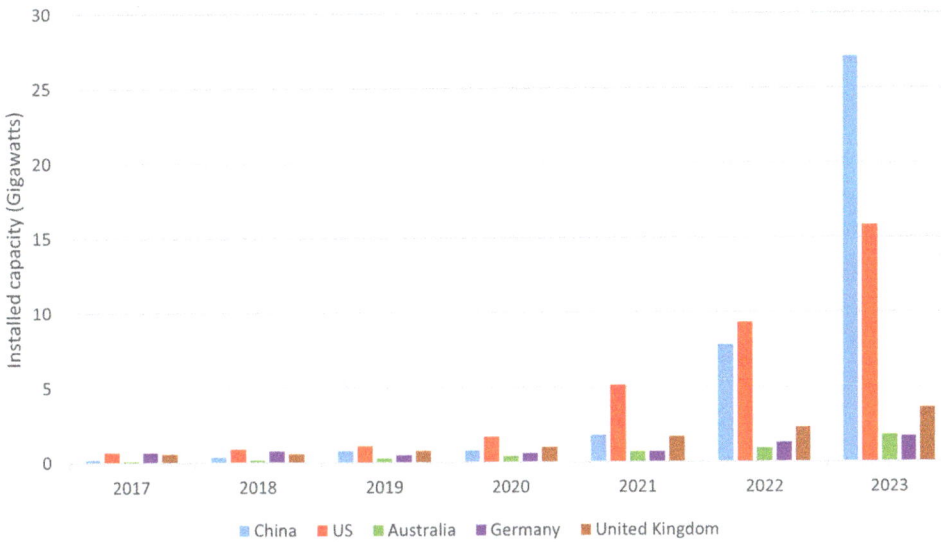

Figure 1.5: Installed energy storage capacity by country
Source: Authors' construction using data from the Energy Institute
(www.energyinst.org/statistical-review).

These contributions have made renewable energy technologies more accessible and affordable worldwide, particularly benefiting emerging and developing economies seeking to expand electricity access. In 2023, China led global renewable energy employment, with 7.4 million jobs, or 46 per cent of the global total, highlighting the country's role in anchoring the industrial base for the green transition (IRENA 2024).

Beyond its export volume, China's prominence in global supply chains is increasingly accompanied by its influence in shaping international standards, especially in energy storage, battery manufacturing and solar technology. With a 48.6 per cent share of the global energy storage market in 2023, and domestic storage capacity reaching 27.1 gigawatts (GW), China is now able to help define global norms and technical specifications in this fast-evolving sector (Figure 1.5).

The projection that China's cumulative new energy storage capacity could reach between 221 gigawatts and 300 gigawatts by 2030 indicates even stronger growth in the sector over the coming years (Yang 2025). As technologies become increasingly commercialised and scale-driven cost reductions persist, China is well positioned to remain a leading supplier of next-generation low-carbon solutions, ranging from grid-scale batteries and smart inverters to electric mobility platforms.

Enabling the global energy transition

Beyond manufacturing, China is actively shaping green infrastructure development across developing countries. Through initiatives such as the BRI and bilateral energy cooperation agreements, Chinese enterprises have constructed landmark renewable projects, including solar parks, wind farms and hydro plants, in more than 100 countries. For example, Chinese companies installed a record 24 gigawatts of energy capacity in BRI countries in 2024, doubling the installations recorded in 2023 (Asian Business Review 2024). These efforts have extended access to clean energy in countries such as Cuba, Ghana, Burkina Faso and Pakistan, where Chinese firms have financed, constructed and operate major solar and energy storage projects. In Cuba, for instance, China is assisting with the construction of more than 20 new solar farms as part of the country's plan to add more than 1,000 megawatts of solar capacity in 2025 (Reuters 2025). In Pakistan, a 20.7-megawatt energy storage battery system is under development with the support of Chinese supplier CATL. This project is expected to help address grid reliability and meet peak electricity demand (Jilani 2025).

China's growing leadership is increasingly evident in its innovation across integrated energy systems. The country has demonstrated how to effectively combine renewable energy generation with grid infrastructure, digital technologies and electrified end-use applications. Its strategic development of EVs, lithium-ion batteries and solar PV illustrates a successful model of building a vertically integrated and export-oriented green industrial ecosystem. These industries now form core components of China's external trade and play a vital role in supporting the global transition to low-carbon energy systems.

A platform for global cooperation

Given the scientific consensus that global greenhouse gas (GHG) emissions must peak before 2025 and decline by 43 per cent by 2030 to limit global warming to 1.5°C (IPCC 2022), China has consistently positioned itself as a partner for international cooperation on climate action. It has eliminated foreign investment restrictions in most energy sectors (except nuclear) and actively participates in global green energy governance (Reuters 2020). China's support for South–South cooperation, including technology transfer and training initiatives, reflects its broader commitment to inclusive energy development. Research has shown that China plays a significant role in transferring clean energy technologies to developing countries, especially across sub-Saharan Africa, where Chinese firms have supported solar, hydro and grid electrification projects under concessional finance and turn-key infrastructure arrangements (Urban 2018). These initiatives are often implemented in collaboration with local governments and aim to address both energy access and capacity-building objectives. Importantly, China's model of technology transfer places emphasis not only on hardware delivery, but also on training, institutional support and technical knowledge-sharing. This integrated approach enhances recipient countries' ability to maintain and adapt clean energy systems independently, offering an alternative to traditional North–South development paradigms.

Headwinds and geopolitical challenges

China's rise as a central player in the global green energy economy has coincided with mounting geopolitical tensions and structural challenges that threaten to fragment the international energy landscape. While clean energy development was once largely framed as a technical and economic issue, it has increasingly become enmeshed in strategic competition, trade disputes and questions of industrial policy.

At the heart of these tensions lies a growing wave of trade protectionism and deglobalisation. Several advanced economies have introduced policies to re-shore clean energy manufacturing and reduce reliance on foreign suppliers, particularly in technology-intensive sectors such as solar PV, wind turbines and battery production. In response to domestic political pressures and concerns about supply chain resilience, some countries have imposed tariffs, introduced subsidy schemes with localisation requirements and tightened controls on outbound investment and technology transfers. A prominent example is the European Union's decision in June 2024 to impose countervailing duties on imports of battery electric vehicles from China. Provisional duties ranging from 17.4 per cent to 38.1 per cent were levied on selected Chinese manufacturers (EC 2024). In parallel, the United States has escalated trade restrictions targeting China's green technology exports. In 2024, the US Government imposed steep tariffs on Chinese-made EVs, batteries, solar cells and critical minerals as part of a broader strategy to counter what it described as 'unfair trade practices' and bolster domestic clean energy manufacturing capacity (Sevastopulo and White 2024). These measures have significantly reshaped the competitive dynamics of the global renewable energy sector, disrupting the global supply chains that have historically underpinned rapid technology diffusion and cost reductions in the clean energy sector.

Beyond trade tensions, rising restrictions on cross-border research and development (R&D) collaboration and the tightening of intellectual property protections are also beginning to constrain the pace of global clean energy innovation. Technological advancement in renewables relies heavily on open channels for knowledge exchange, cross-border partnerships and shared standards. However, increasing geopolitical frictions, particularly between major innovation hubs such as the United States and China, have contributed to the fragmentation of global research ecosystems. According to recent studies, technological decoupling between the world's two largest economies could significantly reduce innovation efficiency, limit spillover benefits and raise the costs of technology development globally (Jinji and Ozawa 2024; Tyers and Zhou 2024). These dynamics are especially concerning in emerging fields such as grid-scale energy storage and green hydrogen, where international collaboration remains critical to scaling deployment and reducing costs.

Policy outlook for energy transformation

China's energy transformation stands at a pivotal moment. Having achieved remarkable progress in decarbonising its power system and advancing renewable technologies, the strategic challenge now lies in ensuring that this momentum contributes meaningfully to a broader global energy transition. As the largest producer and installer of clean energy technologies, China's decisions will shape not only its own energy future but also the development trajectories of emerging economies and the stability of global energy markets.

As the global green transition accelerates, China faces both new responsibilities and new opportunities. Building on its domestic achievements in clean energy deployment and industrial upgrading, China is now positioned to help reshape the international landscape of sustainable trade and low-carbon development. To achieve this, a strategic pivot towards proactive, rules-based cooperation and institutional innovation will be essential.

A priority is to formulate sustainable trade policies and optimise the structure of green product exports. China's global leadership in solar panels, wind turbines and battery technologies provides an opportunity to support energy transitions in other countries, particularly in the Global South. At the same time, it must diversify the composition of green exports beyond hardware to include digital energy services, low-carbon logistics and integrated system solutions. This would enable more value-added participation in global green value chains and reduce vulnerability to trade concentration risks.

To manage the growing complexity of global energy politics, China must also deepen multilateral, bilateral and regional cooperation frameworks. Engagement through platforms such as the G20, BRICS (Brazil, Russia, India, China and South Africa), the Association of Southeast Asian Nations Plus 3 (ASEAN+3) and the Regional Comprehensive Economic Partnership (RCEP) will be essential to safeguard open markets, align regulatory standards and ensure fair access to emerging green technologies. Strengthened diplomatic and institutional efforts will also support cooperation and coordination among major global economies, particularly on issues such as carbon border adjustments, energy security and sustainable supply chains. As noted by Song and Agarwal (2023), China's evolving role as a leading trading nation places it in a pivotal position to help revitalise multilateral institutions and promote more inclusive, rules-based global governance, particularly in areas where trade, climate and development agendas intersect.

Domestically, further progress in advancing low-carbon regulations and aligning them with international norms will enhance China's credibility as a partner in global climate governance. This involves strengthening emission accounting systems, developing carbon footprint methodologies and continuing the expansion of China's national Emissions Trading System. Transparent and robust regulatory alignment will not only support climate goals but also help ease trade frictions and facilitate the flow of low-carbon goods and services. As part of this broader effort, China can also expand its imports of green products and promote market diversification. By actively sourcing high-efficiency technologies and sustainable inputs from abroad, China can accelerate domestic decarbonisation while mitigating trade tensions with key partners. A more diversified and internationally integrated green trade structure will strengthen economic resilience and reduce exposure to geopolitical risks.

A coordinated global transition will require China to strengthen international cooperation across green industrial value chains, finance and standards. This includes supporting collaborative research, joint standard setting and coordinated investment in upstream materials, logistics and recycling systems—all of which are essential for scaling next-generation technologies such as hydrogen, carbon capture and energy storage. China should deepen cooperation in green finance by expanding its leadership in green bonds, climate-related disclosure standards and blended finance mechanisms. Aligning domestic financial initiatives with global taxonomies and participating in cross-border finance platforms will help mobilise capital for green infrastructure, particularly in developing countries. In parallel, China can take a leading role in promoting international mutual recognition of green product certification and labelling. Harmonising standards with key export destinations would reduce compliance costs, facilitate trade and strengthen the credibility of Chinese green exports. Mutual recognition frameworks can also build consumer trust and lower barriers to entry for firms operating across jurisdictions, further supporting the integration of global low-carbon trade.

In the context of global development, China should continue to promote South–South cooperation for the joint development of sustainable trade. This includes sharing best practices, supporting infrastructure investment and building production capacity in low-income countries. Initiatives under the BRI framework can be further aligned with green development goals to ensure they deliver long-term sustainability benefits. In addition, strengthening capacity-building for sustainable trade and promoting coordinated development should remain long-term priorities. China can play a key role by supporting technical training, data-sharing initiatives and policy dialogue with partner countries, thereby enabling more inclusive participation in green globalisation. Coordinated development strategies that integrate trade, climate and development objectives will be essential to achieving a just and effective global energy transition.

Together, these efforts form the basis for a comprehensive international strategy that matches China's domestic progress in green transformation. By aligning trade, regulation, finance and diplomacy with sustainability objectives, China can help shape a global economic architecture that supports low-carbon growth, technological inclusion and shared prosperity.

Structure of the book

In this volume, the authors explore the developments in China's energy transition and thus help provide a holistic view of this issue. The book also covers some of the macroeconomic issues, such as fiscal sustainability from rising debt at the local government level, internal and external imbalances in national savings, new patterns of economic growth, trade and global value chains (GVCs) and agricultural development.

In Chapter 2, Wang, Li and Kong present an integrated framework to assess China's energy transition towards carbon neutrality, emphasising its economic, environmental and public health impacts, as well as international spillover effects. They argue that China's dual carbon goals—carbon peaking by 2030 and neutrality by 2060—require systemic shifts in energy production and consumption. The study introduces a two-module integrated assessment model combining macroeconomic, energy, pollution and health submodels, allowing multidimensional evaluation under different policy scenarios: a current policy scenario (CPS) and an enhanced action scenario (EAS).

Their findings suggest that the energy transition can decouple economic growth from emissions, significantly reduce emissions of carbon dioxide and non–carbon dioxide GHGs and generate substantial health co-benefits. Under the EAS, non-fossil energy could account for 82 per cent of total consumption by 2060, while cumulative net social benefits may exceed US$24.8 trillion. Export-driven spillovers from photovoltaic and wind technologies are projected to cut more than 185 gigatonnes of carbon dioxide globally by 2060. Despite high upfront costs, the long-term gains—including emission reductions, better air quality and avoided deaths—justify sustained investment. The authors urge policymakers to strengthen low-carbon pathways, enhance international cooperation and develop holistic strategies to maximise both the domestic and the global benefits of China's green transition.

Chapter 3 explores China's transition to a renewable electricity system in pursuit of its dual carbon goals. The authors, Yin, Liu and Wang, find that the primary hurdle is no longer the production of renewable energy but the integration of its intermittent supply—especially wind and solar—into a historically centralised, fossil-fuel–based grid. It is suggested that strategies could target three domains: supply, demand and distribution. On the supply side, energy storage technologies (especially lithium-ion batteries and hydrogen storage) and hydrogen-based fuels are critical for smoothing out generation volatility. On the demand side, smart microgrids, demand-side response, EV integration and virtual power plants enable localised energy balancing and better renewable utilisation. On the distribution side, infrastructure flexibility, digitalisation and power market reforms (for example, spot markets, green electricity trading) aim to enhance grid responsiveness. Policy efforts include a focus on improving grid capacity, system flexibility and renewable consumption rates. The authors conclude that China's energy transition hinges on systemic coordination, digital intelligence and market innovation to absorb growing renewable output while ensuring security, economic efficiency and environmental sustainability.

In Chapter 4, Zhao, Sun, Zhang and Deng examine mechanisms for inter-industry collaboration to support China's energy transformation, focusing on the integration of renewable energy in the transport and heating sectors. The first case study explores photovoltaic energy storage charging stations (PV–ES–CS) and analyses their economic and environmental benefits across various building types in Beijing. The findings show that hospitals offer the highest return on investment and carbon emission reductions due to consistent electricity demand that matches solar generation profiles, while residences perform the worst. The study confirms that integrated PV–ES–CS systems outperform stand-alone components in both efficiency and sustainability.

The second case study models the decarbonisation of Beijing's heating system under three scenarios: business as usual, electricity substitution and synergistic development (SD). The SD scenario uses combined heat and power (CHP) retrofitted with renewable energy and heat pumps and achieves 100 per cent renewable energy supply for electricity and heating by 2050, eliminating fossil fuel use and carbon dioxide emissions. Although the SD scenario has the highest investment cost, it provides the best environmental performance, reducing emissions by more than 10 megatonnes through the export of excess renewable electricity. The study concludes that future system costs will be driven more by renewable infrastructure investment than by fossil fuel costs. Their policy recommendations include integrated planning, location-based incentives for PV–ES–CS and differentiated subsidies to enhance system value and accelerate the energy transition.

In Chapter 5 of the book, Li, Pham, Wang and Yang examine the dramatic rise of China's EV industry, which is now the largest globally by production and sales. Supported by strong government policy, China surpassed 12 million EVs produced in 2024, representing 65 per cent of global EV sales. Major manufacturers such as BYD, NIO and Geely have driven innovation and international expansion, with BYD emerging as the global leader. China's advantages lie in vertically integrated supply chains, dominance in battery production and cost efficiencies supported by widespread use of lithium iron phosphate (LFP) batteries. Charging infrastructure has rapidly expanded as well, with more than 2.7 million public charging stations across the country by 2023.

However, regional disparities remain, particularly in rural areas. Policy tools, including the Dual Credit Policy, financial subsidies, tax exemptions and non-financial incentives (such as licence plate privileges), were critical in scaling adoption. Local industrial policies fostered supply chain localisation and competition.

The authors find that, despite success, challenges include global trade barriers, technology bottlenecks (for example, battery performance, recycling) and uneven infrastructure. Rising international tariffs and the European Union's Carbon Border Adjustment Mechanism pose additional risks. Looking forward, integrating EVs into energy systems, developing global standards and ensuring equitable access will be key. The chapter calls for targeted policy reforms to maintain momentum while addressing emerging economic, technological and social challenges.

Zhang in Chapter 6 analyses the European Union's imposition of countervailing duties on Chinese-made EVs based on concerns about unfair subsidies and market distortion. It is revealed that China's rapid EV export growth, driven by subsidies, industrial policy and scale advantages, has triggered trade tensions with the European Union and the United States. The European Union launched an anti-subsidy investigation in 2023, culminating in definitive duties being applied in October 2024, ranging from 7.8 per cent to 35.3 per cent depending on the manufacturer. China has disputed the findings, appealing to the World Trade Organization (WTO) and exploring negotiated solutions, including minimum pricing.

The European Union remains open to compromise, balancing industrial protection with climate goals. However, complexities in enforcing minimum pricing—due to product diversity and past failures with solar panels—make resolution difficult. Chinese automakers are responding by investing in overseas production, though challenges such as local content rules, technology transfer demands and geopolitical risks persist.

The chapter concludes that while tariffs may protect EU industries in the short term, they risk hindering the green transition and consumer access to affordable EVs. A hybrid solution—combining tariffs and price commitments—may offer a pragmatic path forward. The evolving geopolitical landscape, including potential shifts under the Trump presidency, could further influence the trajectory of EU–China EV trade relations.

China's green finance system has evolved significantly over the past decade to support its dual carbon goals. The system, rooted in a top-down governance model, includes green credit, bonds, insurance and a national carbon market. Green loans reached RMB36.6 trillion in 2024, while green bond issuance totalled more than US$120 billion. Despite progress, challenges remain, including limited private sector participation, greenwashing risks and underdeveloped environmental, social and governance (ESG) disclosure standards.

Yue and Nedopil in Chapter 7 focus on China's green finance system including green credit, bonds, insurance and a national carbon market. Internationally, China promotes green finance through the BRI, aligning with global standards via partnerships such as the China–EU Common Ground Taxonomy (CGT) and leadership roles in the G20 and the International Platform on Sustainable Finance (IPSF).

Future priorities include enhancing disclosure, integrating climate risk into financial regulation and leveraging digital tools such as blockchain. However, China's simultaneous expansion of fossil fuel infrastructure raises questions about the consistency of its green finance agenda. The authors call for further research into the effectiveness of China's green finance in achieving real emission reductions and its replicability as a global model.

In Chapter 8, Stern explores whether China is on track to peak its carbon emissions by 2030, as pledged under the Paris Agreement. By comparing data from 2019 (pre-pandemic) and 2023 (post-pandemic reopening), the analysis reveals that China's total carbon emissions rose by 8 per cent, with power sector emissions increasing by 18 per cent.

Despite significant growth in renewable energy, the use of fossil fuel—especially coal—also surged, with coal production up 26 per cent since 2019. While solar and wind power expanded, thermal power still dominated, making up more than 70 per cent of electricity generation in 2023. The slight emissions decline in early 2024 is attributed to increased hydropower from wet weather and sluggish economic activity, not structural changes. Thus, the data suggest China is not yet on a sustainable path to emissions peaking.

Stern also discusses geopolitical and economic factors, such as energy security concerns and strained international relations, which could hinder China's climate ambitions. Although China has strong incentives to lead on climate action, including high vulnerability to climate impacts, recent trends raise doubts about meeting its 2030 and 2060 targets. Stern argues that although China has made progress in green energy, its post-pandemic trajectory indicates a more carbon-intensive path than previously expected, challenging optimistic projections of early emissions peaking.

In Chapter 9, Yao and Yu examine the rapid rise of China's local government debt, identifying its institutional causes, economic impacts and governance challenges. Since 2009, local governments have increasingly relied on off-budget borrowing through local government financing vehicles (LGFVs) to fund infrastructure and development projects. This trend stems from fiscal decentralisation, urbanisation pressures, a unitary political system, frequent official turnover and regulatory gaps in the Budget Law.

While local debt has supported infrastructure growth and economic development, it has also led to significant fiscal risks, including repayment pressures, inefficient investments and systemic financial vulnerabilities. By 2023, off-budget debt reached RMB61.56 trillion—more than 150 per cent of official on-budget debt. The central government has responded with multiple debt-swap programs, converting risky off-budget liabilities into more transparent local government bonds.

Despite these efforts, debt continues to grow, with many LGFVs using new bonds primarily to refinance old ones. The authors propose a three-step resolution: 1) fiscal relief via central government bailouts, 2) transferring LGFV debt to asset management companies for restructuring, and 3) preventing new debt using asset balance sheets and stronger oversight. The chapter emphasises the need for transparency, accountability and structural reforms to ensure fiscal sustainability and economic stability.

Wong takes a detailed look at the growing fiscal crisis facing China's local governments in Chapter 10. Local governments in China are responsible for more than 85 per cent of public spending and infrastructure investment. Local finances are under pressure from three key sources: long-term erosion of tax revenues, the collapse of land-based revenues and a mounting debt crisis. Although these issues appear local, they stem from systemic structural problems. Since 2015, general budget revenues have declined from 22 per cent to 16 per cent of GDP, while social spending obligations have continued to grow. Local governments have increasingly relied on land sales and off-budget borrowing through LGFVs to fund infrastructure and services. However, with the property sector downturn, land revenues have plummeted, exacerbating debt risks.

Despite reforms to legalise local borrowing and curb off-budget debt, LGFV debt has surged, reaching an estimated 84 per cent of GDP by 2024. Recent central government bailouts aim to restructure this debt but offer no relief, leaving local governments with unsustainable burdens. Wong argues that resolving the crisis requires systemic reform, including tax restructuring, rebalancing central–local fiscal responsibilities and ending reliance on infrastructure-led stimulus. Without bold action, China risks prolonged economic stagnation and declining public service quality.

Chapter 11 of the book identifies the structural causes and implications of China's persistently high national savings and current account surpluses. Liu, Song and Zhou in this chapter argue that, since the 2000s, China's macroeconomic imbalances have been driven by low consumption, high savings and investment-heavy growth. This growth was underpinned by demographic shifts, underdeveloped social welfare, financial repression and institutional distortions. Household savings surged due to population ageing, income uncertainty, housing costs and limited welfare coverage. Corporate savings, especially from state-owned enterprises (SOEs), rose due to monopolistic advantages and low dividend payouts. Government savings fluctuated with fiscal reforms and stimulus spending.

Investment, while high, has not matched savings, sustaining external surpluses. SOEs dominate strategic sectors, while private firms face financing constraints. Institutional factors—such as the *hukou* (household registration) system, factor market distortions and limited financial liberalisation—have suppressed consumption and skewed capital allocation. Exchange rate and trade policies also contributed to trade surpluses.

Looking ahead, the authors argue that population ageing, changing consumption patterns and a shrinking real estate sector are expected to reduce savings and narrow external imbalances. Rebalancing towards consumption-led growth requires structural reforms in social welfare, taxation and financial systems to enhance household welfare and ensure sustainable, inclusive development.

Chen, Ai, Wang and Sheng review in Chapter 12 China's rural transformation over the past 50 years and its strategic relevance for developing countries in the final chapter of the book. China has achieved remarkable gains in agricultural productivity, off-farm employment and non-agricultural GDP through market-oriented reforms, infrastructure investment and technological innovation. Agricultural labour productivity rose from US$1,000 in 1970 to more than US$13,000 in 2022, while off-farm employment increased from 20 per cent to 80 per cent. These shifts reflect China's structural transformation from an agriculture-based to a diversified economy.

In Chapter 13, Wang, Xue and Zhou ask whether China can sustain high economic growth amid structural shifts and global value chain (GVC) restructuring. After decades of rapid expansion driven by industrialisation, exports and investment, China's growth has slowed due to demographic ageing, diminishing investment returns, rising costs and geopolitical tensions. The country is transitioning from export-led to consumption-driven growth, supported by policies such as the 'dual circulation' strategy. China's role in global production has evolved from a low-cost assembler to a regional manufacturing hub. However, rising trade frictions and 'de-Sinicisation' efforts by the United States and Japan have reduced China's share in their intermediate goods imports and foreign direct investment (FDI). In response, China has redirected exports through third countries such as Vietnam and Mexico and increased outward FDI to acquire advanced technologies.

The 'new three' industries—EVs, lithium-ion batteries and solar panels—have emerged as key growth drivers, offsetting declines in traditional sectors. These industries also foster global supply chain integration and development in resource-rich countries. However, challenges remain, including overcapacity, trade restrictions and reliance on foreign technologies. The authors conclude that, to sustain growth, China must adapt to GVC shifts, promote innovation and mitigate external risks through strategic industrial and trade policies.

The authors find that China's success stems from market reforms supported by government investment in irrigation, R&D and infrastructure. However, challenges such as climate change, resource degradation and environmental pollution now demand a shift towards 'new quality' agricultural productivity, emphasising inclusiveness, fairness and sustainability. The chapter also compares China's experience with that of other developing regions, showing that Africa lags in productivity and structural transformation. The authors argue that China's experience, characterised by capital deepening, sustainable practices and region-specific policies, offers valuable lessons. Future reforms should focus on science and technology, market-oriented governance and tailored support for small farmers to ensure equitable and resilient rural development.

References

Afshan, Sahar, Younes Ben Zaied, Tanzeela Yaqoob, and Shunsuke Managi. 2024. 'Insights into the Efficiency of China's Green Energy Policies.' *Journal of Cleaner Production* 434: 139913. doi.org/10.1016/j.jclepro.2023.139913.

Asian Business Review. 2024. 'Chinese Firms Install 24 GW Energy Capacity in Belt & Road Countries.' *Asian Business Review*, 7 February. asianbusinessreview.com/in-focus/chinese-firms-install-24-gw-energy-capacity-in-belt-road-countries.

European Commission (EC). 2024. 'EU Commission Imposes Countervailing Duties on Imports of Battery Electric Vehicles (BEVs) from China.' Online post, 12 December. Brussels: European Commission. trade.ec.europa.eu/access-to-markets/en/news/eu-commission-imposes-countervailing-duties-imports-battery-electric-vehicles-bevs-china.

Intergovernmental Panel on Climate Change (IPCC). 2022. 'The Evidence is Clear: The Time for Action Is Now. We Can Halve Emissions by 2030.' Online post, 4 April. Geneva: IPCC. www.ipcc.ch/2022/04/04/ipcc-ar6-wgiii-pressrelease/.

International Energy Agency (IEA). 2022. *Special Report on Solar PV Global Supply Chains*. Paris: IEA. iea.blob.core.windows.net/assets/d2ee601d-6b1a-4cd2-a0e8-db02dc64332c/SpecialReportonSolarPVGlobalSupplyChains.pdf.

International Renewable Energy Agency (IRENA). 2024. *Renewable Energy and Jobs: Annual Review 2024*. Abu Dhabi: IRENA. www.irena.org/Publications/2024/Oct/Renewable-energy-and-jobs-Annual-review-2024.

Jilani, Humza. 2025. 'Chinese Battery Glut Plugs into Solar Boom to Power Pakistan.' *Financial Times*, [London], 2 June. www.ft.com/content/2b4c598e-a4b3-4c6e-9c38-97e46357f819.

Jinji, Naoto, and Shunya Ozawa. 2024. 'Economic Consequences of US–China Technological Decoupling: An Illustrative Quantitative Analysis.' *VoxEU*, 4 October. London: Centre for Economic Policy Research. cepr.org/voxeu/columns/economic-consequences-us-china-technological-decoupling-illustrative-quantitative.

Reuters. 2020. 'China to Abolish Access Restrictions on Foreign Investment in Energy Sector: White Paper.' *Reuters*, 21 December. www.reuters.com/business/energy/china-abolish-access-restrictions-foreign-investment-energy-sector-white-paper-2020-12-21/.

Reuters. 2025. 'Cuba On Track to Install 50 Solar Parks This Year, Ministry Says.' *Reuters*, 20 March. www.reuters.com/world/americas/cuba-track-install-50-solar-parks-this-year-ministry-says-2025-03-20/.

Sevastopulo, Demetri, and Edward White. 2024. 'US Sharply Raises Tariffs on Chinese EVs and Semiconductor Imports.' *Financial Times*, [London], 14 May. www.ft.com/content/972cabfb-f587-4cb3-ab21-ec3380b049da.

Song, Ligang, and Vishesh Agarwal. 2023. 'The Future of Multilateralism in the Post-Pandemic World.' *China & World Economy* 31, no. 1: 62–87. doi.org/10.1111/cwe.12459.

State Council of the People's Republic of China (State Council). 2024. 'China Establishes Internationally Competitive New Energy Industry Chain.' *Xinhua*, 23 October. english.www.gov.cn/news/202410/23/content_WS6718fbc9c6d0868f4e8ec3a1.html.

State Council. 2025. 'Renewable Energy Accounts for 56 pct of China's Total Installed Capacity.' *Xinhua*, 28 January. english.www.gov.cn/archive/statistics/202501/28/content_WS6798de96c6d0868f4e8ef410.html.

Tyers, Rod, and Yixiao Zhou. 2024. 'Tech Wars: Distributional Consequences of Global Tech Rivalry.' *Asian Economic Journal* 38, no. 3: 289–340. doi.org/10.1111/asej.12335.

Urban, Frauke. 2018. 'China's Rise: Challenging the North–South Technology Transfer Paradigm for Climate Change Mitigation and Low Carbon Energy.' *Energy Policy* 113: 320–30. doi.org/10.1016/j.enpol.2017.11.007.

Yang, Anita. 2025. 'China New Energy Storage Capacity to Surge by 2030.' Online post, 14 April. www.icis.com/explore/resources/news/2025/04/14/11092303/insight-china-new-energy-storage-capacity-to-surge-by-2030/.

Zhang, Ming, Nan Cong, Yan Song, and Qing Xia. 2024. 'Cost Analysis of Onshore Wind Power in China Based on Learning Curve.' *Energy* 291: 130459. doi.org/10.1016/j.energy.2024.130459.

2

China's energy transition towards carbon neutrality: A multidimensional framework

Zhen Wang, Boshu Li and Yinghao Kong

Introduction

Climate change is reshaping human existence with unprecedented breadth and depth. According to the latest data from the Sixth Assessment Report of the United Nations Intergovernmental Panel on Climate Change (IPCC), global surface temperatures have already risen by 1.15ºC compared with pre-industrial levels. If current emission trajectories persist, the critical threshold of 1.5ºC is projected to be breached between 2030 and 2035 (IPCC 2023). The ecological and environmental challenges posed by climate change may represent the greatest uncertainty confronting human society.

In this context, China, as the world's largest developing country, has pledged to achieve peak carbon emissions by 2030 and carbon neutrality by 2060. This commitment not only underscores China's responsibility as a major global power but also necessitates the establishment of a comprehensive low-carbon transformation system encompassing the entire industrial chain. Within this transformative process, the energy sector—recognised as the 'main battlefield' for achieving the dual carbon goals—has been undergoing a significant transition from 'light green' to 'deep green' development. Guided by the new energy security strategy of 'Four Revolutions and One Cooperation',[1] substantial progress has been made in transforming energy production and consumption patterns, comprehensively enhancing energy supply security and

---

[1]   Since the Eighteenth National Congress of the Communist Party of China (CPC), in the face of changes in energy supply and demand patterns and new trends in international energy development, the CPC Central Committee with President Xi Jinping at its core has proposed a new energy security strategy of 'four revolutions and one cooperation' (a revolution in energy consumption, supply, technology and systems, and full-scale international cooperation) to provide a guide for high-quality energy development in the new era.

achieving historic breakthroughs in green and low-carbon energy development. By the end of 2023, China's energy consumption per unit of GDP had declined by nearly 26 per cent compared with 2013 levels, while carbon dioxide emissions per unit of GDP had decreased by more than 35 per cent. Simultaneously, the share of clean energy consumption rose to 26.4 per cent and the proportion of coal consumption fell cumulatively by 12.1 percentage points. The installed capacity of renewable energy generation reached 1.38 billion kilowatts, accounting for 51.9 per cent of the nation's total power generation capacity—marking the first time renewable energy capacity surpassed fossil fuel–based power generation capacity.

China's ongoing energy transition is fundamentally reshaping its energy production and consumption landscape. This transformation not only exerts multidimensional impacts on domestic economic growth, industrial transformation, environmental quality and public health, but also has transcended national boundaries, generating significant international spillover effects through global supply chains, technology diffusion and other channels. From 2013 to 2023, the global share of non-fossil energy consumption increased from 13.6 per cent to 18.5 per cent, with China contributing 45.2 per cent of the global growth in non-fossil energy consumption. In 2023 alone, China accounted for more than half the world's newly added renewable energy capacity. Driven by stable policy expectations for the energy transition, China has capitalised on its dominant position in clean energy manufacturing to actively promote the international dissemination of green technologies and equipment. Exports of photovoltaic (PV) cells and wind turbines exceeded US$49.76 billion in 2023, with an export volume of approximately 255 gigawatts (GW). By providing economically accessible clean energy solutions to the global market, China has made substantial contributions to global efforts to combat climate change and achieve sustainable development.

Current research has increasingly focused on China's energy transition and the development trends of its energy sector. Existing studies have examined the latest progress in the energy transition from various perspectives, including industry (He et al. 2022), policy (Li and Taeihagh 2020), technology (Duan et al. 2021) and governance mechanisms (Hepburn et al. 2021). These studies explore the prospects and pathways for transforming energy production and consumption. Notably, regular energy outlook reports published by international organisations (IEA 2024), research institutions (ERI 2024) and energy enterprises (CNEEI 2024) are particularly representative. Based on their respective modelling assumptions and scenario designs, these reports offer diverse pathways and policy recommendations for advancing China's low-carbon energy transition. However, most of these studies focus primarily on demonstrating the necessity of the energy transition, evaluating its achievements and assessing the feasibility of implementation strategies. They often lack a comprehensive analysis that captures the full spectrum of impacts on the economy, society, environment and climate change (Zhang et al. 2024). Furthermore, forward-looking and systematic research on the spillover effects of China's exports of new-energy products, such as PV cells and wind turbines, remains limited. As a result, developing an integrated assessment model to evaluate the impacts and spillover effects of China's energy transition has become critical.

The remainder of this chapter is organised as follows. Section two introduces the integrated assessment modelling framework designed to evaluate the multidimensional impacts and spillover effects of China's energy transition. This framework integrates multidisciplinary methodologies, addressing the limitations of single-model analyses by employing complex and holistic analytical approaches. Section three examines the cascading effects of China's energy transition on macroeconomic performance, industrial structure, carbon reduction, pollution mitigation and public health under various scenarios. This section also quantifies the spillover effects of global emission reductions driven by the export of China's wind power and PV products. Section four concludes by summarising key findings and offering insights and policy recommendations.

Methodology and framework

Comprehensive assessment methodologies typically involve simulating and evaluating the interactions between multiple factors under varying assumptions within a unified framework. This is achieved by linking complex system models across different domains. These methodologies are generally categorised into three types of model: top-down, bottom-up and hybrid. Building on the theoretical foundations of cost–benefit analysis, this study develops a novel integrated assessment model for evaluating the multidimensional impacts and spillover effects of China's energy transition (Figure 2.1). The framework comprises two modules.

Module I integrates four submodels: a macroeconomic model, an energy system optimisation model, a co-benefits approach and a public health benefit evaluation model. This framework is designed to evaluate the impacts of the energy transition across multiple dimensions under various policy constraint scenarios. The macroeconomic model is a multi-sector, recursive dynamic computable general equilibrium (CGE) model. It simulates and evaluates the effects of economic, social and environmental constraints on overall economic growth, industrial development and labour market dynamics. The energy system optimisation (ESO) model, built on extensive technical data, encompasses the entire energy lifecycle—from extraction, processing and conversion to transportation and end use. It focuses on modelling key energy-consuming and emission-intensive sectors, such as industry, building operations and transportation, to optimise energy transition pathways and estimate the costs of transition under different policy scenarios. Additionally, the framework establishes soft linkages with the greenhouse gas and air pollution interaction and synergies (GAINS) model developed by the International Institute for Applied Systems Analysis (Kilmont 2025) and the public health benefits evaluation (HEL) model. These linkages enable the quantification of the co-benefits of carbon and pollution reduction resulting from the energy transition. Furthermore, the framework provides monetised benefits of public health improvements, offering a more intuitive representation of the synergies between pollution mitigation and health benefits.

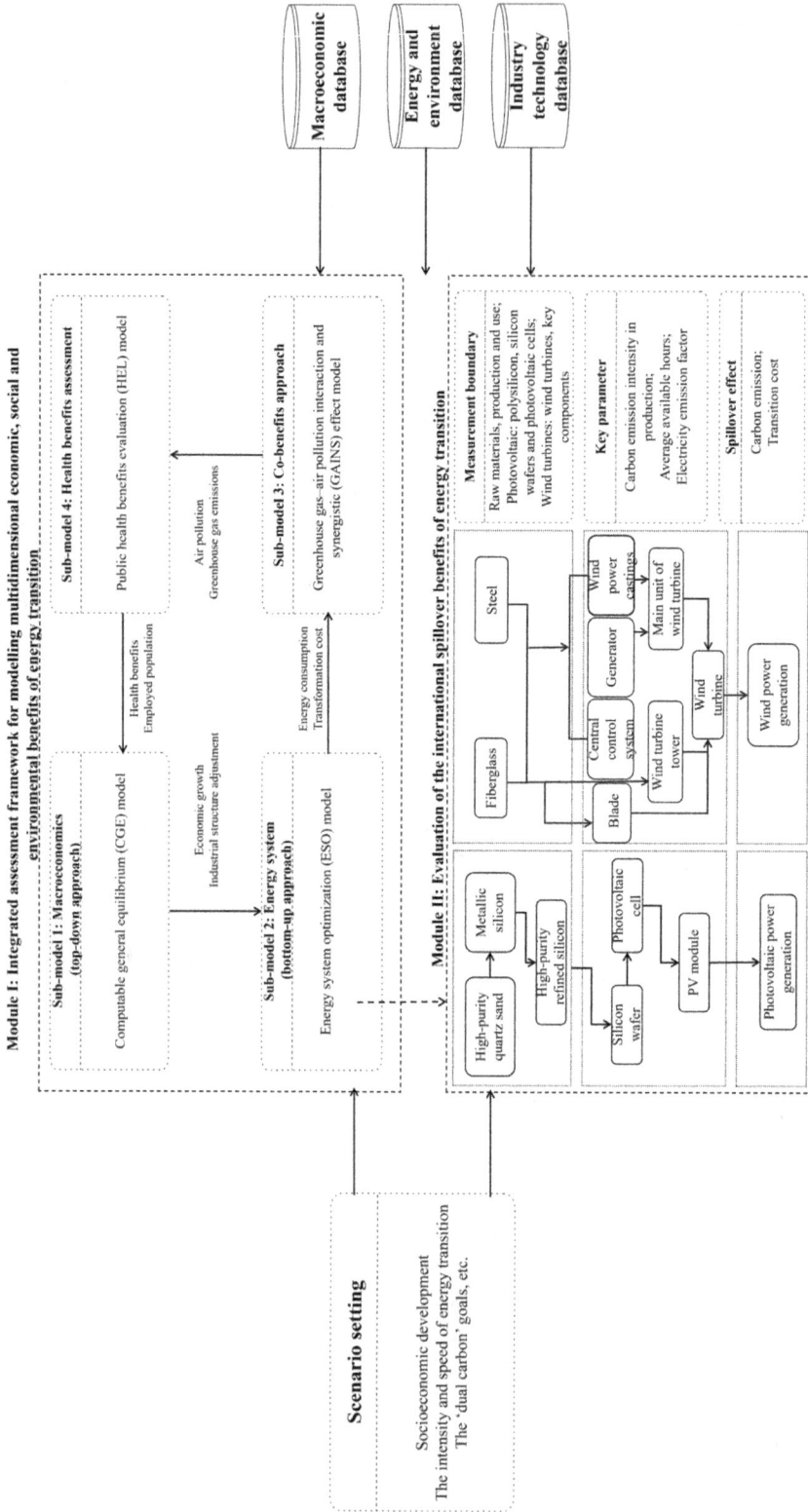

Figure 2.1: Integrated assessment framework for modelling multidimensional economic, social and environmental benefits of China's energy transition

Figure 2.1 illustrates the coupling relationships among the four submodels in Module I. Under the constraints of economic and social development goals, carbon reduction and energy policy, the CGE model generates projections of economic growth and structural adjustments, which are subsequently transmitted to the ESO model as constraints on energy service demand forecasting. The ESO model, optimised to minimise energy system costs, determines long-term transition trends and technological pathways for key energy-consuming and carbon-emitting sectors. Simultaneously, the ESO model provides national energy consumption trajectories to the GAINS model, which simulates and optimises changes in three key air pollutant emissions and particulate matter (PM) 2.5 concentrations. Leveraging the linear and nonlinear exposure-response functions embedded in the HEL model, health impacts—including morbidity and mortality cases—are quantified. These results are then fed back into the CGE model, enabling further analysis of the corresponding economic variation and changes in employment. This iterative process allows policymakers to evaluate the comprehensive costs and benefits of the energy transition from a social perspective. By integrating cross-domain interactions, this framework overcomes the limitations of single-model approaches, offering a more holistic and interdisciplinary perspective on the energy transition and its multidimensional impacts.

Module II defines a country's export contribution to global emission reductions as the sum of emissions generated in the domestic production process and the emission reductions realised during the use phase in foreign countries. The measurement boundaries and key parameters are illustrated in Figure 2.1. For PV products, production-related emissions can be calculated as the product of net export volumes and carbon emission intensity, as expressed in Equation 2.1.

Equation 2.1

$$P_j = \frac{\sum_{i=1}^{n} x_{ij}}{p_j} \times \theta_j$$

In Equation 2.1, P_j denotes the production-related carbon emissions from the net exports of photovoltaic products in year j, reflecting China's contribution to global carbon emission reductions during the production phase; i denotes the product type, where $i = 1$ corresponds to high-purity silicon, $i = 2$ corresponds to silicon wafers and $i = 3$ corresponds to PV cells (Liu et al. 2019); j denotes the year, where $j = 1$ corresponds to 2015 and $j = 9$ corresponds to 2023. X_{ij} represents the net export value of product i from China to the world in year j; P_j denotes the price of PV modules in year j; and θ_j denotes the carbon emission intensity of PV modules in year j, with the unit of tonnes of carbon dioxide per megawatt (tCO_2/MW).

To calculate the emission reductions during the use phase, it is essential to determine the amount of electricity generated by all exported products. Since polysilicon and silicon wafers are used in the production of PV cells, their export values must be converted into the equivalent quantity of PV cells. As trade data in commodity trade statistics databases are typically reported in monetary terms, the average price of the products is used to convert monetary values into the corresponding quantity of PV modules. Consequently, the electricity generation capacity of PV modules over their full lifecycle can be calculated using Equation 2.2.

Equation 2.2

$$C_j = \frac{\sum_{i=1}^{n} x_{ij}}{p_j} \times T \times n \times \beta_j$$

In Equation 2.2, T denotes the average available hours of the power-generating equipment in photovoltaic power plants; n denotes the service life of PV power plants; β_j denotes the electricity emission factor in year j; and C_j denotes the potential emission reductions during the use phase of net exported PV products in year j, reflecting China's contribution to global carbon emission reductions during the use phase.

The total contribution to emission reductions is calculated by summing the emission reduction contributions from the production phase and the use phase, as shown in Equation 2.3. The emission reduction contributions of wind turbine exports are analogous to those of PV products and will not be elaborated further here.

Equation 2.3

$$T_j = P_j + C_j$$

Population growth and urbanisation rates are the key factors influencing economic development and energy consumption patterns. The social cost of carbon, which quantifies the economic loss caused per unit of carbon emissions, serves as a critical tool for conducting cost–benefit analyses of the energy transition and climate policies. In this study, we comprehensively compared projections of population growth and urbanisation rates from major institutions, including the United Nations (UNDESA 2024; UN-Habitat 2024) and the National Bureau of Statistics of China. Furthermore, the latest assessments of the social cost of carbon are systematically reviewed (Wang et al. 2022; Ricke et al. 2018). To facilitate cross-sectoral comparisons of the costs and benefits associated with the energy transition and to provide policymakers with intuitive insights, we monetised the health benefits using environmental valuation, such as the value of statistical life (Jin and Zhang 2018). The specific assumptions for these key parameters are presented in Table 2.1. In addition, the data inputs, including input–output tables, energy balances, industrial statistics and carbon intensity data, were sourced from government publications, industry associations and relevant literature (Zheng et al. 2021; Liu et al. 2021; Pan et al. 2022). Emission factors, product lifespans and average annual operating hours were estimated based on industry averages. Air pollutant and carbon emission factors were determined using the latest IPCC Guidelines for National Greenhouse Gas Inventory (IPCC 2019) and data from China's Carbon Emission Accounts and Datasets (CEADs).[2]

---

2   See details at the CEADs website: www.ceads.net.cn/data/emission_factors/.

Table 2.1: Pre-set value of China's population, urbanisation rate, social cost of carbon and value of statistical life

Parameters	2025	2030	2040	2050	2060
Population (billion)	1.41	1.39	1.36	1.30	1.20
Urbanisation rate (%)	67.2	70.6	75.0	80.0	85.0
Social cost of carbon (US$/tCO$_2$)	26.8	33.2	41.7	74.9	88.2
Value of statistical life (US$ million)	–	–	–	–	1.34

Note: To eliminate the impact of price fluctuations and reflect the real changes in monetised results such as GDP, unless otherwise specified, all monetised values in this study have been adjusted to constant prices using the 2010 price level as the base year.

Source: Authors' calculations and forecasts.

In addition to examining economic and social impacts, this study focuses on the synergistic emission reductions of three key air pollutants—sulphur dioxide, nitrogen oxides and PM2.5—and their associated public health benefits. Given that PM2.5 pollution is the leading contributor to both mortality and morbidity, the health benefit valuation in this study exclusively targets PM2.5 pollution, drawing on epidemiological principles and insights from the Global Burden of Disease studies (Murray et al. 2020; Xu et al. 2021). The health endpoints and exposure-response coefficients related to PM2.5 pollution are explained in detail in Table 2.2.

Table 2.2: Exposure-response functions for health endpoints related to PM2.5 pollution

Endpoint	Exposure-response functions		
	Confidence interval (95%) low	Medium	Confidence interval (95%) high
All causes (China)[a]	–0.0003	0.0009	0.0018
Chronic obstructive pulmonary diseases[b]	Nonlinear function		
Lung cancer[b]			
Ischemic heart disease (25y–65y)[b]			
Cardiovascular disease (25y–65y)[b]			
Lower respiratory infection[b]			

Sources: [a] Cao et al. (2011); [b] Pope et al. (2002).

The primary global spillover effects of China's energy transition are the impacts on carbon dioxide mitigation in other countries from China's exports of PV modules and wind turbines. While the term 'spillover effects' can encompass a broader range of phenomena, such as international knowledge transfers or shifts in global fossil fuel markets, the scope of this study is focused on the direct impact of technology exports due to limitations in data availability. These spillover effects encompass both the production phase (Yang et al. 2015; Wang 2023) and the use phase. Key parameters influencing these effects include global electricity emission factors, product service lifespans and average annual operating hours, as summarised in Table 2.3.

Table 2.3: Pre-set values of global electricity emission factors, product lifespans and average annual operating hours

Product	Product lifespan (years)	Average annual operating hours (hours/year)	Global electricity emission factor (tCO$_2$/GWh)
Photovoltaic	30	1,700	671
Wind turbines	30	1,900	671

Source: Authors' calculations and forecasts.

Based on variations in the depth and pace of energy transition, this study develops two scenarios: the current policy scenario (CPS) and the enhanced action scenario (EAS). The divergence between these two scenarios underscores the 'action gap' that China must address to accelerate its energy transition and achieve its dual goals of 'carbon peaking and carbon neutrality'.

The CPS is grounded in China's 2021 Nationally Determined Contributions (NDCs) under the Paris Agreement and reflects the policy intensity and energy system transformation efforts outlined in sectoral development plans, energy conservation objectives and carbon reduction targets established under the '1+N' policy framework for dual carbon goals. This scenario assumes that emission reduction efforts after 2030 will largely follow the current trajectory. Furthermore, it assumes that the global energy transition will maintain its present momentum, while China's exports of PV modules and wind turbines will continue to grow at their current rates.

The EAS is designed to capture the policy intensity and energy system transformation efforts necessary to address the constraints imposed by China's socioeconomic and ecological development goals. Building on the CPS, the EAS further strengthens and updates more ambitious NDC targets and action requirements, while fully leveraging the potential for energy conservation and emission reductions. It also intensifies the stringency and scale of dual carbon-control measures. This scenario is characterised by a substantial increase in the adoption of advanced energy technologies, the penetration of low-carbon technologies and the electrification of end-use energy consumption. Concurrently, the costs of low-carbon energy sources, including wind and solar power, as well as energy storage, continue to decline. These advances collectively facilitate the establishment of a new energy system, thereby supporting the realisation of China's 'carbon peak' and 'carbon neutrality' goals. Compared with the CPS, the EAS envisions an accelerated global energy transition, alongside faster growth in China's PV and wind turbine exports.

Results and discussion

The advancement of the energy transition will play a pivotal role in supporting Chinese-style modernisation and driving industrial transformation. At present, China is in the mid-to-late stages of industrialisation and urbanisation, steadily transitioning towards a high-quality development phase under the guidance of its new development philosophy. In 2024, China's GDP was projected to reach approximately US$18.9 trillion, reflecting a 5 per cent increase from the previous year. Under both the CPS and the EAS, China's macroeconomic trajectory is expected to exhibit sustained expansion in scale, accompanied by a gradual deceleration in growth rates (Table 2.4). By about 2035, China's GDP is anticipated to surpass that of the United States, positioning it as the world's largest economy.

Table 2.4: Changes in China's average GDP growth rate

Scenario	2026–30	2031–35	2036–40	2041–50	2051–60
CPS	5.0	4.5	4.5	3.4	2.4
EAS	4.5	3.5	3.5	2.5	1.5

Source: Authors' calculation.

Against the backdrop of China's 'Two Centenary Goals'[3] and historical trends in the industrial structure of developed countries, China's future economic development can be divided into three distinct phases (Figures 2.2a and 2.2b).

Now until 2035

The growth rate and drivers of the economy will enter a critical period of transformation. The share of secondary industry is expected to stabilise and gradually decline, with a steady reduction in the scale of energy-intensive manufacturing industries. Despite this, China will retain its position as a major manufacturing powerhouse, while continuously optimising and upgrading its industrial structure. High-tech manufacturing sectors, including new-energy vehicles and strategic emerging industries, are projected to maintain rapid growth. Meanwhile, tertiary industry will steadily increase its share in the national economy and its contribution to overall economic growth. By 2035, per capita GDP is anticipated to reach the level of moderately developed countries, marking the basic realisation of socialist modernisation and the 'Beautiful China' vision.

---

3   The CPC first proposed the 'Two Centenary Goals' in 1997. The first centenary goal was, by 2021—the centenary of the CPC—to build China into a moderately prosperous society in all respects and doubling GDP and per capita income against 2010. The second centenary goal is, by 2049—the centenary of the People's Republic of China—to build a modern socialist country that is prosperous, strong, democratic, culturally advanced, harmonious and beautiful.

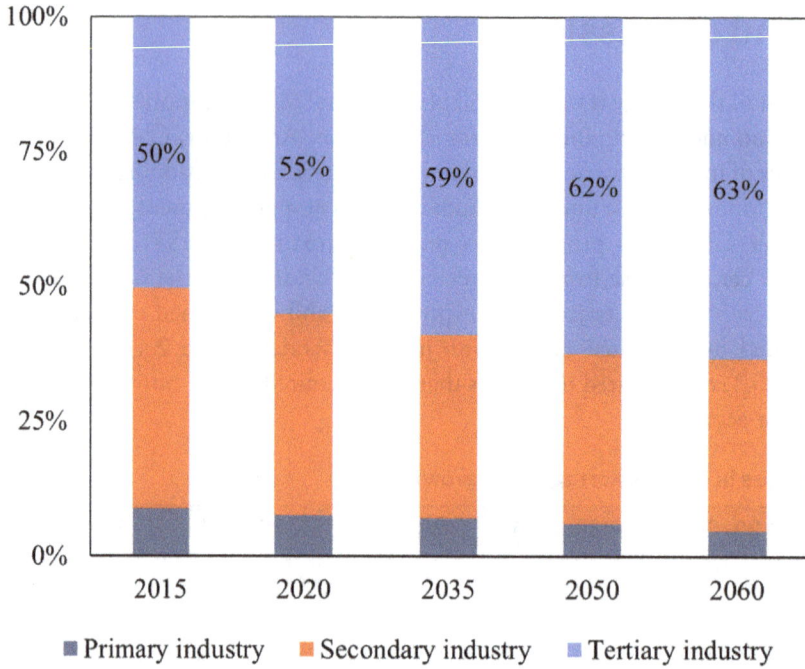

Figure 2.2a: Changes in the proportions of China's industrial structure under the CPS

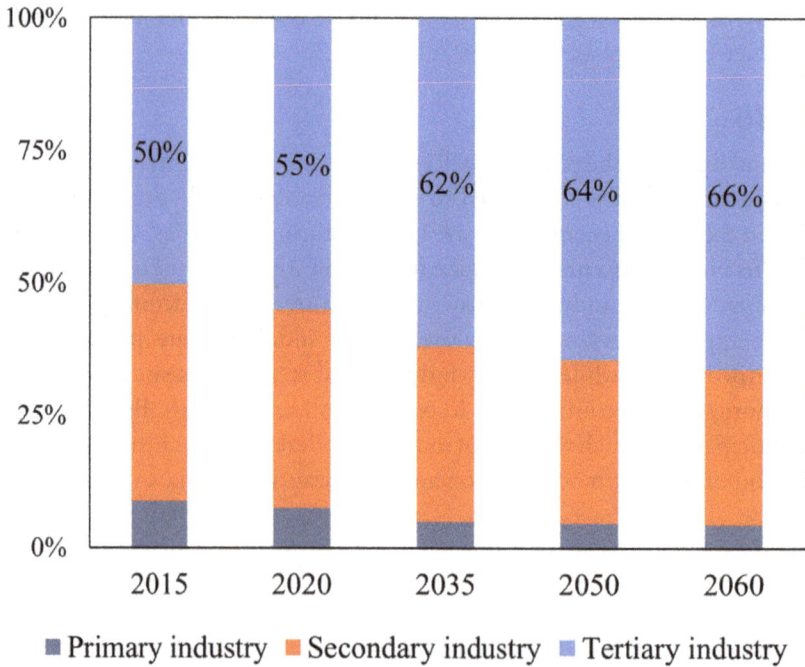

Figure 2.2b: Changes in the proportions of China's industrial structure under the EAS

Source: Authors' calculation.

2035–50

During this second phase, economic growth will be predominantly driven by the service sector and domestic consumption. Tertiary industry will consolidate its dominant position, accounting for more than 62 per cent of GDP. Concurrently, manufacturing will undergo a transformation characterised by digitalisation, high-end innovation and green development, elevating China to the ranks of the leading global manufacturing powers.

2050–60

In the third phase, China's economy is expected to play an increasingly prominent role in driving global development, fostering the emergence of numerous new business formats and innovations. A green economy will take centre-stage globally, with China positioned as a key engine and stabiliser of global economic growth.

The energy transition will catalyse a transformative leap forward in clean energy development. China's total primary energy consumption (TPEC) is expected to follow a trajectory characterised by sustained growth, followed by a gradual decline. This transition can be attributed to four primary factors: 1) the earlier than anticipated peak in population imposes an upper limit on the expansion of domestic aggregate demand, thereby diminishing its supportive effect on energy consumption; 2) the stabilisation and gradual slowdown of economic growth rates create favourable conditions for controlling total energy consumption; 3) industrial restructuring and upgrading contribute to a progressive reduction in energy intensity across sectors; and 4) technological advancements, coupled with the deepening of demand-side energy conservation behaviours, enhance energy efficiency. Taken together, these factors will significantly decelerate the growth rate of energy consumption, leading to a peak, after which consumption will steadily decline.

Under the CPS, TPEC is projected to rise from 5.97 billion tonnes of standard coal equivalent (btce) in 2024 to 6.21 btce by 2035, after which it will plateau. During this period, energy consumption will gradually decouple from economic growth, declining to 5.42 btce by 2060 (Figure 2.3a). Under the EAS, with the implementation of stricter carbon constraints, improvements in energy efficiency and more intensive demand-side conservation measures, TPEC is expected to peak earlier, in 2030, at a lower level of approximately 5.85 btce. By 2060, it is projected to decrease further, to 4.98 btce (Figure 2.3b).

With the continued advance of the energy transition and the implementation of the 'carbon peak and carbon neutrality' goals, China's energy consumption structure is expected to undergo significant transformations, characterised by three key trends: 1) a gradual, stepwise decline in the shares of coal and oil consumption; 2) natural gas consumption initially increasing before declining; and 3) a rapid and substantial rise in the share of non-fossil energy consumption.

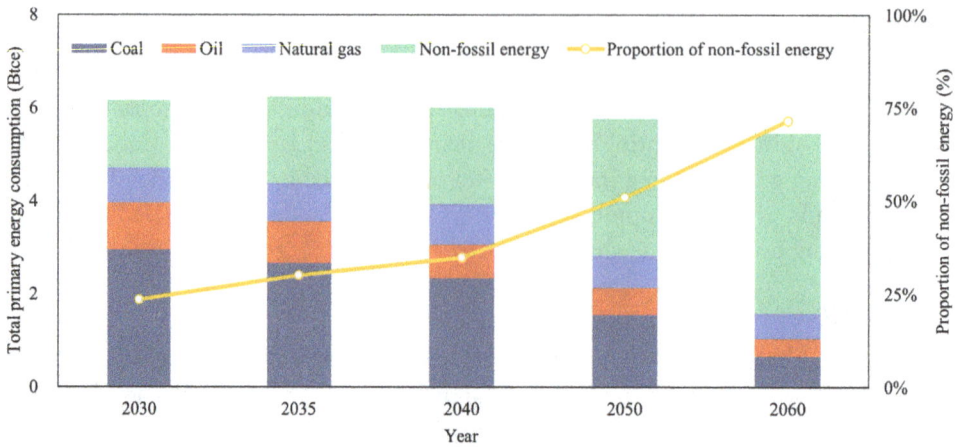

Figure 2.3a: Changes in China's total primary energy consumption under the CPS

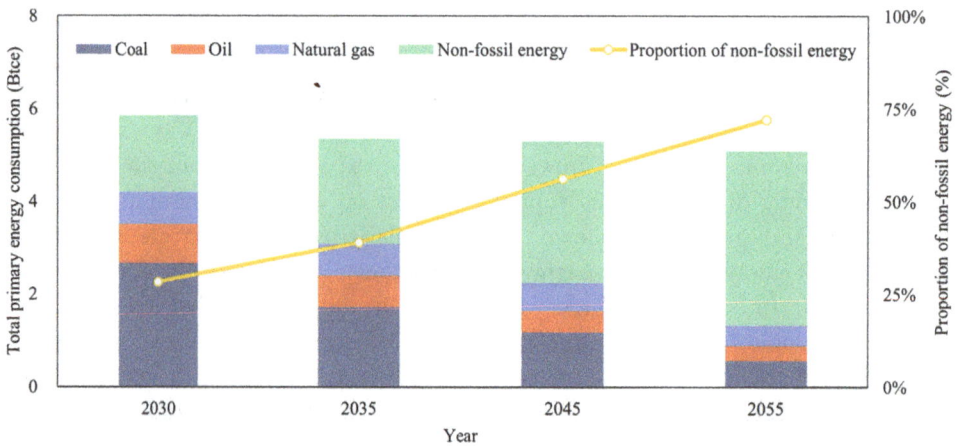

Figure 2.3b: Changes in China's total primary energy consumption under the EAS
Source: Authors' calculation.

Under the EAS, the share of non-fossil energy is projected to reach 82 per cent of total energy consumption by 2060, surpassing the policy target of 80 per cent outlined in China's energy and climate strategies. Meanwhile, considering the dual imperatives of energy security and cost constraints, the phased withdrawal of fossil fuels will adhere to the principle of 'establishing new energy systems before dismantling old ones'. Coal and oil consumption are expected to peak by about 2025. For natural gas, peak consumption is projected to occur about 2040 under the CPS, while under the EAS, the peak is anticipated to arrive earlier, by approximately 2035. Overall, fossil fuels will continue to play a vital role as a cornerstone of energy security and maintaining a stable energy supply during the transition period.

The energy transition will significantly contribute to the substantial reduction of carbon dioxide emissions while further unlocking the potential for synergistic mitigation of other greenhouse gases (GHGs). The CPS aligns with China's current stage and characteristics of development, recognising the need for a certain level of carbon emissions to ensure the low-carbon transition of the economy and society before peaking about 2030. Under this scenario, carbon dioxide emissions are expected to peak at approximately 11.4 billion tonnes. From 2030 to 2060, the low-carbon transformation of the energy structure is expected to accelerate under the EAS, with greater efforts directed towards achieving carbon neutrality through enhanced technological efficiency and industrial restructuring. These efforts will contribute significantly to emission reductions, with economic growth increasingly decoupling from energy consumption and carbon emissions. By 2060, carbon dioxide emissions are projected to decline to 830 million tonnes, representing a reduction of 1.02 billion tonnes compared with the CPS (Figure 2.4a).

Carbon dioxide, as the most dominant GHG by emission share, remains the primary driver of global climate change. However, numerous studies underscore the critical role of GHGs other than carbon dioxide, which cannot be overlooked. For instance, nearly one-third of emission reductions in developed countries have been achieved through non–carbon dioxide GHG mitigation (Montzka et al. 2011; Rogelj and Lamboll 2024). Without effective control, non–carbon dioxide GHG emissions are projected to continue rising, potentially offsetting the progress made in reducing carbon dioxide emissions. Under the CPS, non–carbon dioxide GHG emissions are expected to peak about 2035, plateauing before gradually declining, with a peak level of approximately 2.89 gigatonnes of GHG carbon dioxide equivalent. By 2060, these emissions are projected to decrease slowly, to 2.23 gigatonnes of carbon dioxide equivalent. In contrast, the EAS envisions non–carbon dioxide GHGs peaking concurrently with carbon dioxide, achieving a peak reduction of approximately 0.33 gigatonnes of carbon dioxide equivalent compared with the CPS and declining further, to approximately 1.39 gigatonnes of carbon dioxide equivalent by 2060 (Figure 2.4b).

To underscore the indispensable role of non–carbon dioxide GHGs in national climate strategies, China's updated NDCs should include explicit peak targets for non–carbon dioxide GHGs as soon as possible. Furthermore, carbon neutrality goals should explicitly account for these gases to highlight their integral role in the country's overall strategy for addressing climate change.

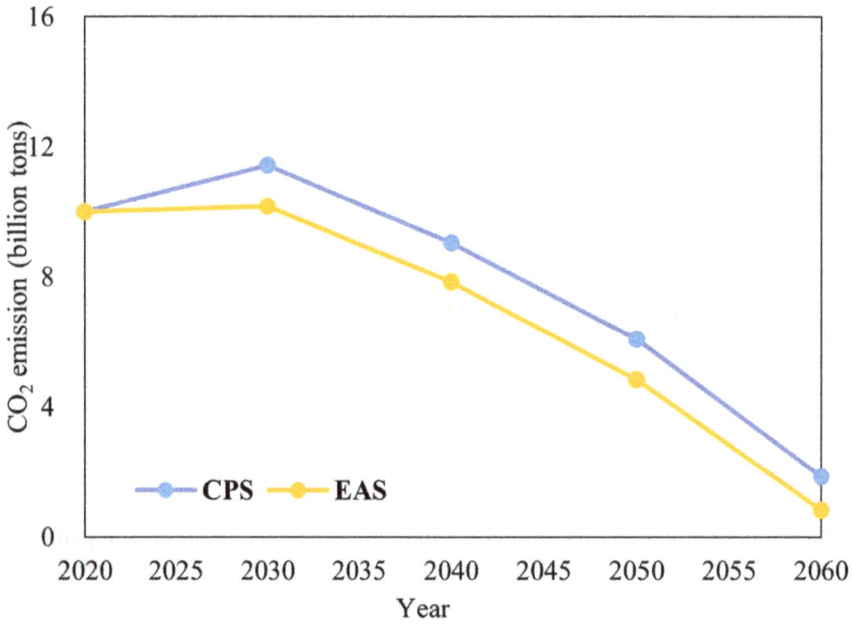

Figure 2.4a: Changes in China's carbon dioxide emissions under different scenarios

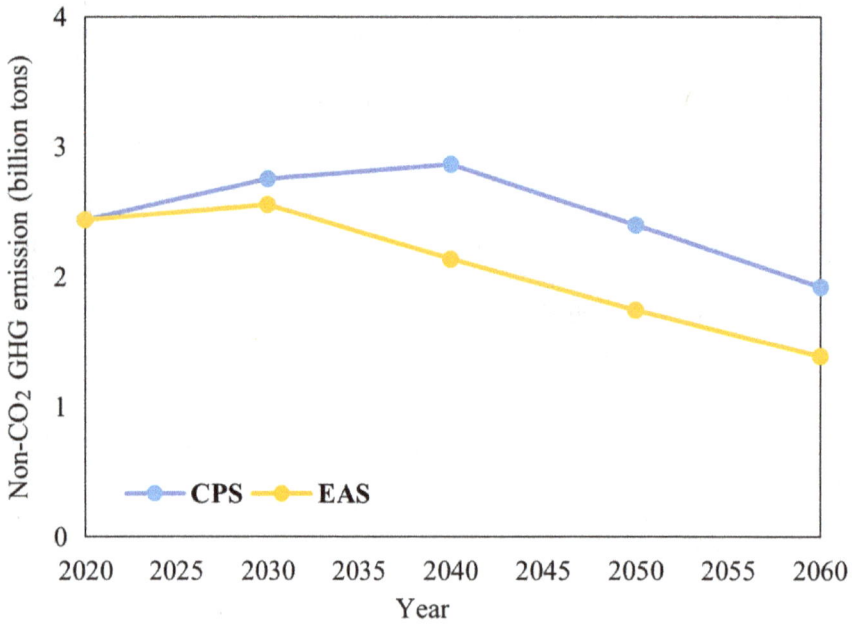

Figure 2.4b: Changes in China's non–carbon dioxide greenhouse gas emissions under different scenarios

Source: Authors' calculation.

Concurrent with decarbonisation, the energy transition has also yielded substantial improvements in air quality. Historical evidence indicates that GHG emissions and air pollution challenges in developed countries did not emerge simultaneously. As a result, these two issues have traditionally been treated as separate problems. However, in contemporary China, they are deeply interconnected, representing critical domains for integrated governance. China is currently one of the world's largest emitters of sulphur dioxide, nitrogen oxides and particulate matter, with ambient concentrations of these pollutants exceeding the safe levels recommended by the World Health Organization (WHO). This has led to a range of environmental and public health issues (Li et al. 2020).

Based on the energy transition scenarios analysed in this study, China's energy consumption structure is expected to undergo significant transformations in the coming decades. By continuously strengthening source control and implementing comprehensive pollution control measures across key industries, emissions of air pollutants are projected to decline substantially. Under the EAS, greater reductions are achieved than under the CPS, with sulphur dioxide, nitrogen oxides and PM2.5 emissions declining to 920,000 tonnes, 2.19 million tonnes and 630,000 tonnes, respectively, by 2060 (Figure 2.5a). Using these emission reductions as inputs, simulations conducted with the EMEP-CTM atmospheric transport model, integrated into the GAINS framework (Amann et al. 2020), it is projected that annual average PM2.5 concentrations nationally will decrease to 35.4 micrograms per cubic metre ($\mu g/m^3$) by 2030 and 17.4 $\mu g/m^3$ by 2060 under the CPS. The EAS reduces PM2.5 concentrations by an additional 2.8 $\mu g/m^3$ in 2060 compared with the CPS (Figure 2.5b).

Accelerating the energy transition is expected to enable most regions in China to meet the national secondary air quality standard for PM2.5 (≤ 35 $\mu g/m^3$) by approximately 2030, marking significant progress towards achieving the 'Beautiful China' initiative. However, after the initial substantial reductions, the rate of decline in annual average PM2.5 concentrations is expected to slow, making further reductions increasingly challenging. Even under the EAS, China's PM2.5 concentrations in 2060 are projected to remain well above the revised 2021 WHO guideline of 5 $\mu g/m^3$. This underscores the need for a scientifically designed, integrated approach that leverages the shared origins of air pollutants and carbon dioxide emissions. By adopting coordinated strategies for long-term decarbonisation and pollution mitigation, China can achieve greater synergies, facilitating fundamental improvements in air quality while advancing its climate goals.

Accelerating the energy transition can significantly enhance public health. The cumulative impacts of air pollution pose pressing challenges to society and public health. In China, approximately 1 million premature deaths annually are attributed to atmospheric pollutant emissions (Wang et al. 2024). This study further estimates the number of premature deaths caused by five PM2.5-related diseases under different scenarios: chronic obstructive pulmonary disease, lung cancer, ischemic heart disease, cerebrovascular disease and lower respiratory infections.

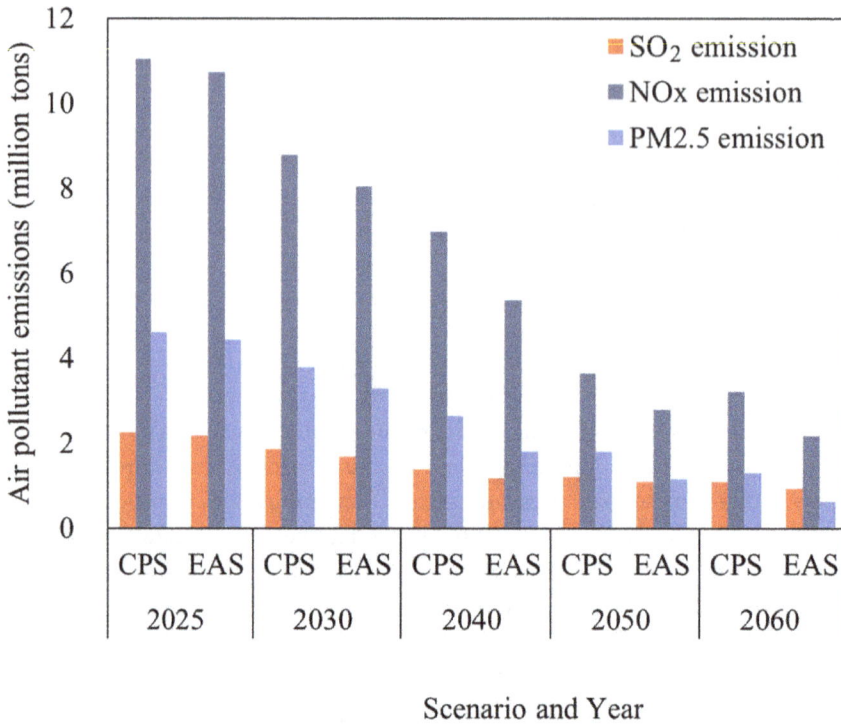

Figure 2.5a: Changes in emissions of sulphur dioxide, nitrogen oxides and PM2.5

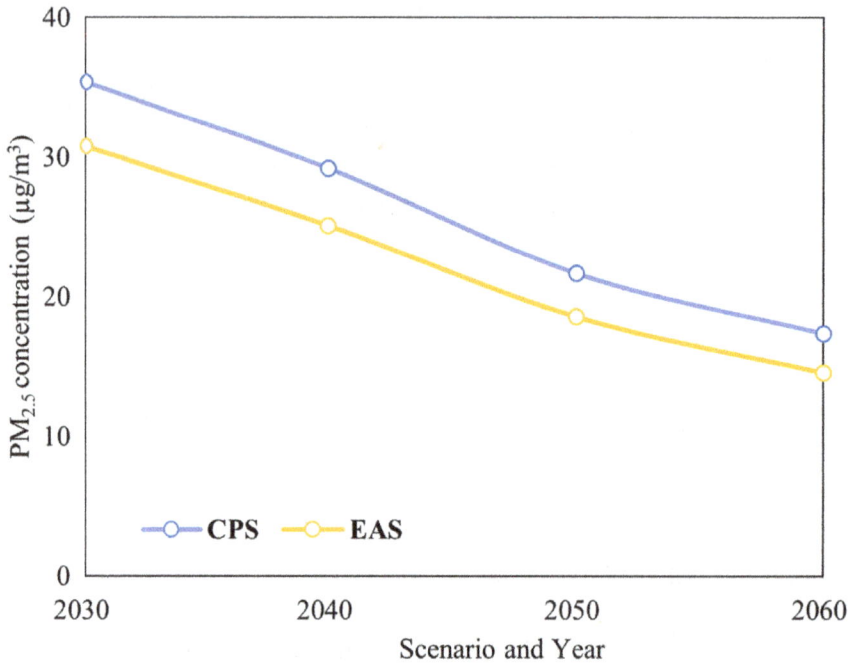

Figure 2.5b: Changes in PM2.5 concentration

Source: Authors' calculation.

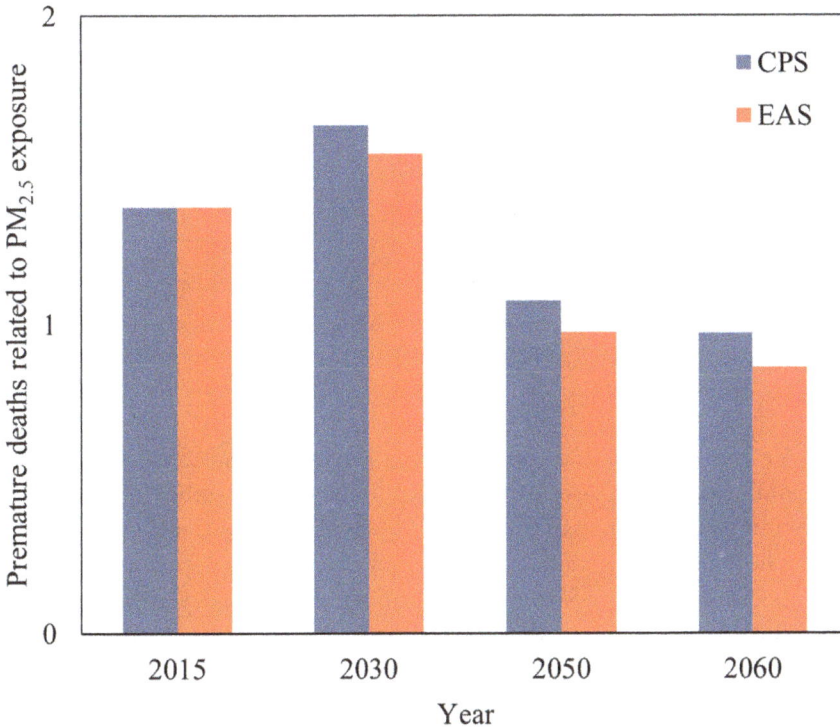

Figure 2.6: Changes in premature deaths related to PM2.5 exposure
Source: Authors' calculation.

Under the EAS, accelerated reductions in PM2.5 concentrations are projected to result in greater improvements in public health. By 2030, the total number of premature deaths is estimated to reach 1.55 million, with 91,000 premature deaths avoided compared with the CPS, underscoring the substantial health benefits of an accelerated transition. By 2060, the health co-benefits of coordinated emission reductions are further amplified, as PM2.5-related premature deaths decrease to 970,000 (Figure 2.6). These findings demonstrate that deep transformations in the energy structure, coupled with stricter end-of-pipe pollution control measures, can generate significant health benefits. Such improvements in air quality can offset much of the employment loss associated with population ageing, highlighting the broader socioeconomic advantages of pursuing an accelerated energy transition.

While the substantial costs of transformative change cannot be overlooked, China's energy transition offers significant long-term net social benefits. Adopting a social perspective, this study systematically evaluates the potential impacts of China's energy transition and establishes a cost–benefit analysis framework (Table 2.5). This framework integrates energy, economic, climate and health-related costs and benefits, providing a quantitative assessment of each dimension.

Table 2.5: Cost–benefit analysis framework for China's low-carbon energy transition

Classification	Composition	Definition
Benefits	Climate benefits	Benefits derived from carbon emission reductions induced by energy transition.
	Health benefits	Monetisation of avoided premature deaths and illnesses resulting from marginal reductions in long-term PM2.5 exposure, due to synergistic effects of carbon reduction and pollution control.
Costs	Energy costs	Changes in energy costs incurred by various technologies and production processes.
	Economic costs	Initial investment and operational maintenance costs associated with deploying energy transition–related technologies.

Source: Authors' compilation.

Total costs under both the CPS and the EAS are projected to increase steadily, driven by higher economic output, large-scale deployment of cleaner low-carbon technologies and substantial investments in transformative and disruptive innovations. These trends highlight that the energy transition is a sustained, high-investment process, with cumulative costs estimated at US$13.2 trillion under the CPS and US$18.8 trillion under the EAS by 2060 (Figure 2.7). Consequently, policymakers and industry stockholders should act promptly to define future industrial structural layouts, prioritise key directions for breakthroughs in emerging technologies and enhance research and innovation capacities. These measures will help mitigate future cost burdens and maximise potential benefits.

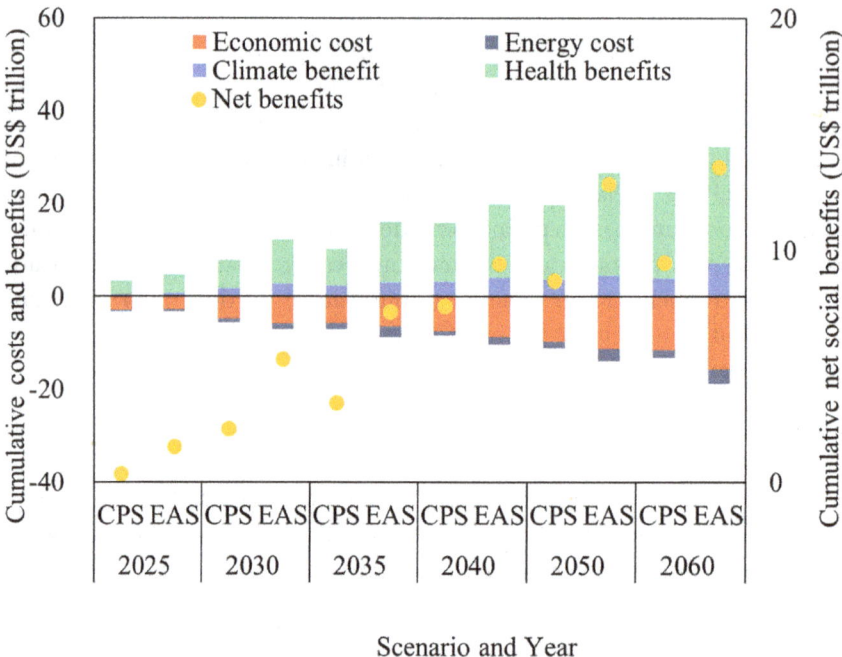

Scenario and Year

Figure 2.7: Cumulative costs and benefits of China's energy transition
Source: Authors' calculation.

Nevertheless, the benefits—particularly those related to climate and public health—are expected to far outweigh the associated costs. Both scenarios yield positive net social benefits, with the EAS demonstrating significantly greater net benefits compared with the CPS. This underscores the importance of accelerating the energy transition and should strengthen policymakers' confidence and determination to pursue this path. Net social benefits are projected to grow over time: under the EAS, cumulative net social benefits reach US$5.3 trillion in 2030, US$13.6 trillion in 2050 and US$24.8 trillion by 2060. These findings highlight the long-term necessity of implementing comprehensive policy measures to support an accelerated energy transition.

Moreover, a fully realised energy transition is expected to substantially improve socioeconomic footprints, enhance household welfare and boost employment. However, due to data limitations, this study does not quantify these additional economic benefits, suggesting that the net social benefits of the energy transition may be even greater than estimated.

China's exports of PV and wind turbine technologies have generated significant global emission reduction spillover effects. By 2023, the net exports of Chinese PV modules and critical components had contributed to potential carbon reductions exceeding 20 gigatonnes in other countries over their full lifecycle. Similarly, exports of wind turbines and critical components contributed more than 3 gigatonnes of potential carbon reductions. With the continued advancement of the global energy transition, China's PV and wind exports are expected to maintain robust growth. Under the CPS, the cumulative carbon reductions from Chinese PV and wind exports are projected to surpass 140 gigatonnes by 2060, including approximately 128 gigatonnes from PV exports (Figure 2.8a) and 12 gigatonnes from wind exports (Figure 2.8b). Under the EAS, these cumulative reductions are expected to exceed 185 gigatonnes by 2060, with PV exports contributing around 170 gigatonnes and wind exports approximately 16 gigatonnes.

More importantly, a combination of factors has driven continuous cost reductions in China's PV and wind industries, significantly lowering the global costs of the energy transition. According to the International Renewable Energy Agency (IRENA 2024), the global average levelised cost of electricity for solar PV has declined by more than 80 per cent in the past decade, with China playing a pivotal role. Similarly, Wood Mackenzie (Forbes-Cable 2023) reports that the cost of producing one megawatt (MW) of PV electricity in China is 11–64 per cent lower than in other markets. The cost declines in China's PV and wind sectors can be attributed to three key factors. First, policy initiatives and public awareness campaigns have fostered a broad social consensus about and stable expectations of the energy transition. This has incentivised corporate investment in new-energy technologies. Since 2021, China has consistently ranked first globally in clean energy investment. Second, stable transition expectations have facilitated rapid deployment of new-energy infrastructure, such as wind and solar power. This has created vast opportunities for the adoption of advanced technologies and the upgrading of energy products, effectively reducing costs and promoting economies of scale. Third, China's robust manufacturing capabilities have enhanced its competitiveness in equipment production, technological research and development (R&D) and industrialisation related to the energy transition. This has enabled the efficient production of high-quality products essential for global energy transition efforts.

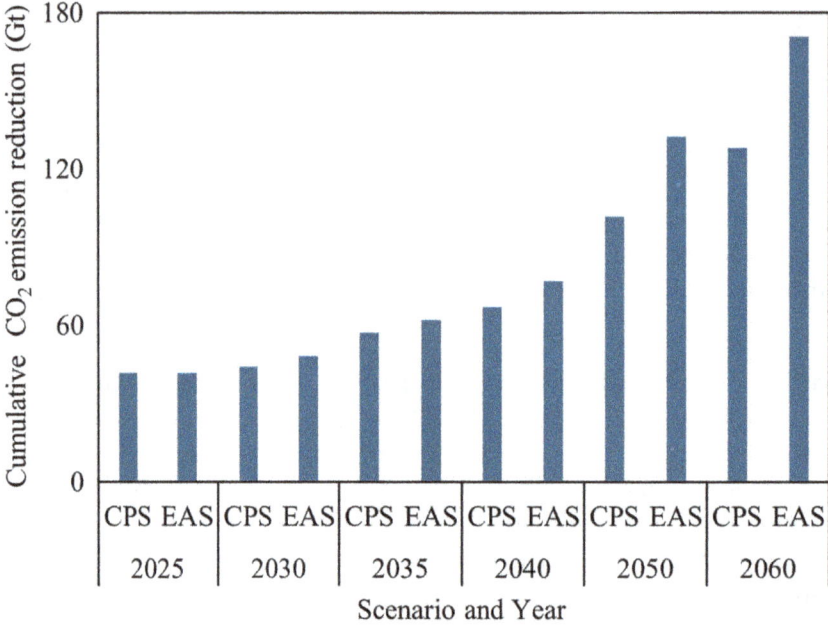

Figure 2.8a: Contribution of China's PV exports to global carbon emission reductions

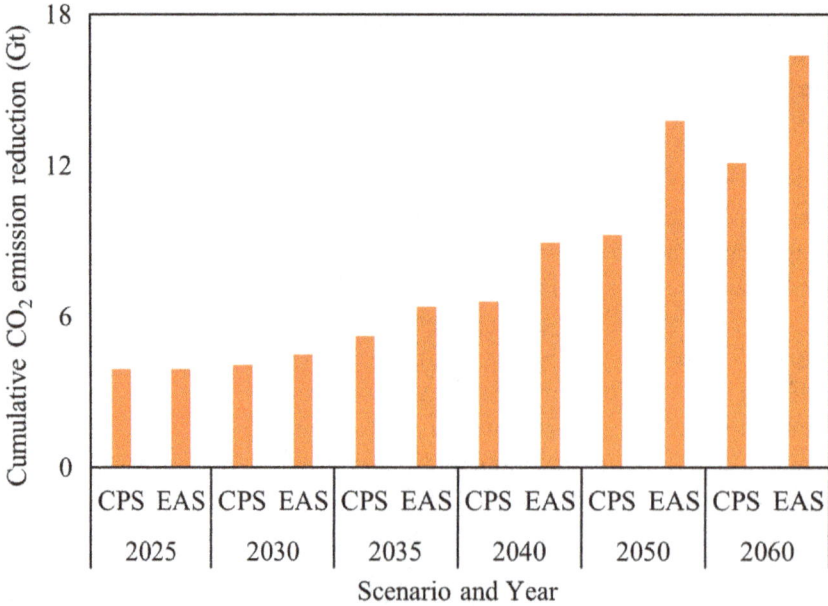

Figure 2.8b: Contribution of China's wind turbine exports to global carbon emission reductions

Source: Authors' calculation.

Conclusions and policy implications

This study develops an innovative integrated assessment modelling framework to evaluate the comprehensive impacts and spillover effects of China's energy transition. The framework systematically simulates and analyses the multidimensional impacts of varying depths and paces of energy transition on energy and economic systems, climate change, environmental outcomes and public health. A particular focus is placed on assessing the global emission reduction spillover effects of China's PV and wind turbine exports, enabling a comprehensive evaluation of both the integrated impacts and the spillover effects of China's energy transition.

Given the substantial net social benefits and significant global carbon emission reduction effects derived from accelerating the energy transition, this study argues that policymakers should maintain strategic resolve in advancing the energy transition and achieving the dual carbon goals (carbon peak and carbon neutrality). In the face of escalating global climate crises and increasing domestic structural emission reduction pressures, China should use its dual carbon goals as a guiding framework to reinforce its strategic commitment to the low-carbon restructuring of its energy system.

Although the energy transition entails incremental investment pressures, the comprehensive benefits—including climate change mitigation, health improvements, job creation and enhanced energy security—are significantly positive and continue to grow over time. To overcome barriers related to low-carbon financing mechanisms, technology diffusion and regulatory systems, China should plan clear and actionable long-term deep decarbonisation pathways for the 2030–60 period. These pathways should align with current energy conservation and carbon reduction policies, the forthcoming 2035 NDC targets and their implementation plans, while also accelerating absolute carbon dioxide emission reductions after 2030. Additionally, China should strengthen the management and control of non–carbon dioxide GHGs, such as methane, nitrous oxide and fluorinated gases, while prioritising R&D of breakthrough technologies for deep emission reductions. Efforts should also be directed towards enhancing agricultural and forestry carbon sinks to achieve economy-wide deep decarbonisation and reduction of all GHGs.

China's green transformation is globally significant. Leveraging its world-leading industrial scale and technological innovation capabilities, China has established substantial cost advantages in new-energy sectors such as PV module and wind turbine manufacturing. Its export trade not only directly alleviates global financial pressures for the energy transition but also contributes positively to global carbon dioxide emission reductions. To further advance global energy transition efforts, it is recommended to deepen international collaboration along the green industry value chain. Technologically advanced nations could focus on core technology R&D, while emerging economies capitalise on their manufacturing strengths to establish a tiered collaborative network encompassing R&D, manufacturing and application. At the same time, caution is needed to prevent unilateral trade protectionism from driving up transition costs.

A tariff preference framework for new-energy products should be developed to transform the production capacity advantages of manufacturing powerhouses such as China into accelerators for global carbon neutrality.

References

Amann, Markus, Gregor Kiesewetter, Wolfgang Schöpp, Zbigniew Klimont, Wilfried Winiwarter, Janusz Cofala, Peter Rafaj, Lena Höglund-Isaksson, Adriana Gomez-Sabriana, Chris Heyes, Pallav Purohit, Jens Borken-Kleefeld, Fabian Wagner, Robert Sander, Hilde Fagerli, Agnes Nyiri, Laura Cozzi, and Claudia Pavarini. 2020. 'Reducing Global Air Pollution: The Scope for Further Policy Interventions.' *Philosophical Transactions of the Royal Society A: Mathematical, Physical and Engineering Sciences* 378, no. 2183: 20190331. doi.org/10.1098/rsta.2019.0331.

Cao, Jie, Chunxue Yang, Jianxin Li, Renjie Chen, Bingheng Chen, Dongfeng Gu, and Haidong Kan. 2011. 'Association between Long-Term Exposure to Outdoor Air Pollution and Mortality in China: A Cohort Study.' *Journal of Hazardous Materials* 186, nos 2–3: 1594–600. doi.org/10.1016/j.jhazmat.2010.12.036.

China National Offshore Oil Corporation (CNOOC) Energy Economics Institute (CNEEI). 2024. *Energy Outlook 2060.* Beijing: CNOOC Energy Economics Institute.

Duan, Hongbo, Sheng Zhou, Kejun Jiang, Christoph Bertram, Mathijs Harmsen, Elmar Kriegler, Detlef P. van Vuuren, Shouyang Wang, Shinichiro Fujimori, Massimo Tavoni, Xi Ming, Kimon Keramidas, Gokul Iyer, and James Edmonds. 2021. 'Assessing China's Efforts to Pursue the 1.5°C Warming Limit.' *Science* 372, no. 6540: 378–85. doi.org/10.1126/science.aba8767.

Energy Research Institute (ERI). 2024. *China Energy Transformation Outlook 2024.* Beijing: Energy Research Institute of the Chinese Academy of Macroeconomic Research.

Forbes-Cable, Malcolm. 2023. *Top of the Charts: Five Low-Carbon Tech Trends Worth Tracking.* Online post, December. Edinburgh: Wood Mackenzie. www.woodmac.com/horizons/five-low-carbon-tech-trends/.

Global Burden of Disease (GBD) 2019 Risk Factors Collaborators. 2020. 'Global Burden of 87 Risk Factors in 204 Countries and Territories, 1990–2019: A Systematic Analysis for the Global Burden of Disease Study 2019.' *The Lancet* 396, no. 10258: 1223–49. doi.org/10.1016/s0140-6736(20)30752-2.

He, Jiankun, Zheng Li, Xiliang Zhang, Hailin Wang, Wenjuan Dong, Ershun Du, Shiyan Chang, Xunmin Ou, Siyue Guo, Zhiyu Tian, Alun Gu, Fei Teng, Bin Hu, Xiu Yang, Siyuan Chen, Mingtao Yao, Zhiyi Yuan, Li Zhou, Xiaofan Zhao, Ying Li, and Danwei Zhang. 2022. 'Towards Carbon Neutrality: A Study on China's Long-Term Low-Carbon Transition Pathways and Strategies.' *Environmental Science and Ecotechnology* 9: 100134. doi.org/10.1016/j.ese.2021.100134.

Hepburn, Cameron, Ye Qi, Nicholas Stern, Bob Ward, Chunping Xie, and Dimitri Zenghelis. 2021. 'Towards Carbon Neutrality and China's 14th Five-Year Plan: Clean Energy Transition, Sustainable Urban Development, and Investment Priorities.' *Environmental Science and Ecotechnology* 8: 100130. doi.org/10.1016/j.ese.2021.100130.

Intergovernmental Panel on Climate Change (IPCC). 2019. *2019 Refinement to the 2006 IPCC Guidelines for National Greenhouse Gas Inventory.* Geneva: IPCC.

Intergovernmental Panel on Climate Change (IPCC). 2023. *Climate Change 2023: Synthesis Report.* Geneva: IPCC. doi.org/10.59327/ipcc/ar6-9789291691647.

International Energy Agency (IEA). 2024. *World Energy Outlook 2024.* Paris: International Energy Agency. www.iea.org/reports/world-energy-outlook-2024.

International Renewable Energy Agency (IRENA). 2024. *Renewable Power Generation Costs in 2023.* Abu Dhabi: International Renewable Energy Agency. www.irena.org/Publications/2024/Sep/Renewable-Power-Generation-Costs-in-2023.

Jin, Yana, and Shiqiu Zhang. 2018. 'An Economic Evaluation of the Health Effects of Reducing Fine Particulate Pollution in Chinese Cities.' *Asian Development Review* 35, no. 2: 58–84. doi.org/10.1162/adev_a_00114.

Kilmont, Zbigniew. 2025. *Greenhouse Gas and Air Pollution Interactions and Synergies (GAINS).* Online post. Austria: International Institute for Applied Systems Analysis. iiasa.ac.at/models-tools-data/gains.

Li, Bo-Shu, Yan Chen, Shaohui Zhang, Zheru Wu, Janusz Cofala, and Hancheng Dai. 2020. 'Climate and Health Benefits of Phasing Out Iron & Steel Production Capacity in China: Findings from the IMED Model.' *Climate Change Economics* 11, no. 3: 2041008. doi.org/10.1142/s2010007820410080.

Li, Lili, and Araz Taeihagh. 2020. 'An In-Depth Analysis of the Evolution of the Policy Mix for the Sustainable Energy Transition in China from 1981 to 2020.' *Applied Energy* 263: 114611. doi.org/10.1016/j.apenergy.2020.114611.

Liu, Da, Jinchen Liu, Shoukai Wang, Ming Xu, and Syed Junaid Akbar. 2019. 'Contribution of International Photovoltaic Trade to Global Greenhouse Gas Emission Reduction: The Example of China.' *Resources, Conservation and Recycling* 143: 114–18. doi.org/10.1016/j.resconrec.2018.12.015.

Liu, Zhu, Zhu Deng, Gang He, Hailin Wang, Xian Zhang, Jiang Lin, Ye Qi, and Xi Liang. 2021. 'Challenges and Opportunities for Carbon Neutrality in China.' *Nature Reviews Earth & Environment* 3: 141–55. doi.org/10.1038/s43017-021-00244-x.

Montzka, S.A., E.J. Dlugokencky, and J.H. Butler. 2011. 'Non-CO_2 Greenhouse Gases and Climate Change.' *Nature* 476: 43–50. doi.org/10.1038/nature10322.

Pan, Xiongfeng, Shucen Guo, Haitao Xu, Mengyuan Tian, Xianyou Pan, and Junhui Chu. 2022. 'China's Carbon Intensity Factor Decomposition and Carbon Emission Decoupling Analysis.' *Energy* 239, Part C: 122175. doi.org/10.1016/j.energy.2021.122175.

Pope, C. Arden, III, Richard T. Burnett, Michael J. Thun, Eugenia E. Calle, Daniel Krewski, Kazuhiko Ito, and George D. Thurston. 2002. 'Lung Cancer, Cardiopulmonary Mortality, and Long-Term Exposure to Fine Particulate Air Pollution.' *JAMA* 287, no. 9: 1132–41. doi.org/10.1001/jama.287.9.1132.

Ricke, Katharine, Laurent Drouet, Ken Caldeira, and Massimo Tavoni. 2018. 'Country-Level Social Cost of Carbon.' *Nature Climate Change* 8: 895–900. doi.org/10.1038/s41558-018-0282-y.

Rogelj, Joeri, and Robin D. Lamboll. 2024. 'Substantial Reductions in Non-CO_2 Greenhouse Gas Emissions Reductions Implied by IPCC Estimates of the Remaining Carbon Budget.' *Communications Earth & Environment* 5: 35. doi.org/10.1038/s43247-023-01168-8.

United Nations Department of Economic and Social Affairs, Population Division (UNDESA). 2024. *World Population Prospects 2024: Summary of Results*. UN DESA/POP/2024/TR/No. 9. New York: UNDESA. desapublications.un.org/publications/world-population-prospects-2024-summary-results.

United Nations Human Settlements Programme (UN-Habitat). 2024. *World Cities Report 2024: Cities and Climate Action*. Nairobi: United Nations Human Settlements Programme. unhabitat.org/wcr/.

Wang, Huaibin. 2023. 'Life Cycle Carbon Footprint and Carbon Emission Reduction Potential of Photovoltaic Power Generation.' *Energy of China* 45, no. 8: 34–44.

Wang, Mengya, Yingying Liu, Sumei Li, Parham Azimi, Sha Chen, and Steve Hung Lam Yim. 2024. 'Air Quality and Health Benefits of Achieving Carbon-Neutrality in Building Sector over Beijing, China.' *Journal of Environmental Management* 370: 122652. doi.org/10.1016/j.jenvman.2024.122652.

Wang, Tianpeng, Fei Teng, Xu Deng, and Jun Xie. 2022. 'Climate Module Disparities Explain Inconsistent Estimates of the Social Cost of Carbon in Integrated Assessment Models.' *One Earth* 5, no. 7: 767–78. doi.org/10.1016/j.oneear.2022.06.005.

Xu, Meng, Zhongfeng Qin, and Shaohui Zhang. 2021. 'Carbon Dioxide Mitigation Co-Effect Analysis of Clean Air Policies: Lessons and Perspectives in China's Beijing–Tianjin–Hebei Region.' *Environmental Research Letters* 16, no. 1: 015006. doi.org/10.1088/1748-9326/abd215.

Yang, Dong, J. Liu, Jianxin Yang, and Ding Ning. 2015. 'Carbon Footprint of Wind Turbine By Life Cycle Assessment.' *Huanjing Kexue Xuebao/Acta Scientiae Circumstantiae* 35, no. 3: 927–34. www.researchgate.net/publication/281961611_Carbon_footprint_of_wind_turbine_by_life_cycle_assessment.

Zhang, Shu, Wenying Chen, Qiang Zhang, Volker Krey, Edward Byers, Peter Rafaj, Binh Nguyen, Muhammad Awais, and Keywan Riahi. 2024. 'Targeting Net-Zero Emissions While Advancing Other Sustainable Development Goals in China.' *Nature Sustainability* 7: 1107–19. doi.org/10.1038/s41893-024-01400-z.

Zheng, Heran, Yangchun Bai, Wendong Wei, Jing Meng, Zhengkai Zhang, Malin Song, and Dabo Guan. 2021. 'Chinese Provincial Multi-Regional Input–Output Database for 2012, 2015, and 2017.' *Scientific Data* 8: 244. doi.org/10.1038/s41597-021-01023-5.

3

Striding over the next hurdle in China's transition to renewable electricity

Haitao Yin, Boyu Liu and Feng Wang

China is determined to achieve its carbon peak in 2030 and carbon neutrality by 2060. As energy consumption is the predominant source of carbon emissions, this ambitious goal means China must make a fundamental transition—from a fossil fuel–dominated system to one dominated by renewal energy. In January 2024, at the eleventh collective study session of the twentieth Political Bureau of the CPC, President Xi Jinping stated that 'green development is a defining feature of high-quality development, and new-quality productive forces are essentially green productive forces' (Xinhua 2024a). He emphasised the necessity to accelerate the transition to a green development model to support the goals of carbon peak and carbon neutrality, highlighting the importance of strengthening the green manufacturing industry, developing green services and expanding the green energy sector (Xinhua 2024a).

Following his important statement, in February 2024, the twelfth collective study session of the twentieth Political Bureau delved into the issue of electricity system restructuring, which is required to support China's green development. President Xi stressed:

> [E]nergy security is crucial to the overall development of the economy and society. Actively developing clean energy and promoting the green and low-carbon transformation of the economy and society have become a global consensus in addressing climate change. We must seize the momentum and redouble our efforts to drive the high-quality development of China's new energy sector, provide a reliable energy guarantee for China's modernisation and make greater contributions to building a clean and beautiful world. (Xinhua 2024b)

Electricity is the cornerstone of modern development. China's electricity system must undergo a profound transformation to clean and low-carbon sources while ensuring safety and efficiency. The challenge has been much discussed. Traditionally, China's electricity system is highly centralised, with grid operators carefully balancing supply and demand, which is possible because these can be reasonably predicted and therefore planned for. The penetration of renewable energy makes this quite challenging as electricity supply becomes much more volatile because of the instability of wind and solar power generation. This demands stronger capabilities to manage supply-side volatility and, at the same time, greater flexibility on the demand side, as well as enhanced resilience and efficiency in electricity distribution. To achieve this, a diversified energy supply structure, an efficient electricity market and the support of digital and intelligent technologies are indispensable.

This trend, while posing challenges, also provides great opportunities for innovation. First, on the supply side, breakthroughs in new energy and storage technologies are imperative to enhance the flexibility and resilience of the energy supply. Second, energy demand must become softer and more flexible. For this purpose, digital and intelligent technologies are required to improve demand-side management. Future intelligent energy management, enabled by advanced information technologies, focusing on data-based optimisation, scheduling and dispatch, rather than digital displays of energy consumption data, will become a core driver and safeguard of the future energy system. Third, an efficient and open electricity market is the core bridge to balance energy supply and consumption, while also providing additional business cases for investment in renewable energy and energy storage. Only through coordination of supply, configuration and demand can a clean and reliable energy system be sustainable. This will further facilitate China's transition to a low-carbon society and economy—driving the formation of new productive forces and fostering high-quality development.

The rest of this chapter proceeds as follows. The next section will discuss the main hurdles to China's transition to a renewable electricity system and therefore the key tasks on which it must work in the next five to 10 years. Section three provides an overview of the strategies that have emerged to help China stride over these hurdles, while section four concludes with a summary.

The key challenges in China's transition to green electricity

A defining feature of the new electricity system is the gradual replacement of fossil fuels with non-fossil energy sources, which is essential for China to achieve its carbon goals. Zhang et al. (2022) and the International Energy Agency (IEA) (Lim and Hart 2021) investigated the electricity structure change required to ensure carbon neutrality by 2060. They agree that wind and solar power must provide 60–70 per cent of China's total electricity generation by 2060—a shift from an electricity structure dominated by thermal power (nearly 70 per cent in 2020). Zhang et al. (2022) also estimated that total electricity production would increase from 7,511 terawatt hours (TWh) in 2020 to 15,100 TWh in 2060. Consequently, by 2060, the combined generation from wind and solar power must reach 8,700 TWh—far exceeding today's current total electricity consumption. This grand transition must occur within 40 years.

This transition is further fuelled by the need to ensure national energy security. China has few reserves and little production of fossil fuels. This, and the fact that China's economy is heavily reliant on fossil fuels for its energy needs, leads to an increasing dependence on international energy markets. In 2023, China's crude oil consumption was 772 million tonnes, with imports accounting for 564 million tonnes, resulting in an external dependency rate of 73 per cent (NBS 2023, 2024; Fan et al. 2024). In 2021, President Xi emphasised that 'as a major manufacturing country, China must hold its own energy bowl firmly in its own hands' (Bai and Zhao 2024). Hence, fully utilising China's abundant wind and solar resources to replace traditional fossil fuels is vital for enhancing China's national energy security. Therefore, the transition to low carbon is necessary not only for sustainable development but also for ensuring energy security.

With strong political determination and policy support, China's new-energy industry has become globally significant, with its wind and photovoltaic (PV) industries rapidly progressing from 'followers' to 'leaders'. China now holds global leadership in new-energy technology and equipment manufacturing, having established the world's largest and most comprehensive new-energy industrial chain and occupying a dominant position in the global market. According to *China's Energy Transition White Paper* (NEA 2024a), by the end of 2023, the combined grid-connected installed capacity of wind and PV power had reached 1.05 terawatts (Figure 3.1), accounting for 36 per cent of the national total installed capacity. By the end of November 2024, China's cumulative power generation capacity had reached 3.23 terawatts—a 14.4 per cent increase year-on-year, with photovoltaic and wind power capacities reaching 820 gigawatts and 490 gigawatts and year-on-year growth rates of 46.7 per cent and 19.2 per cent, respectively (NEA 2024a). China's global leadership in PV and wind industries is rooted in its continuous innovations in new-energy technology, from the introduction of advanced foreign technologies to independent innovation. China's remarkable achievements in new-energy technological innovation and industrial expansion have provided a strong impetus for the global green and low-carbon energy transition.

As a result of continued technological innovation and expansion of production capacity, the cost of wind and solar electricity generation has quickly declined. For instance, in 2024, the levelised cost of electricity generation from solar power was only RMB0.15–0.25 per kWh in northern China—a 90 per cent drop from 2010, and already cheaper than coal-fired electricity generation, which often ranges from RMB0.3 to RMB0.4 per kWh (Soochow Securities Research Institute 2024). As a result, the key hurdles to the transition to renewable electricity generation are no longer the manufacturing and operation of solar panels.

Installed Capacity / Gigawatts

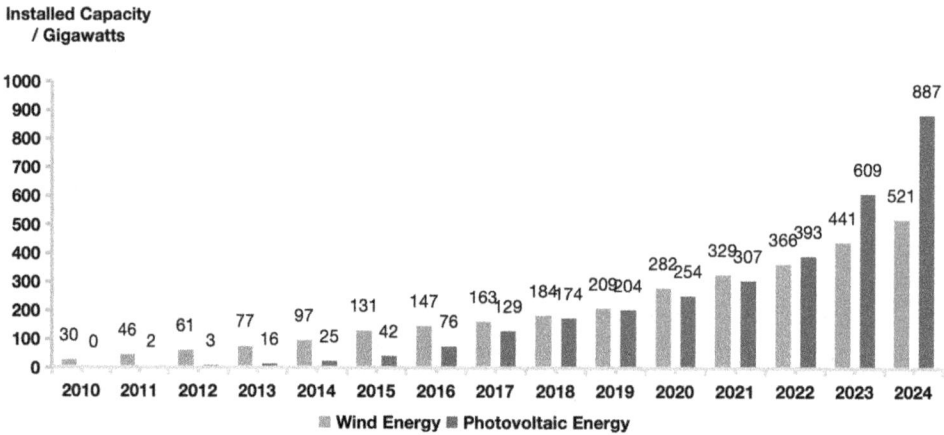

Figure 3.1: Installed capacity of wind and photovoltaic energy in China, 2010–24
Source: *China Statistical Yearbooks* (2010–24) (data.stats.gov.cn/english/).

The key hurdles for the future will be the integration of renewable electricity into the existing energy system. In China, the installed capacity of wind and PV power is growing at an unprecedented rate, placing significant stress on the grid. The inherent intermittency and volatility of renewable sources such as wind and solar exacerbate supply instability and regulatory pressure on the grid. For instance, wind power generation peaks at night when low-level airflow is stable and wind speeds are relatively high, while PV generation peaks at midday during summer, and therefore is often misaligned with the periods of peak electricity demand, leading to curtailment of wind and solar power. In 2023, provinces with substantial PV capacity, such as Henan and Shandong, frequently restricted PV grid connection during sunny periods, highlighting the absorption bottleneck. According to a report on the national new-energy grid connection and consumption in February 2024, the national PV utilisation rate dropped to 93.4 per cent—the first time it had fallen below 95 per cent (NEA 2024f). Data from the National New Energy Consumption Monitoring and Early Warning Centre (2024) show that, in the first four months of 2024, regions such as Gansu, Qinghai, Ningxia, Xinjiang and Tibet repeatedly saw renewable energy utilisation rates falling below 95 per cent, with PV utilisation rates in these regions dropping below 90 per cent in October. Notably, the PV utilisation rate in Tibet was only 71.8 per cent from January to April, further declining, to 65.5 per cent, in October—significantly lower than the national average of 90 per cent (National New Energy Consumption Monitoring and Early Warning Centre 2024). These figures indicate that the grid's capacity to integrate renewable electricity is approaching its limit. As installed renewable generation capacity further increases, the power system must manage significantly larger generation fluctuations, increasing difficulty in balancing the system and risks to safe and stable operation (Guo 2023).

Under current circumstances, the primary challenge for the energy transition is no longer the production and installation of new-energy equipment, but the mismatch between the rapid expansion of new-energy generation capacity and the limited capacity of the grid to integrate it. If the mismatch persists or intensifies, further development of renewable generation capacity will be futile. This challenge was underscored in the CPC Central Committee's decision on 'Further Comprehensively Deepening Reform and Promoting Chinese-Style Modernisation' (Xinhua 2024c), which emphasised the urgency of planning and building a new energy system that facilitates the utilisation of renewable electricity, and improving policies and measures for renewable energy consumption and regulation.

The Chinese Government has implemented several policy measures to ensure the efficient and sustainable development of the new-energy system, maintain safe and stable operations and enhance utilisation capacity. On 28 May 2024, the National Energy Administration of China (NEA 2024d) issued a notice that highlighted that addressing consumption is not only a critical link in the planning and construction of a new-energy system but also of great significance for increasing the proportion of new energy in consumption and achieving the 'dual carbon' goals. The challenge has become how to efficiently utilise and integrate renewable energy into the existing system. In response, the NEA outlined several measures, including accelerating the construction of new-energy–supporting grid projects, actively promoting system flexibility to enhance the coordinated development of load–network–source integration and leveraging the role of the power grid as a resource allocation platform. These measures aim to enhance the power system's capacity to consume renewable energy, ensuring its effective integration and optimal distribution (NEA 2024d).

On 6 January 2025, the National Development and Reform Commission (NDRC) released the 'Special Action Plan to Optimise Power System Regulation Capacity'. The plan set a goal for the period 2025–27 of the grid supporting an average annual increase of more than 200 gigawatts in new-energy consumption through capacity construction and optimisation, and ensuring a national new-energy utilisation rate of at least 90 per cent. This is the first time the state has set a clear target for the annual utilisation rate for renewable energy consumption (NDRC 2025). Considering the newly installed renewable generation capacity reached 373 gigawatts in 2024 and a 95 per cent utilisation rate, this action plan reveals a more prudent attitude to renewable electricity development, in better alignment with the grid's capacity to accommodate it.

Strategies to enhance the utilisation of renewable energy

As the previous section revealed, the greatest hurdle for China's energy transition is integrating renewables into the existing supply and consumption network. For this purpose, the coordinated efforts of the supply side, the demand side and the distribution network are essential. Energy system reform is a comprehensive and integrated project.

It necessitates optimising the energy configuration side by restructuring the energy mix, while enhancing the adaptability of both the demand and the supply sides to develop a system that is cleaner, safer, more efficient and more resilient.

Supply side: Reducing the volatility of renewable electricity supply

As previously noted, on the supply side, China's energy structure is undergoing a profound transformation. In recent years, great leaps have been made in the installed capacity of renewable energy. In 2024, China's newly installed renewable energy capacity reached 373 gigawatts, representing a year-on-year increase of 23 per cent from 2023 and accounting for 86 per cent of the country's total newly installed power capacity. The composition of the newly installed capacity was as follows: hydropower, 13.78 gigawatts; wind power, 79.82 gigawatts; solar power, 278 gigawatts; and biomass-generated power, 1.85 gigawatts (NEA 2024c).

However, despite the continued expansion of renewable energy installations, the intermittent and volatile nature of wind and solar generation has led to output instability, making it difficult to closely align with demand in the power system and thereby posing greater challenges for grid integration and consumption. To improve the controllability of electricity supply, the importance of energy storage systems is becoming increasingly evident. These can store excess energy during periods of surplus renewable generation and release it during shortages, effectively smoothing the output curve of renewable energy and significantly enhancing its scheduling and reliability. Thus, energy storage technology, as a core solution to this challenge, has garnered widespread attention from both industry and society.

Currently, the development of such technologies is primarily concentrated in the field of electrochemical energy storage. Among these, lithium-ion batteries, with their mature technology and broad range of applications, have emerged as the leading solution. By the end of 2023, lithium-ion battery energy storage constituted approximately 97.4 per cent of the newly installed energy storage capacity in China (NEA 2024b). However, there are several challenges to this type of storage. First, safety remains a critical issue for lithium-ion batteries. Under abnormal conditions such as overcharging or short-circuiting, lithium-ion batteries may experience thermal runaway, potentially leading to fires or even explosions. These incidents not only threaten equipment safety but also pose significant risks to the energy grid. Second, the rapid growth of electric vehicles and energy storage has increased global demand for lithium, cobalt and other heavy metals, exacerbating resource shortages and driving up costs. Finally, the issue of recycling decommissioned batteries is becoming increasingly urgent. The rapid advancement of new-energy vehicle technology has accelerated model turnover, and battery performance degrades with repeated charging. If retired batteries are not properly managed, they can cause serious environmental pollution. Currently, there are two main approaches for recycling decommissioned batteries: cascade utilisation and raw material recovery. Cascade utilisation involves disassembling and reassembling decommissioned batteries for use in energy storage applications with lower energy density requirements, while raw material recovery focuses on extracting valuable metals such as cobalt, nickel and lithium from scrapped batteries for recycling.

Electrochemical technology is more applicable to short-term energy storage. If the volatility of electricity generation must be smoothed out over months or seasons, or across regions, hydrogen storage seems to be more appealing. It enables energy storage and release through the production, storage and use of hydrogen. Electrolysis is used to convert surplus or low-cost renewable electricity into hydrogen, which can then be transformed back into electricity or heat via fuel cells when needed or when demand and electricity prices are high. This process alleviates supply-side pressures associated with intermittent renewable energy. A notable example is the power-to-gas project in Prenzlau, Germany, where excess wind power is electrolysed into hydrogen for storage, power generation, heating and powering fuel cell vehicles.

The development of hydrogen energy storage technology has created a new pathway for the broad application of clean energy, enabling renewables to be utilised in various sectors. This not only accelerates the transition from traditional fossil fuels to clean energy but also offers several core advantages. Hydrogen energy storage is suitable for both short-term and long-term storage, has minimal dependence on external environmental conditions and allows for flexible site selection with a small environmental impact. Compared with conventional technologies such as pumped storage and compressed air storage, hydrogen is more competitive and can meet the needs of large-scale, long-duration storage, making it particularly suitable for energy storage that spans seasons. It also supports a variety of flexible storage and transportation methods, such as pipeline transport and blending with natural gas, and can be closely integrated with clean energy generation systems to form a complementary energy supply network. By storing hydrogen during periods of excess renewable energy generation and generating electricity via fuel cells during peak demand, hydrogen storage significantly enhances the flexibility and stability of the energy system.

However, hydrogen storage has its own significant challenges. First, the high costs of large-scale electrolysis make its widespread adoption difficult. Meanwhile, hydrogen's low density, leakage risks and flammability complicate storage and transportation, with current methods including high-pressure gas, liquid hydrogen and solid-state storage each having limitations. Moreover, hydrogen storage currently exhibits lower energy conversion efficiency and higher costs than other storage technologies, posing barriers to commercial viability.

To address these challenges, the Chinese Government has implemented several supportive policies. The NDRC and the NEA released a plan for the development of a hydrogen energy industry in March 2022, underscoring hydrogen's role as a crucial component of the future national energy system. The plan proposes the establishment of a supply system based on hydrogen as an industrial by-product and locally utilised renewable hydrogen production during the Fourteenth Five-Year Plan period (NDRC

2022). This provides policy support and a clear direction for the development of hydrogen energy storage technology. On 31 December 2024, the General Office of the Ministry of Industry and Information Technology, the General Office of the NDRC and the General Department of the NEA issued 'The Implementation Plan for Accelerating the Application of Clean and Low-Carbon Hydrogen in the Industrial Field' (State Council 2024). This plan highlights that accelerating the application of clean hydrogen, such as industrial by-product hydrogen and hydrogen produced from renewable electricity, is a vital for promoting high-quality development and productivity in the hydrogen energy industry. It is also a core pathway for promoting energy conservation, carbon emission reduction and new industrialisation. These policy initiatives have provided robust support and clear guidance for the advancement of hydrogen energy storage technology.

The propeller for hydrogen energy development is twofold. First, China needs a large-scale energy storage approach to smooth the seasonal fluctuations in wind and solar power. Second, it needs a large-scale energy storage approach that can turn otherwise wasted wind and solar power into useable resources. The momentum for hydrogen development will not dwindle unless other technologies that can meet these two needs emerge and mature. Of course, the challenges with hydrogen energy must be addressed. The US Department of Energy (DOE 2022) published its *National Clean Hydrogen Strategy and Roadmap* in 2022, setting a goal of reducing the production cost of renewable hydrogen to US\$1 per kilogram. In 2024, China had already witnessed a sharp decline in this cost.

Meanwhile, the challenges associated with hydrogen storage and transportation have spurred the development of hydrogen-based fuel synthesis technology, which involves combining hydrogen with carbon dioxide captured from the atmosphere to produce methanol, ammonia and ether-based fuels. This not only addresses hydrogen storage and transportation issues but also enables carbon recycling, offering substantial environmental and economic benefits. On 9 September 2024, the NEA Party Leadership Group published a signed article in the *People's Daily* that proposed satisfying terminal energy consumption with diversified clean alternatives, including electricity, hydrogen and ammonia (People's Daily 2024). Previously, only electricity had been mentioned. The addition of hydrogen and ammonia was new. This statement not only affirms the development trend towards hydrogen-based fuel synthesis but also outlines the direction for future energy transformation.

Hydrogen-based fuel synthesis has made notable progress. For example, the Carbfix project in Iceland has successfully produced methanol by capturing atmospheric carbon dioxide and combining it with hydrogen generated through water electrolysis (using geothermal power) for local transportation and industrial use. Additionally, Australia's Fortescue Future Industries is developing several green ammonia projects, which plan to use renewable energy to produce hydrogen and combine it with atmospheric nitrogen to synthesise ammonia. 'The Implementation Plan for Accelerating the Application of Clean and Low-Carbon Hydrogen in the Industrial Field' (State Council 2024) states that, by 2027, significant progress will be made developing the equipment and technology for clean and low-carbon hydrogen in industries such as metallurgy, ammonia synthesis, methanol synthesis and refining. Demonstrations will be carried out of industrial green microgrids, shipbuilding, aviation and rail transit, forming several commercial application models for hydrogen energy in transportation, power generation and energy storage. These advancements demonstrate that hydrogen-based fuel synthesis is not only theoretically feasible but also holds prospects for broad practical application.

Energy storage and hydrogen-based fuel synthesis technologies (also as an approach for energy storage) are key to promoting the construction of a clean energy structure on the supply side, consuming excess renewable electricity that would otherwise be wasted and fostering sustainable development. Energy storage technology effectively mitigates the intermittency and volatility of renewable energy generation, enhancing supply stability and reliability. Hydrogen-based fuel synthesis addresses hydrogen storage and transportation challenges while enabling carbon recycling, offering significant environmental benefits. With continuous technological progress and policy support, clean energy industries such as energy storage and hydrogen energy will provide solid support for the energy transition, accelerating the shift from traditional fossil fuels to clean energy.

Demand side: Smart energy construction

The demand side plays a crucial role in improving the utilisation of renewable electricity and therefore increasing the penetration of renewables into China's power mix. Energy-consuming enterprises, through energy-saving measures and renewable energy substitutions, can not only effectively reduce their reliance on high-polluting energy sources, but also, more importantly, alleviate the burden that renewables' intermittency places on the grid, and even improve the grid's capacity to accommodate the penetration of renewables through demand-side management. Among these initiatives, the development and application of microgrid technology have emerged as vital means of enhancing renewable energy consumption on the demand side. Demand-side energy transformation, supported by digital and intelligent technologies, not only optimises energy efficiency but also provides robust support for the safe and stable operation of the power system. For instance, when renewable energy generation on the supply side is insufficient due to weather, the demand side can reduce electricity load through flexible responses and even convert energy storage resources (including electric vehicles) into power suppliers. Demand-side smart energy systems are a core strategy to enhance the utilisation of renewable electricity, with four main directions warranting further exploration.

Intelligent microgrids

The development of microgrids, underpinned by digitalisation and intelligent technologies, will be a cornerstone of smart energy solutions on the demand side. As defined by the NDRC and the NEA, a microgrid is a localised power distribution system with energy storage and monitoring and protection capabilities. It primarily utilises locally generated renewable energy to directly serve end users (NDRC 2021). On the demand side, microgrids can dynamically meet energy supply needs and optimise energy usage through advanced big data analytics and intelligent management. This enhances local voltage stability, reduces energy consumption and mitigates supply shortages. For the broader power grid, microgrids function as agile power units, capable of rapidly addressing internal and external transmission requirements. The exchange of energy between microgrids and the main grid significantly improves the overall stability of the power system. As more enterprises integrate distributed energy resources and transition towards combined energy production and consumption models, microgrids that incorporate local renewables, storage, load equipment and energy management are becoming essential infrastructure. Under the new power system architecture, local electricity needs will be met locally and the reduced power exchange between microgrids and the national grid will ease the burden of renewable electricity integration.

China's microgrid demonstration projects—such as the Xiapu Microgrid Demonstration Project in Ningde, Fujian Province, and the Cixi hydrogen-electric coupled DC microgrid project in Ningbo, Zhejiang Province—have been instrumental in driving the high-quality development of renewable energy and enhancing the power system's capacity to absorb new energy in these areas. Thus, the development and operation of microgrids are critical for renewable energy consumption. On 9 October 2024, the NEA released for public comment the 'Management Measures for the Development and Construction of Distributed Photovoltaic Power Generation'. The draft stipulated that 'the electricity generated from large-scale industrial and commercial distributed solar projects must be consumed locally and are not allowed to be fed into the grid' (NEA 2024e). Under this regulation, large-scale industrial and commercial distributed PV projects exceeding 6 megawatts will no longer be able to operate under the traditional model in which surplus electricity is fed into the grid. The '6-megawatt' threshold is expected to be lowered in the future. This sends a signal that future industrial and commercial distributed PV projects must operate self-sufficiently, precluding the option of feeding surplus electricity into the grid. While this will help to mitigate the impact of distributed generation on the power grid and is a significant measure to address renewable energy consumption challenges, it also means that project owners must establish an intelligent operating system that covers PV power generation capacity, storage, load adjustment and the charging/discharging of electric vehicles to ensure system stability and cost optimisation. In essence, this will drive the development of demand-side microgrids, which face numerous management issues (Xu et al. 2021).

From a technical perspective, intelligent microgrids must be supported by advanced information, communication and control technologies. In its opinions on 'Accelerating the Development of Digital and Intelligent Energy', the NEA clarified the development direction for microgrids. It proposed using digital and intelligent technologies to accelerate the clean energy transition, expedite the development of new-energy microgrids and high-reliability digital distribution systems and enhance the configuration and operation of demand-side distributed power supplies and new-energy storage resources (Energy Bureau 2023). This indicates that the government is placing significant emphasis on the intelligent and digital development of microgrids, aiming to achieve modernisation and upgrading of the energy industry through technological progress.

Microgrids, state grid ancillary services and virtual power plants

The significance of smart microgrids could go beyond realising local self-sufficiency in electricity needs and reducing dependence on the central grid. Leveraging agile operational control and energy management capabilities, smart microgrids can rapidly respond to supply fluctuations and emergencies on the energy production side, ensuring continuous and reliable power supply and improving grid capacity that integrates renewables.

In essence, as a critical component of the energy transition, microgrids can significantly bolster the grid's ancillary service capabilities through demand-side management, including frequency regulation and peak–valley load balancing, thereby enhancing grid stability and reliability. In demand-side management, microgrids can utilise technologies such as artificial intelligence (AI) to optimise the allocation and consumption of power and improve energy efficiency. This includes the possibility of maintaining supply–demand balance through the precise control of demand-side electricity needs. Based on the principle of minimising net load, the power system can align the load curve with the renewable energy generation curve through demand-response technology, utilising load and energy storage resources to consume excess renewable energy when it is abundant and reduce electricity consumption when renewable generation is insufficient, or even supply electricity back to the grid. This interaction between microgrids and distribution networks will support the operational optimisation of the power system and play a vital role in enhancing renewable energy utilisation capacity, the energy structure and energy efficiency.

As a distributed energy aggregator, microgrids can become a core component of virtual power plants in the future, participating in grid scheduling and enhancing the grid's renewable energy utilisation capacity. With this consideration, the *Fourteenth Five-Year Plan for Modern Energy System* (State Council 2022) underscores the importance of actively developing smart microgrids, highlighting their role in innovations in grid structure and operation to advance the construction of new power systems. The plan emphasises the core role of microgrids in diversifying and evolving energy systems, particularly in improving renewable energy efficiency and reducing costs. The NDRC and the NEA published a guide on 'Accelerating the Development of Virtual Power

Plants', which sets a goal for the dispatch capacity of virtual power plants to reach 20 gigawatts by 2027 and 50 gigawatts by 2030. To enhance the grid's capacity to accommodate unstable renewables generation, the development of virtual power plants has been fast-tracked (NDRC and NEA 2025).

Participation in grid ancillary services and virtual power plants necessitates robust digital and intelligent technology capabilities. In its 'Opinions on Accelerating the Development of Digital and Intelligent Energy', the NEA emphasised the pivotal role of these technologies in upgrading the energy industry's foundations and modernising the industrial chain (Energy Bureau 2023). Through the application of digital and intelligent technologies, microgrids can more effectively participate in grid ancillary services and virtual power plant development, enhancing overall grid performance and efficiency and providing solid support for the development of a clean, low-carbon, safe and efficient energy system.

Microgrids engaging in the power spot market through demand-side optimisation

In 2023, the NDRC and the NEA issued a notice on 'Accelerating the Construction of the Power Spot Market', which required regions with a relatively high installed capacity of distributed renewable energy to encourage generators to participate in the electricity market. Time-of-use electricity would send market price signals to guide renewable generation and encourage new entities such as energy storage systems, virtual power plants and load aggregators to participate in peak shaving and valley filling, optimising supply quality and exploring innovative models such as 'renewable energy plus energy storage' (NDRC 2023a).

In the electricity spot market, as illustrated in Figure 3.2, electricity prices fluctuate in real time according to market supply and demand. With the increasing proportion of renewable energy, favourable wind and solar conditions can lead to an excess of renewables generation, resulting in negative electricity prices. Conversely, unfavourable wind and solar conditions can lead to insufficient generation, causing electricity prices to soar. If a demand-side entity based on a distributed microgrid has developed renewable energy generation forecasts and flexible load adjustment capabilities, it can capitalise on these price fluctuations by purchasing electricity during negative price periods and selling it during high price periods, thereby achieving arbitrage. This not only diversifies the profit models for distributed energy owners, but, more importantly, also creates market incentives to enhance the utilisation of renewables and support the stable operation of the power grid as the penetration rate of renewables continues to increase.

Nodal price (Yuan/MWh)

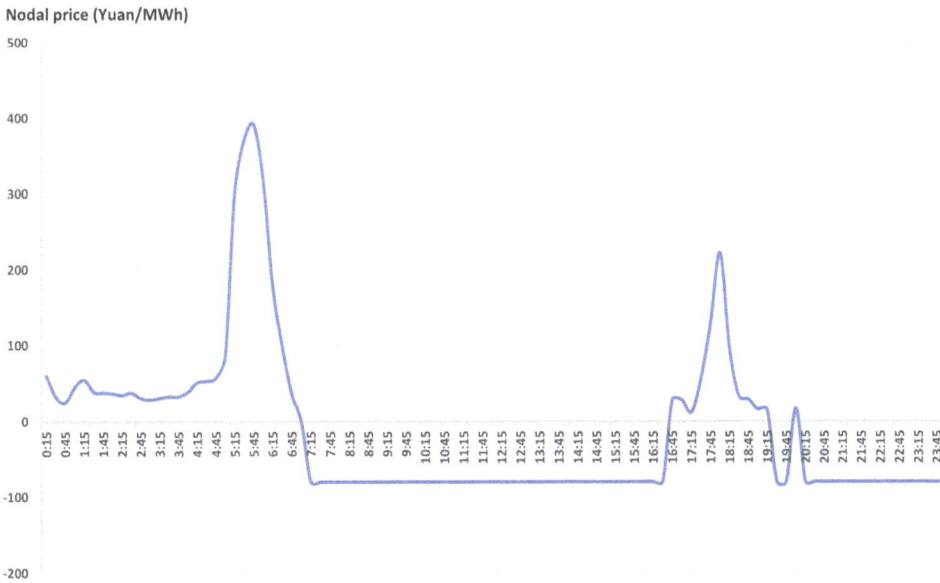

Figure 3.2: Real-time trading price data for Shandong electricity spot market on 1 May 2023

Source: Shandong Electricity Spot Market.

As a platform for demand-side energy optimisation, microgrids can enhance their intelligence levels through digital technology and participate in the power spot market, thereby improving market competitiveness and achieving efficient energy allocation. This model has been successfully implemented in several regions, particularly in the integrated smart energy projects of the State Power Investment Corporation. The demand-side optimisation in these projects has demonstrated the potential for participation in the power spot market. For instance, 12 of the State Power Investment Corporation's smart energy projects were selected for the 2023 'Smart Energy Projects Excellent Project Case Collection' (CCPITEP 2023), including industrial parks and cluster building models. These projects showcase the corporation's demand-side optimisation and participation in the electricity spot market. For example, the integrated smart energy demonstration project in Beijing's Future Science City business park has achieved an 18 per cent self-sufficiency power supply rate for enterprises through the installation of rooftop PV systems, ground-mounted PV arrays, small wind turbines and energy storage systems. This not only improves energy efficiency but also maximises the economic benefits through participation in the electricity spot market. These projects highlight the importance of demand-side optimisation in spot markets and the dual value of microgrids in promoting the energy transition and enhancing economic efficiency. Of course, efficient participation in the electricity spot market by intelligent microgrids, supported by advanced information, communication and control technologies, is also crucial.

Integration of electric vehicle charging infrastructure with the power grid

In recent years, there has been rapid growth in the uptake of electric vehicles (EVs) in China (Figure 3.3). The onboard batteries of EVs, acting as a form of energy storage, can participate in grid operations through intelligent charging and discharging technologies. Based on vehicle-to-grid charging strategies, EV charging networks can enhance renewable energy absorption, effectively coordinate and optimise EV–grid interactions and optimise grid loads. Studies show that, by flexibly regulating EV charging loads, the power system can alter the spatiotemporal distribution of those loads while fully meeting demand. This process can shift some electricity originally dependent on thermal power to renewable energy generation, thereby increasing local renewable energy consumption and reducing overall carbon emissions of the coupled network (Ye et al. 2023). This optimisation not only improves energy efficiency but also contributes positively to carbon emission reduction.

In response to these developments, in December 2023, the NDRC issued its opinions on strengthening the integration of and interactions between new-energy vehicles and the power grid (NDRC 2023b). It described the future development of the charging network, aiming to tap into the regulatory potential of batteries as load control or mobile energy storage. By establishing an information and energy exchange system between new-energy vehicles and the power supply network, the policy provides strong support for the efficient and economical operation of new power systems: charging vehicle batteries when renewable generation is abundant while feeding the battery power to the grid when electricity on the grid is in short supply. According to statistics from the NEA, from January to September 2024, the number of EV charging facilities in China increased by approximately 2.837 million units. By the end of September 2024, the total number of EV charging facilities in China reached 11.433 million units—a year-on-year increase of 49.6 per cent. Among these, 3.329 million were public and 8.104 million were private charging facilities (Dai 2024). This trend indicates the significant potential of the EV charging network to help adjust power loads, helping smooth the fluctuation of renewable generation.

Building on this foundation, some leading charging station enterprises, such as TELD, are using charging networks, microgrids and energy storage networks to build virtual power plant platforms. By aggregating resources such as EV charging, PV microgrids, mobile energy storage and cascade energy storage, these platforms can achieve functions such as frequency modulation, peak regulation, demand-side response, aggregated electricity sales, green electricity consumption and carbon trading. These platforms directly provide ancillary services for the power grid, thereby enhancing the grid's capacity to accommodate renewable energy consumption.

**Electric Vehicle (EV)
Sales Volume
/ 10000 Vehicles**

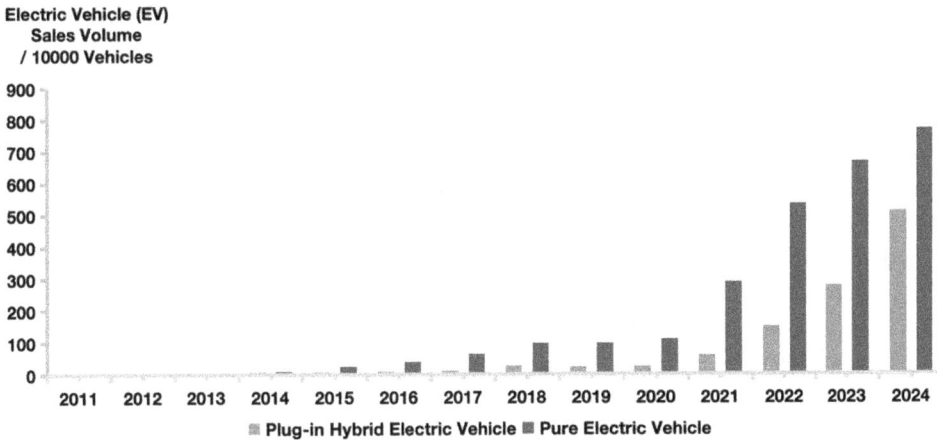

Plug-in Hybrid Electric Vehicle Pure Electric Vehicle

Figure 3.3: Sales volumes of electric vehicles in China, 2011–24
Source: China Association of Automobile Manufacturers (en.caam.org.cn/Index/lists/catid/78.html).

These four components form a comprehensive development strategy that uses digitalisation and intelligent operations of the demand-side energy system to advance the power system's capacity to accommodate renewable electricity. This is instrumental for a cleaner, safer, more efficient and resilient electricity system.

Distribution side: Smart grid development

Energy distribution is the complex and critical task of efficiently and safely transporting electrical energy from the source to the load. Traditionally, grid operators rely on a centralised planning and dispatch system to make sure supply meets demand. This is achievable when electricity demand and supply are predictable; however, it is difficult, if not impossible, in a power system that is predominantly sustained with renewable energy. The reason, again, is the intermittency and fluctuation in solar and wind generation.

As the linkage between power generation and consumption, distribution methods and grid structure play a decisive role in the grid's capacity to accommodate renewable generation. The key challenge is to enhance the flexibility and resilience of the entire energy system through optimal scheduling and intelligent management. There are three major development trends on the distribution side: flexible and adjustable infrastructure, digital and intelligent capability and the development of an open and diversified power market.

The infrastructure component focuses on adjustable and flexible loads

Flexible loads are those that can adjust their use patterns according to system requirements and market signals. For example, due to their high adjustability, heating, ventilation and airconditioning systems and charging piles allow the power grid to flexibly adjust load power without affecting user experience, thereby balancing electricity supply and demand and optimising energy costs. Additionally, the application of flexible load technology can mitigate the peak–valley differences in the power grid and enhance its operational efficiency and reliability. With continuous technological advancements, the application scenarios for flexible loads continue to expand. For instance, charging piles can flexibly adjust charging power based on the state of the grid and optimise charging times. Distributed energy sources, such as small-scale solar and wind power, can generate electricity locally, reducing dependence on the grid. The Chinese Government is actively promoting the development of flexible loads through various policies. On 17 January 2025, the NEA issued new administrative measures to regulate the 'Development and Construction of Distributed Photovoltaic Power Generation', solve grid accommodation issues, protect user rights and foster the sustainable development of the distributed PV industry (NEA 2025).

Energy digitalisation is also a core technological trend on the distribution side

Within the smart grid framework, an optimised scheduling system, built on big data and AI technologies, can accurately forecast power loads and achieve intelligent power scheduling. This system effectively links the supply side with the demand side, ensuring the efficient and stable operation of the power grid. Using real-time collection and analysis of grid operation data, the system can automatically adjust power distribution according to demand and supply, thereby maintaining the stability of the entire grid. Device status monitoring and predictive scheduling functions are also crucial. By monitoring equipment status in real time and predicting operational demand, energy management systems can pre-emptively adjust energy distribution, optimise equipment operation, reduce failures and enhance the reliability of energy supply. An intelligent energy management system can help enterprises monitor their energy use in real time, control costs, improve energy utilisation efficiency and provide strong support for the stable operation of the power grid.

Moreover, the digitalisation of energy also provides vast potential for the development of virtual power plants. As a novel type of management system, virtual power plants achieve coordination and optimisation of the power supply, energy storage, load and charging piles through the Internet of Things and other technologies. They integrate distributed energy resources and use advanced algorithms to optimise energy allocation and scheduling. On 24 September 2024, the NEA's 'Basic Rules for the Registration of the Electricity Market' clarified the regulations for market access for virtual power plants (NEA 2024g). Policy support at both national and local levels provides a clear direction and strong support for the development of virtual power plants as part of smart grids. In the future, with the continuous maturation of technology and policy, virtual power plants are expected to become vital in promoting the energy transition and achieving China's dual carbon goals.

The electricity market is a crucial development on the distribution side

The electricity market functions as an economic system that allocates power resources through market mechanisms. Within this framework, market entities such as power generation companies, users, grid enterprises and trading institutions engage in trading activities for electricity and related products through negotiation, bidding and other means to determine prices and transaction volumes. As the electricity market gradually opens and diversifies, both the supply side and the consumption side can participate more flexibly in a variety of market activities, including medium and long-term power trading, power spot trading, demand-side response ancillary services, green electricity trading and carbon trading. This flexibility not only enhances the overall efficiency of the energy market but also creates more opportunities for the consumption of renewable energy.

On 12 October 2023, the NDRC and the NEA issued their notice on 'Further Accelerating the Construction of the Power Spot Market'. This notice clarified five core tasks: accelerating the construction of new-energy projects supporting the power grid, promoting the system's adjustment capacity and coordinated development of network sources, fully leveraging the role of power grid resource allocation platforms, scientifically optimising new-energy utilisation targets and ensuring the statistical management of new-energy consumption data (NDRC 2023a). On 28 May 2024, the NEA issued a notice on 'Doing a Good Job in the Absorption of New Energy to Ensure the High-Quality Development of New Energy', re-emphasising the urgency of these five tasks and highlighting the pivotal role of the power market in building a new-energy system and addressing the challenge of consumption (NEA 2024d).

These three development strategies are essential to improve energy consumption from the distribution side. Flexible and adjustable infrastructure adapts to the intermittency and volatility of renewable energy generation by creating a flexible grid architecture and equipment. Intelligent digital construction, supported by advanced information technology, achieves precise regulation and management of the energy system. The construction of an open and diversified power market introduces market competition mechanisms to promote the efficient allocation of resources and the consumption of renewable energy. Only through continuous efforts and improvements in these three areas can the energy distribution side effectively link the demand side and the supply side, build a new-energy consumption system and drive the entire energy system towards being clean, efficient, green and safe.

Conclusion

The most significant challenge in the transition to renewable electricity in China is not the production and installation of solar panels or wind turbines, but the mismatch between the rapid growth of renewable energy generation capacity and the limited capacity of the grid to accommodate it. Because of the intermittency and volatility of wind and solar generation, the rapid expansion of capacity has posed critical challenges to the safe operation of the grid. Therefore, future efforts must centre on improving the utilisation of renewable electricity through the supply, distribution and demand sides of the electricity market and delivering on security, economic and environmental values.

From a security perspective, the future energy system could achieve precise load management and risk prediction through a multi-source structure, virtual power plants and smart energy systems, even with intermittent and unstable renewable electricity as the predominant energy source. This will ensure the stability of energy supply and enhance system security. On the supply side, the large-scale adoption of energy storage technology would help address the intermittency and volatility of renewable energy generation, ensuring stable supply. On the distribution side, virtual power plants leverage the Internet of Things and big-data technologies to monitor and optimise the scheduling of distributed energy in real time, quickly identifying and responding to potential risks. On the demand side, the use of big data and algorithms would advance demand-side management: forecasting energy demand and achieving accurate load regulation. Precise load management can help match fluctuating electricity supply. Stored energy can serve as baseload power when renewable generation cannot meet demand, thereby ensuring the stability of energy supply and enhancing the security of the entire energy system.

In terms of economic significance, the renewable energy consumption system enhances energy utilisation efficiency, reduces energy waste and costs through energy complementarity, intelligent scheduling and precise management, and secures economic benefits by engaging in market transactions. On the supply side, the multi-source system integrates diverse energy resources to achieve complementarity, reduce waste of renewable generation capacity and cut supply costs. The virtual power plant on the distribution side boosts energy efficiency and cuts consumption via intelligent scheduling and optimised management. On the demand side, smart energy systems identify and reduce energy waste, improve efficiency and lower costs through meticulous management. Meanwhile, accurate measurement and control empower users to actively participate in renewable energy market trading, secure additional economic benefits through arbitrage and enhance the overall economic performance of the energy system.

From an environmental perspective, the improved utilisation of renewable electricity helps reduce dependence on traditional fossil fuels, lower carbon emissions and combat global climate change. On the supply side, the multi-source structure optimises the energy mix by integrating more renewable energy sources, thereby reducing carbon emissions. The virtual power plant on the distribution side enhances the utilisation of renewable energy through intelligent management and optimal scheduling, reducing the curtailment of wind and solar power. On the demand side, smart energy systems further reduce carbon emissions by optimising energy use and minimising unnecessary consumption. The goal of all the discussed efforts is to ensure renewable electricity, as much as possible, is used to meet energy needs. This system would provide strong support for achieving carbon neutrality and promote the development of a green, low-carbon energy system.

This chapter reviews how the supply, distribution and demand sides of China's energy system could evolve in the future. Developments will centre on how to stride over the key hurdle to China's transition to renewable electricity—that is, making sure that renewable generation capacity can be utilised to the largest extent to meet energy demand. We identify several areas that hold high potential: energy storage, including electrochemical battery storage, hydrogen and hydrogen-based fuels, microgrid development and its roles in providing grid ancillary services; virtual power plants and their role in providing grid ancillary services; development of electricity spot markets; and digital and intelligent infrastructure for precise load management and optimal scheduling of power generation and dispatch. The success of China's transition to renewable electricity hinges on the development of these areas, and technological and managerial breakthroughs in these areas will provide social and economic rewards.

Acknowledgement

Financial supports from the National Social Science Foundation of China (23&ZD095 22VRC180) are greatly appreciated.

References

Bai, Yuejing, and Yu Zhao. 2024. 'Victory Answer Sheet: Looking at China's Oil and Gas Industry to Increase Storage and Produce New Kinetic Energy from Shengli Oilfield.' *China Electric Power News*, 24 May. baijiahao.baidu.com/s?id=17999219114956856635&wfr=spider&for=pc.

China Council for the Promotion of International Trade Electric Power Industry Committee (CCPITEP). 2023. '12 Projects of State Power Investment Corporation Were Selected in the "2023 Comprehensive Smart Energy Excellent Project Case Collection".' Press release, 31 July. www.ccpitep.org.cn/cnt_18184.html.

Dai, Xiaohe. 2024. 'An Increase of 49.6%! The Year-On-Year Growth Rate of the Number of Electric Vehicle Charging Facilities in the First Three Quarters Reflects Three Major Trends.' *Xinhua*, 2 November. www.gov.cn/lianbo/bumen/202411/content_6984504.htm.

Energy Bureau. 2023. *Several Opinions of the National Energy Administration on Accelerating the Development of Digital and Intelligent Energy*. 28 March. Beijing: General Office of the State Council of the People's Republic of China. www.gov.cn/zhengce/zhengceku/2023-04/02/content_5749758.htm.

Fan, Dalei, Wang Zongli, Li Jian, and Wang Yuyan. 2024. 'Analysis and Outlook of Global Oil and Gas Resources Situation in 2023.' *China Mining Magazine* 33, no. 1: 30–37. doi.org/10.12075/j.issn.1004-4051.20240076.

Guo, Jianbo. 2023. 'Understanding the New Power System Evolution Trend.' *China Energy News*, 5 June: 2. paper.people.com.cn/zgnyb/html/2023-06/05/content_25996391.htm.

Lim, Jinsun, and Craig Hart. 2021. *Climate Resilience: Electricity Security*. Paris: International Energy Agency. iea.blob.core.windows.net/assets/a7d3273e-0d09-4384-8837-29d121b545 ef/ClimateResilience_JinsunLIMandCraigHart.pdf.

National Bureau of Statistics (NBS). 2023. *2023 China Statistical Yearbook*. Beijing: National Bureau of Statistics of China. www.stats.gov.cn/sj/ndsj/2023/indexch.htm.

NBS. 2024. *Energy Production in December 2023*. Report [2024], 17 January. Beijing: National Bureau of Statistics of China. www.stats.gov.cn/sj/zxfb/202401/t20240116_1946618.html.

National Development and Reform Commission (NDRC). 2021. 'Glossary of Terms in the Outline of the 14th Five-Year Plan 64: Intelligent Microgrid.' *National Development Strategy and Planning*. Beijing: NDRC. www.ndrc.gov.cn/fggz/fzzlgh/gjfzgh/202112/t20211224_1309317.html.

NDRC. 2022. 'The National Development and Reform Commission and the National Energy Administration Jointly Issued "The Medium and Long-Term Plan for the Development of Hydrogen Energy Industry (2021–2035)".' Press release, 24 March. Beijing: NDRC. www.gov.cn/xinwen/2022-03/24/content_5680973.htm.

NDRC. 2023a. 'Notice of the General Office of the National Development and Reform Commission and the General Department of the National Energy Administration on Further Accelerating the Construction of the Power Spot Market.' *Development and Reform Commission System Reform* [2023], no. 813 (12 October). Beijing: General Office of the NDRC. www.gov.cn/zhengce/zhengceku/202311/content_6913560.htm.

NDRC. 2023b. 'Implementation Opinions of the National Development and Reform Commission and Other Departments on Strengthening the Integration and Interaction between New Energy Vehicles and the Power Grid.' *Development and Reform Energy* [2023], no. 1721 (13 December). Beijing: NDRC. www.gov.cn/zhengce/zhengceku/202401/content _6924347.htm.

NDRC. 2025. 'Notice on the Issuance of the Implementation Plan for the Special Action to Optimise Power System Regulation Capacity (2025–2027).' *Development and Reform Energy* [2024], no. 1803 (6 January). Beijing: NDRC. www.ndrc.gov.cn/xwdt/tzgg/202501/ t20250106_1395479.html.

NDRC and National Energy Administration (NEA). 2025. 'Guiding Opinions of the National Development and Reform Commission and the National Energy Administration on Accelerating the Development of Virtual Power Plants.' *Development and Reform Energy* [2025], no. 357 (25 March). Beijing: Planning Department, NDRC. www.nea.gov.cn/ 20250411/94a805e55f064493901cab6196e0448d/c.html.

NEA. 2024a. *China's Energy Transition White Paper (Full Text)*. Beijing: Information Office of the State Council of the People's Republic of China. www.nea.gov.cn/2024-08/29/c_ 1310785406.htm.

NEA. 2024b. 'Transcript of the National Energy Administration's Press Conference in the First Quarter of 2024.' 25 January. Beijing: NEA. www.nea.gov.cn/2024-01/25/c_1310762019.htm.

NEA. 2024c. 'The National Energy Administration Held a Press Conference to Release the Operation of Renewable Energy Connected to the Grid in the First Quarter.' Press release, 29 April. Beijing: NEA. www.nea.gov.cn/2024-04/29/c_1212357856.htm.

NEA. 2024d. 'Notice of the National Energy Administration on Doing a Good Job in the Absorption of New Energy to Ensure the High-Quality Development of New Energy.' *National Energy Power Generation* [2024], no. 44 (28 May). Beijing: NEA. www.gov.cn/zhengce/zhengceku/202406/content_6956401.htm.

NEA. 2024e. 'Notice of the Comprehensive Department of the National Energy Administration on Publicly Soliciting Opinions on the "Management Measures for the Development and Construction of Distributed Photovoltaic Power Generation (Draft for Comment)".' 9 October. Beijing: NEA. zfxxgk.nea.gov.cn/2024-10/09/c_1212404143.htm.

NEA. 2024f. 'National New Energy Grid Connection and Consumption Situation in October 2024.' *New Energy Information*, 29 November. Beijing: NEA. www.pvmeng.com/2024/11/29/32442/.

NEA. 2024g. 'Notice of the National Energy Administration on Issuing "The Basic Rules for the Registration of the Electricity Market".' *National Energy Administration Regulation* [2024] No. 76. Beijing: NEA. www.gov.cn/zhengce/zhengceku/202409/content_6976860.htm.

NEA. 2025. 'Notice of the National Energy Administration on Issuing "The Administrative Measures for the Development and Construction of Distributed Photovoltaic Power Generation".' 17 January. Beijing: NEA. www.nea.gov.cn/20250123/112c5b199c5f45dd8e7ac93c9f5e4eaf/c.html.

National New Energy Consumption Monitoring and Early Warning Centre. 2024. 'National New Energy Grid Connection and Consumption Situation in February 2024.' 2 April. mp.weixin.qq.com/s/5DVLaySV1Jq3lyKBFVxUHQ.

People's Daily. 2024. 'Supporting Chinese Modernisation with Energy Transformation Development.' *People's Daily*, 9 September. jl.people.com.cn/n2/2024/0909/c349771-40970463.html.

Soochow Securities Research Institute. 2024. 'Photovoltaic LCOE Offers Strong Cost-Effectiveness and Possible Relaxation of Renewable Absorption Limits—Installation Capacity Expected to Exceed Growth Forecasts.' *Power Equipment Industry Review Report*. Suzhou, China: Soochow Securities. pdf.dfcfw.com/pdf/H3_AP202403111626322177_1.pdf?1710230464000.pdf.

State Council. 2022. 'Notice of the National Development and Reform Commission and the National Energy Administration on Issuing the "14th Five-Year Plan for Modern Energy System".' *NDRC Energy* [2022], no. 210 (29 January). Beijing: Energy Bureau of the NDRC. www.gov.cn/zhengce/zhengceku/2022-03/23/content_5680759.htm.

State Council. 2024. 'Notice of the Ministry of Industry and Information Technology and Two Other Departments on Issuing the "Implementation Plan for Accelerating the Application of Clean and Low-Carbon Hydrogen in the Industrial Field".' Ministry of Industry and Information Technology Joint Letter [2024], no. 499 (30 December). Beijing: State Council of the People's Republic of China. www.gov.cn/zhengce/zhengceku/202412/content_6995692.htm.

US Department of Energy (DOE). 2022. *DOE National Clean Hydrogen Strategy and Roadmap*. Draft, September. Washington, DC: US Department of Energy. www.hydrogen.energy.gov/ docs/hydrogenprogramlibraries/pdfs/clean-hydrogen-strategy-roadmap.pdf?Status=Master.

Xinhua. 2024a. 'Xi Jinping Stressed at the 11th Collective Study Session of the Political Bureau of the CPC Central Committee: Speed Up the Development of New Mass Productivity Promoting High-Quality Development.' *Xinhua*, 1 February. www.gov.cn/yaowen/liebiao/ 202402/content_6929446.htm.

Xinhua. 2024b. 'Xi Jinping Stressed at the 12th Collective Study Session of the Political Bureau of the CPC Central Committee: Vigorously Promote the Development of New Energy with High Quality in Our Country, A Much Larger Contribution to Build Clean, Beautiful World.' *Xinhua*, 1 March. www.gov.cn/yaowen/liebiao/202403/content_6935251.htm.

Xinhua. 2024c. 'Decision of the Central Committee of the Communist Party of China on Further Comprehensively Deepening Reform and Promoting Chinese-Style Modernisation.' *Xinhua*, 21 July. www.gov.cn/zhengce/202407/content_6963770.htm?sid_for_share=80113_2.

Xu, W.D., L. Wei, J. Luo, and H.T. Yin. 2021. 'Energy Revolution Brought About by the Development of Energy Storage Industry and Its Key Management Science Issues.' *Journal of Systems Management* 30, no. 1: 191–97.

Ye, Y.J., Q. Yuan, and Y. Tang. 2023. 'Low Carbon Load Optimization Method for Electric Vehicles in Transportation–Power Grid Coupling Network for Dual Carbon Targets.' *China Electric Power* 56, no. 5: 72–79.

Zhang, X.L., X.D. Huang, Y. Geng, D. Zhang, L.X. Tian, Y. Fan, and W.Y. Chen. 2022. 'Research on the Path and Policy of Energy Economy Transition Under the Goal of Carbon Neutrality.' *Management World* 38, no. 1: 35–51.

4

Mechanisms for inter-industrial collaboration in China's energy transformation

Xiaoli Zhao, Chuyu Sun, Hongjun Zhang and Yarui Deng

Climate change has imposed widespread and profound impacts on ecosystems, human health and socioeconomic systems, becoming one of the most pressing global environmental challenges. Mainstream scientific research has reached a broad consensus that the continuous rise in greenhouse gas (GHG) emissions is the principal driver of global warming (Gu et al. 2009; Lashof and Ahuja 1990; Mohajan 2017). In 2021, Syukuro Manabe and Klaus Hasselmann were awarded the Nobel Prize in Physics for their pioneering work in climate system modelling (von Storch and Heimbach 2022; Navarra 2023). Their research, grounded in physical modelling, quantitatively established the causal relationship between GHG emissions—especially carbon dioxide—and global temperature rise, thereby strengthening the scientific foundation for mitigation-focused climate governance. According to the Intergovernmental Panel on Climate Change (IPCC 2014), 80 per cent of global GHG emissions originate from energy-related activities, including power generation, heating, transportation and industrial processes. Given the dominant role of the energy system in global emissions, its deep structural transformation has become a fundamental pathway to achieving emission reductions. This low-carbon transition requires the coordinated advancement of multiple pathways, including replacing fossil fuels with renewable energy, improving energy system efficiency and optimising end-use energy consumption patterns.

Against this backdrop, China faces a series of complex challenges in its energy transition. In terms of energy structure, coal has long held a dominant position, resulting in carbon emission intensity significantly higher than the global average. At the same time, total energy consumption continues to rise, placing increasing pressure on the scale, dispatch complexity and safe operation of the energy system. China's inter-regional energy supply and demand are seriously mismatched and there are constraints on grid access and load matching. Renewable energy on the supply side is developing rapidly, but stable and large-scale grid connection capacity on the demand side is insufficient, and problems such as wind and solar abandonment continue to exist. By the end of 2024, China's installed capacities of wind and solar power had reached 1.407 terawatts and 886 gigawatts, respectively, accounting for 42.03 per cent of total installed power generation capacity (Ding 2025). However, the average utilisation rates of wind and solar power generation are only 95.3 per cent and 96.8 per cent, respectively (National Renewable Energy Consumption Monitoring and Early Warning Centre 2025). This supply–demand mismatch not only reduces the efficiency of renewable energy utilisation, but also increases the burden on conventional energy sources for peak regulation during the transition period. In September 2021, 21 provinces in China experienced electricity shortages and power restrictions were imposed on residents in the three north-eastern provinces (Peng and Xu 2021). This crisis revealed the lack of an effective mechanism to coordinate the phase-out of fossil fuel energy and the growth of renewables, leading to structural instability risks. Moreover, international experience shows that poorly managed structural adjustment in the energy sector can trigger significant socioeconomic disruptions. From 1985 to 2005, coal industry employment in the United Kingdom fell from 221,000 to 7,000 people, with many affected regions still struggling with structural unemployment, inadequate social support and a lack of alternative job opportunities. In the United States, fossil energy employment dropped by more than 30 per cent after 2008 and, despite government intervention, many regions face persistent economic decline (Sainato 2020). These examples demonstrate that the energy transition is not only a matter of technological substitution, but also a systemic challenge involving multiple sectors and governance levels.

Given these challenges, coordinated planning across power supply and demand sectors is essential to push the low-carbon transition of energy structure (Englberger et al. 2021; Huang et al. 2019; Eldeeb et al. 2018; Beuse et al. 2021). Among China's end-use sectors, transport and heating play a critical role in achieving carbon neutrality. The transport sector accounts for approximately 10 per cent of China's total carbon dioxide emissions, ranking third after the power and industrial sectors (MEE 2022). The heating sector, especially in northern China, also poses significant challenges. The integration of renewable electricity into heating systems—such as through power–heat coupling or advanced combined heat and power (CHP) retrofits—offers a promising pathway but requires systemic coordination across sectors.

Given the high emission intensity and decarbonisation difficulty of transportation and heating, this chapter selects these two sectors as representative domains to investigate coordination mechanisms between terminal energy-consuming sectors and renewable energy. Accordingly, it focuses on mechanisms for inter-industrial collaboration in China's energy transformation and presents two empirical case studies. The first explores the integration of renewable energy into the transport sector through photovoltaic–energy storage–charging station (PV–ES–CS) systems, evaluating their economic and environmental performance considering location and scale. The second builds a power–heat coupling model using Beijing as a case study, assessing how renewable energy can support regional heating system decarbonisation. Through these two cases, this chapter aims to reveal the underlying logic and institutional requirements of inter-industrial coordination and offer policy-oriented recommendations for China's deep energy decarbonisation.

Development pathways for coupled photovoltaic–energy storage–charging station

The coupled PV–ES–CS is an important approach for promoting the transition from fossil energy to low-carbon energy consumption. The use of electric vehicles (EVs) and the installation of distributed rooftop photovoltaics (PV) can form a feedback loop (Kaufmann et al. 2021), which is an efficient approach to integrating distributed PV and EV charging systems (Denholm et al. 2013). In addition, distributed PV is intermittent and volatile, thus requiring energy storage (ES) devices (Omran et al. 2011; Zhang et al. 2019). However, the integrated charging station (CS) is underdeveloped. One of the key reasons for this is that there has been little evaluation of its economic and environmental benefits. Based on the electricity load of different types of buildings and data about EV charging stations in Beijing, we analyse the economic and environmental benefits of integrated charging stations developed over different scales using the capacity optimisation model.

First, we provide a detailed introduction to the PV–ES–CS system, then construct the optimal capacity allocation method of this system and analyse its economic and environmental benefits with the goal of profit maximisation. In the case study, we aggregated several datasets for calculation, including solar radiation data for Beijing and charging data from 21 EV charging stations (covering seven types of buildings such as hospitals and residences). PV–ES–CS locations in different types of buildings determine the different characteristics of a building's electricity load and EV charging rules, which lead to the heterogeneity of the economic and environmental benefits of PV–ES–CS. We analysed the economic and environmental benefits of different scales of PV–ES–CS in different locations.

The PV–ES–CS we discuss comprises distributed PV power generation modules, ES modules, EV charging modules and seven other power user modules such as hospitals and teaching buildings that are connected to the power grid (Figure 4.1). Besides solar PV, the PV–ES–CS system can buy electricity from power grids at the daily valley price. The electricity from the PV–ES–CS system is used not only for EV charging, but also for other uses within buildings. Because the electricity price for building power users is cheaper than for EV charging, the PV–ES–CS gives priority to charging EVs and then to supplying electricity to building users. The grid system could supply electricity to building users (Sabadini and Madlener 2021), but the electricity prices of 'PV–ES–CS to building users' are lower than the prices of 'grid to building users' (through peak–valley spread arbitrage, the PV–ES–CS could provide a cheaper electricity price than the grid system), thus building users would prefer to get electricity from PV–ES–CS than from a grid system.

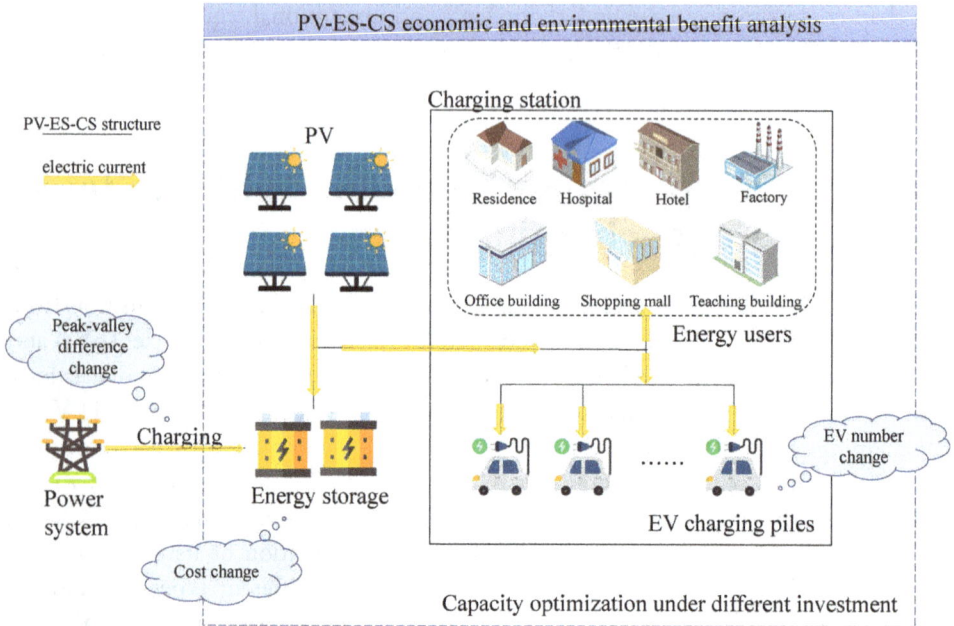

Figure 4.1: Structure of the PV–ES–CS model
Source: Authors' schema and analysis.

Methods

Objective function

Based on the maximum net present value (NPV) model of CS (Yang et al. 2020), the objective function in this chapter is to maximise the NPV of the PV–ES–CS (Equation 4.1). NPV is calculated by Equations 4.2 to 4.4. In Equation 4.2, NPV is maximised with the decision variables *a*, denoting the capacity of distributed PV; *b*, denoting the capacity of ES; and *c*, the initial investment in the CS (including equipment purchase costs, related facility construction costs and land utilisation costs; *c* is a positive integer). The NPV equals the discounted annual profit minus the initial investment in distributed PV (capacity *a* kilowatt), energy storage (capacity *b* kilowatt hour) and *c* charging piles, where P_{pv}, P_s, $P_{evc,c}$ and $P_{evc,l}$ represent the investment costs of distributed PV, ES, each charging pile and land, respectively. The land use of the charging pile is indicated by the symbol *neil*. *Y* is the lifecycle of a PV–ES–CS.

Equation 4.1

$$\max NPV$$

Equation 4.2

$$NPV = \sum_{y=1}^{Y} \frac{AP(y)}{(1+r)^{y-1}} - a \cdot P_{pv} - b \cdot P_s - c \cdot P_{evc,c} - neil(c) \cdot P_{cvc,l}$$

The annual profit is calculated by Equation 4.3 as the sum of the hourly profit for the year, $HP(t, d, y)$, minus the annual distributed PV operation and maintenance cost, C_{pvm}.

Equation 4.3

$$AP(y) = \sum_{d=1}^{D}\sum_{t=1}^{T} HP(t,d,y) - C_{pvm}$$

The annual profit in year y is the summation of hourly profit in year y. As shown in Equation 4.3, the sum of the hourly profit is expressed as $\sum_{y=1}^{Y} \dfrac{AP(y)}{(1+r)^{y-1}}$, which means the sum of hourly profit in year y from t = 1–T and d = 1–D.

Equation 4.4

$$HP(t,d,y) = EVQ(t,d,y) \cdot \left(P_{cv,t} + P_{sv,t}\right) + UQ(t,d,y) \cdot P_t - SCQ(t,d,y) \cdot P_t$$

The hourly profit is shown in Equation 4.4, comprising three parts. The first is charging income, which is calculated by multiplying the charging capacity of the EV by the electricity price per kilowatt hour. The electricity price consists of the electricity price and the service price, denoted as $EVQ(t,d,y) \cdot \left(P_{cv,t} + P_{sv,t}\right)$. The second part is the income from the PV–ES–CS supplying electricity to nearby buildings—that is, multiplying the energy consumption of the building by the commercial electricity price, denoted as $UQ(t,d,y) \cdot P_t$. The third part is the cost of purchasing electricity from the grid for the PV–ES–CS, denoted as $SCQ(t,d,y) \cdot P_t$. The hourly profit is calculated by EV charging income plus nearby buildings' charging income minus the cost of the PV–ES–CS.

Constraints

The constraints of this model include the following four aspects.

1) Investment constraints

Equation 4.5

$$a \cdot P_{pv} + b \cdot P_s + c \cdot P_{evc} \leq MI$$

The construction of the PV–ES–CS requires investment in distributed PV, ES and CS. Equation 4.5 indicates that the total investment in a PV–ES–CS comprises a, kilowatt capacity distributed PV; b, kilowatt hour ES; and c, charging piles; and does not exceed the fund, MI, owned by the investor.

2) Constraints on power balance

Equation 4.6

$$S(t+1,d,y) = S(t,d,y) - EVQ(t,d,y) - UQ(t,d,y)$$
$$+ PVQ(t,d,y) + SCQ(t,d,y)$$

Equation 4.6 indicates that the difference between the next period of electricity of the PV–ES–CS and the current period of electricity must be equal to the difference between the power generation of distributed PV, $PVQ(t,d,y)$; the power generation of large power grids, $SCQ(t,d,y)$; the actual charging capacity of EVs, $EVQ(t,d,y)$; and the actual power consumption of users, $UQ(t,d,y)$.

3) Capacity constraints

Equation 4.7

$$S(t,d,y) \le b$$

Equation 4.8

$$S(t+1,d,y) \le b$$

Equations 4.7 and 4.8 indicate that the ES capacity per hour cannot exceed the ES capacity of the PV–ES–CS (simplified to b).

Equation 4.9

$$PVQ(t,d,y) \le PV(t,d,y)$$

Equation 4.9 indicates that the power generation of the distributed PV used by the PV–ES–CS cannot exceed its power generation in the current period. The power generation of the distributed PV is calculated by Equation 4.10, in which a represents the installed capacity of the distributed PV; $RA(t,d,y)$ represents solar radiation per hour; η_1 represents the photoelectric conversion efficiency; and η_2 represents the conversion coefficient between the installed distributed PV capacity and the area of the distributed PV panel.

Equation 4.10

$$PV(t,d,y) = a \cdot RA(t,d,y) \cdot \eta_1 \cdot \eta_2$$

Equation 4.11

$$EVQ(t,d,y) \le EVC(t,d,y)$$

Equation 4.12

$$EVQ(t,d,y) \le S(t-1,d,y)$$

Equations 4.11 and 4.12 indicate that the charging capacity of EVs in a certain hour cannot exceed the EV power demand, $EVC(t,d,y)$, and the remaining power, $S(t-1,d,y)$, in the ES.

Equation 4.13

$$UQ(t,d,y) \leq U(t,d,y)$$

Equation 4.14

$$UQ(t,d,y) \leq S(t-1,d,y) - EVQ(t,d,y)$$

Under the limited ES capacity of the PV–ES–CS and the charging priority of EVs over large power users, Equations 4.13 and 4.14, respectively, represent the fact that the power purchased by large power users from the PV–ES–CS cannot exceed the power consumption of the large power users, $U(t,d,y)$, and the remaining power in the ES, $S(t-1,d,y) - EVQ(t,d,y)$.

4) Constraints on charging and discharging

Equation 4.15

$$EVQ(t,d,y) \leq SD_{max}$$

Equation 4.16

$$UQ(t,d,y) \leq SD_{max} - EVQ(t,d,y)$$

Equation 4.17

$$SCQ(t,d,y) \leq SC_{max}$$

Equation 4.15 indicates that the actual charge of EVs cannot exceed the maximum discharge of the PV–ES–CS hourly SD_{max}. Equation 4.16 indicates that the charging power of the large power users cannot exceed the remaining discharge power of the PV–ES–CS, $SD_{max} - EVQ(t,d,y)$, as EVs and large power users share the output line of the PV–ES–CS during the given period. Equation 4.17 indicates that the electricity purchased by the PV–ES–CS from the large power grid cannot exceed the maximum charging capacity of the power station, SC_{max}.

Model for calculating economic benefits and carbon dioxide emissions reduction

The rate of return on investment (ROI) in a PV–ES–CS can be calculated by Equation 4.18, based on the average annual profit divided by the total investment (Yang et al. 2020).

Equation 4.18

$$ROI = \frac{\frac{1}{Y}\sum_{y=1}^{Y}AP(y)}{a \cdot P_{pv} + b \cdot P_{s} + c \cdot P_{evc}}$$

Carbon dioxide emissions reduction is used here to measure the environmental benefits of the PV–ES–CS (Pu et al. 2021), which are calculated by multiplying the use of large-scale grid electricity with the carbon dioxide emissions per kilowatt hour of the large grid—that is, the electricity directly used by the PV–ES–CS from distributed photovoltaics multiplied by the large grid's carbon dioxide emissions per kilowatt hour (Equation 4.19).

Equation 4.19

$$CR = EF_{grid} \cdot \sum_{y=1}^{Y}\sum_{d=1}^{D}\sum_{t=1}^{T}PVQ(t,d,y)$$

In Equation 4.19, EF_{grid} represents the carbon dioxide emissions per kilowatt hour of electricity from China's regional power grid and PVQ represents the consumption of the distributed PV power generation.

Equation 4.20 expresses the distributed PV curtailment rate—an index commonly used to measure the utilisation efficiency of distributed PV.

Equation 4.20

$$PV_{curtailment} = 1 - \frac{\sum_{y=1}^{Y}\sum_{d=1}^{D}\sum_{t=1}^{T}PVQ(t,d,y)}{\sum_{y=1}^{Y}\sum_{d=1}^{D}\sum_{t=1}^{T}PV(t,d,y)}$$

In Equation 4.20, PVQ represents the consumption of the distributed PV power generation, PV.

Case study

Study area

The world's largest PV–ES–CS in an urban centre was built in China's capital, Beijing, in 2019 (He 2019). Beijing is both the political and the cultural centre of China (Shang et al. 2021), and building new projects there has greater influence and facilitates the promotion of PV–ES–CS. Therefore, Beijing is chosen for our case study.

In Beijing, the electricity loads of each PV–ES–CS are closely related to their location. PV–ES–CS near different types of buildings have different charging curves for EVs and loads for electricity users, which result in different economic and environmental benefits. We selected typical productive and non-productive buildings from the data of the Ministry of Housing and Urban–Rural Development (2020) to roughly divide installation locations of PV–ES–CS into seven categories: office buildings, factories, teaching buildings, hotels, shopping malls, hospitals and residences.

Data

To calculate the economic and environmental effects of the PV–ES–CS in different locations and at different scales, it is necessary to calculate the distributed PV power generation and EV charging load of the PV–ES–CS.

1) Power generation prediction for distributed PV

The long short-term memory (LSTM) network, a time-series prediction model based on the recurrent neural network (RNN) (Shang et al. 2021), is commonly used to predict distributed PV power generation and performs well in long-term prediction scenarios (Luo et al. 2021b; Ibrahim et al. 2021), as it does not have gradient explosion or disappearance problems over the long-term RNN learning process (Hochreiter and Schmidhuber 1997). The LSTM structure contains four neural network (NN) layers, which are the forgetting gate layer, input gate layer, renew values layer and output gate layer. The four-layer NN helps the LSTM network avoid the long-term dependency problem.

An LSTM model is used in this study. The model is fed with Beijing's hourly solar radiation data for 27,024 hours from 2017 to January 2020 to regress and predict future solar radiation in MATLAB Version 2020a. With an interval of 100 data points, 70 per cent of the data are used for training the model and the other 30 per cent for validation. A comparison of real and predicted solar radiation data is shown in Figure 4.2, which indicates that the predicted results are credible. The hourly solar radiation data predicted by MATLAB for 20 years are recorded as $RA(24, 365, 20)$, then the distributed PV power generation is $PV = a \cdot RA(24,365,20) \cdot \eta_1 \cdot \eta_2$, where a represents the installed capacity of the distributed PV, η_1 represents the conversion efficiency of the distributed PV and η_2 represents the conversion rate between the installed capacity and distributed PV panel area.

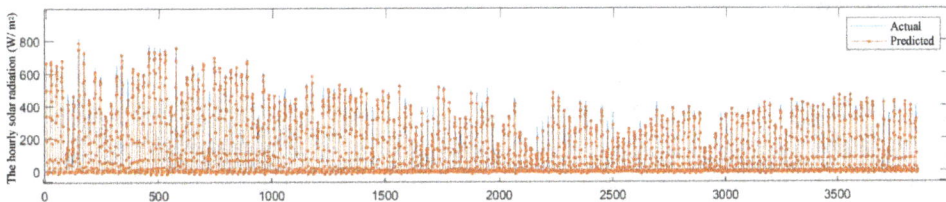

Figure 4.2: Real versus predicted data

Source: Data from the US National Oceanic and Atmospheric Administration (www.noaa.gov).

2) Electricity load prediction of charging pile

For conventional EV charging pile load analysis, the charging and discharging behaviour of EVs is generally simulated through data such as those from a 'Family Travel Survey Report', as the total load of the charging pile is accumulated from the bottom up. However, China lacks such surveys; it is difficult to fully cover the charging and discharging behaviour of EVs in Beijing through questionnaire data. Studies have shown that the power remaining when EVs drive into a charging pile is random (Luo et al. 2011)—that is, the charging power is independent of the charging start time. The electric load model of CS is constructed in this study through a probability analysis of the hourly EV charging pile discharge data obtained for Beijing. The hourly discharge amount of the charging pile when EV charging behaviour occurs is determined, then whether there is an EV charging in that hour is judged by analysing the EV charging start time.

The hourly electricity load of the charging pile of the PV–ES–CS can be obtained using the above method. Let X_1, X_2 ... X_n be the hourly charging capacities of given EVs. The Arena simulation tool is used to fit the probability distribution function of hourly charging capacity, F(x). The charging time of EVs can be analysed from the existing data and represented by $g_n(t,d,y)$, which is a variable from 0 to 1, where $g_n(t,d,y) = 0$ means that there is no EV charging at that moment and $g_n(t,d,y) = 1$ means that there are EVs charging at that moment; $g_n(t,d,y)$ is related to the number of EVs. The frequency of $g_n(t,d,y) = 1$ doubles as the number of EVs doubles. The charging and discharging of each charging pile in the PV–ES–CS is shown in Equation 4.21.

Equation 4.21

$$EVC(t,d,y) = \sum_{n=1}^{c} F_n(x) \cdot g_n(t,d,y)$$

In Equation 4.21, t represents time on the day, d, of the year, y, while $F_n(x)$ represents the probability distribution to which the unit charge of the n charging pile is subject, and c is the number of charging piles that must be optimised.

Hourly charging and discharging data from 21 public EV charging stations in Beijing from 2019 to 2020 are collected with the cooperation of State Grid EV Service Limited, which covers all administrative districts and seven building categories in Beijing. Each charging station has 20 to 30 charging piles. Excluding the abnormalities in the original data—such as EVs that do not charge when plugged in or EVs plugged and unplugged immediately without charging—650,255 valid data points remained. The charging data for each building type are accumulated before calculating the charging capacity per hour; then Arena software is used to generate a distribution function obeyed by the hourly charging capacity. As shown in Figure 4.3f, taking a hospital as an example, bars represent the hourly charge distribution of 8,760 hours and the curve represents the fitted distribution function. The data description item is the descriptive statistics of the CS near the hospital. There are 49,694 effective data points for hospitals in this sample. The average hourly charge value (kilowatts) is 21.5, the maximum value is 58.5 and the minimum value is 0.003. The hourly curve for hospitals is fitted with nine common distributions (for example, β distribution and normal distribution), then the distribution function with the smallest error is selected. The distribution functions of other building types are shown in Figures 4.3a to 4.3e and Figure 4.3g, as well as Table 4.1.

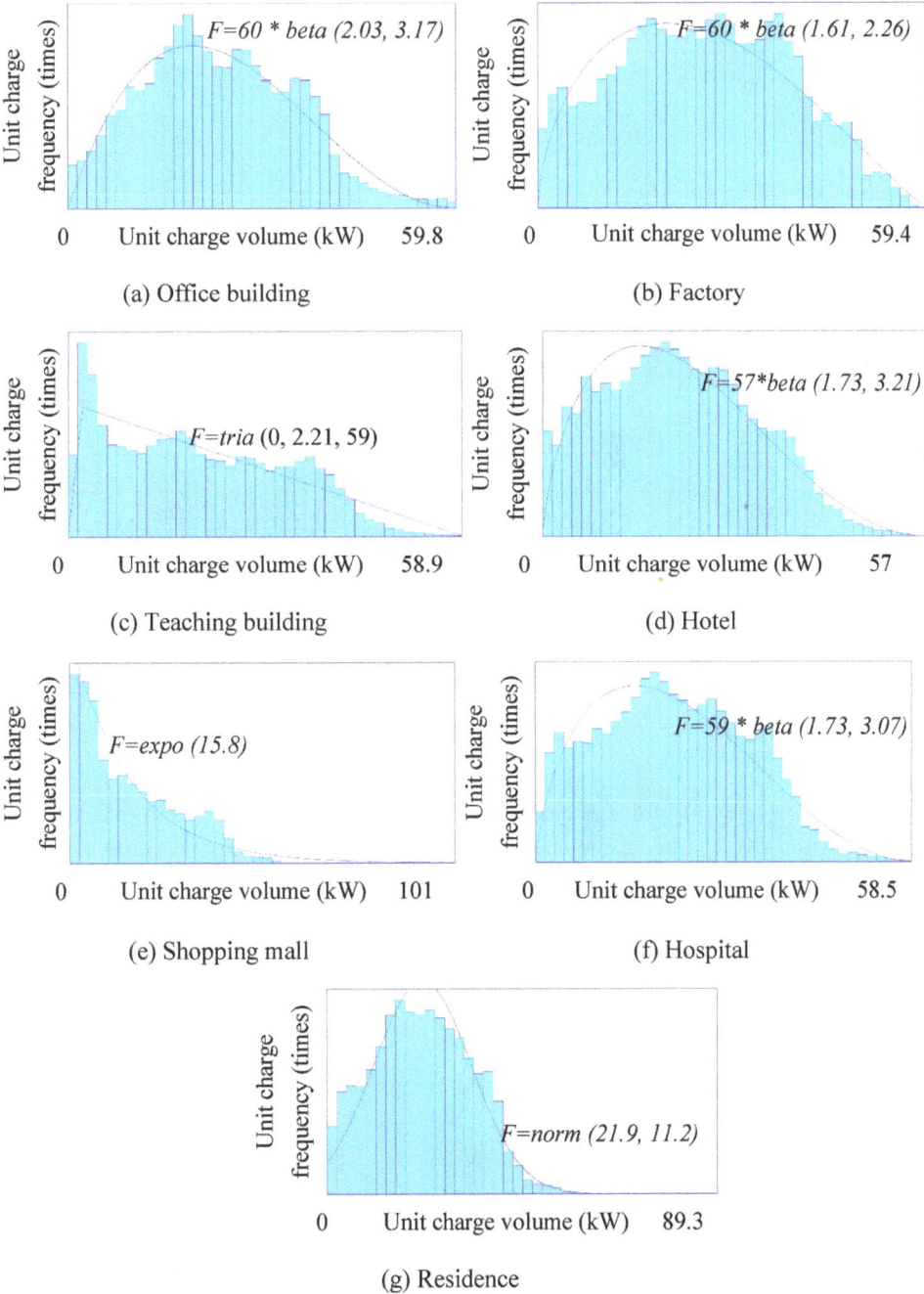

(a) Office building

(b) Factory

(c) Teaching building

(d) Hotel

(e) Shopping mall

(f) Hospital

(g) Residence

Figure 4.3: Hourly charging capacity distribution curve of charging station

Source: Data from State Grid EV Service Limited.

Table 4.1: Hourly charging capacity distribution of various building types (kilowatts)

	Office building	Factory	Teaching building	Hotel	Shopping mall	Hospital	Residence
Data description							
Valid data	43,598	23,826	33,534	62,603	4,226	49,694	95,660
Mean	23.5	25.4	19.3	20	15.8	21.5	21.9
Standard deviation	11.8	13.2	13	11.2	12.8	11.5	11.2
Maximum	59.8	59.4	58.9	57	101	58.5	89.3
Minimum	0.055	0.008	0.002	0.003	0.028	0.003	0.001
Errors under various distribution functions							
Beta	0.001	0.001	0.002	0.001	0.003	0.001	0.003
Normal	0.001	0.001	0.002	0.001	0.021	0.001	0.001
Triangular	0.001	0.003	0.002	0.003	0.026	0.003	0.013
Gamma	0.002	0.003	0.003	0.003	0.003	0.003	0.005
Erlang	–	0.004	0.007	0.007	0.002	0.004	0.005
Uniform	0.011	0.006	0.010	0.010	0.050	0.010	0.023
Lognormal	0.005	0.007	0.011	0.007	0.007	0.008	0.012
Exponential	0.014	0.011	0.010	0.010	0.002	0.012	0.020
Weibull	–	0.011	0.008	0.012	0.160	0.002	0.006
Optimal distribution	Beta	Beta	Triangular	Beta	Exponential	Beta	Normal
Distribution charge	60 × beta (2.03, 6.87)	60 × beta (1.61, 6.28)	Tria (0, 2.21, 59)	57 × beta (1.73, 3.21)	expo (15.8)	59 × beta (1.73, 3.07)	norm (21.9, 11.2)
Hourly	3.17	2.26	–	–	–	–	–

Source: Authors' calculations.

In this study, $g(t, d, y)$ indicates charging (or not charging) at a certain time. The solid line in Figure 4.4f represents the charging frequency of a CS near a hospital in 2019, the dotted line represents the charging situation in 2020, the coloured lines represent the number of charging EVs in an hour for each charging pile and the black line represents the simulated charging number. The simulation curves fit well for all types of buildings analysed in this study, which validates the proposed model. The charging frequency simulations of CS near other building types are shown in Figures 4.4a–e and Figure 4.4g.

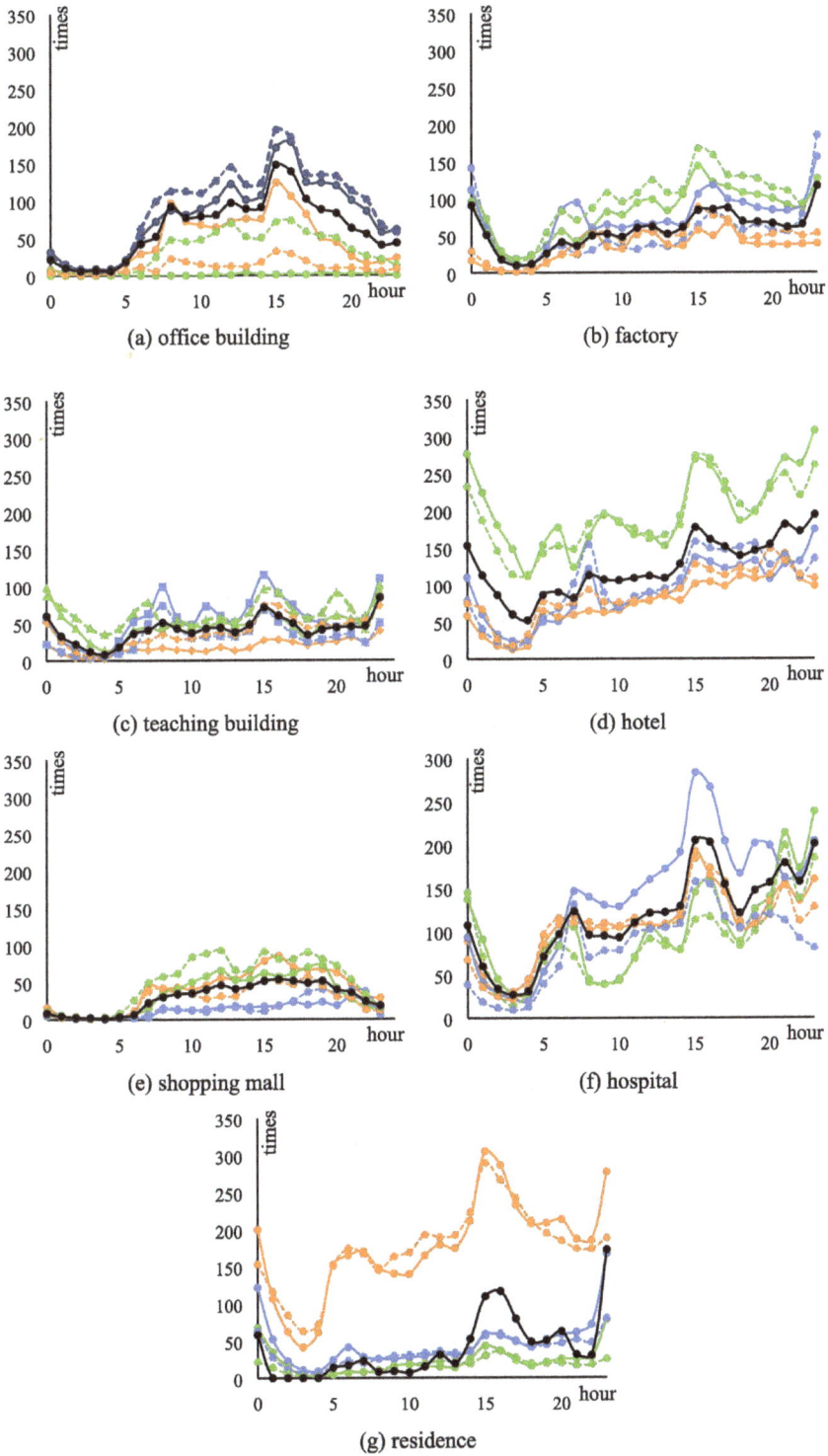

Figure 4.4: Real and simulated charging times for charging station (24 hours)

Source: Data from State Grid EV Service Limited.

3) Other parameters selection

According to the 'Specification for Photovoltaic Power Generation System Performance' issued by China's National Energy Administration (NEB 2020), the usual lifespan of a distributed PV module is 25 years. For ES modules, this chapter selects as the object the lithium iron phosphate battery—a common battery in PV–ES–CS; its configuration costs US\$300 per kilowatt hour and the operation and maintenance cost is US\$0.30 per kilowatt hour. The lithium iron phosphate battery has a lifespan of 10.91 years (Eldeeb et al. 2018). As a new business model, the PV–ES–CS does not have a unified computing lifecycle standard. If the lifecycle is set to 10 or 15 years, compared with 20 years, the distributed PV part of the PV–ES–CS cannot be fully utilised. If the lifecycle is set to 25 years, three more ESs will be needed, among which the service period of the third ES could be about three years only. In this case, the income and rate of return of the PV–ES–CS for 20 years of service are higher than that for 25 years of service. The lifecycle of the PV–ES–CS is set to 20 years accordingly.

The electricity consumption of power users is taken from Ye (2018), with 1 denoted as *U(24, 365, 20)*. The characteristics of users' electricity consumption on a typical day are shown in Figure 4.5. The EV charging fee, $P_{ev,t}$; service fee, $P_{s,t}$; and Beijing peak and valley electricity price, P_t, are shown in Figure 4.6 (Beijing Municipal Commission of Development and Reform 2020b). The exchange rate is from the Bank of China, where the exchange rate was RMB6.5249 to US\$1 on 1 January 2021. Other parameters and sources used in this work are listed in Table 4.2.

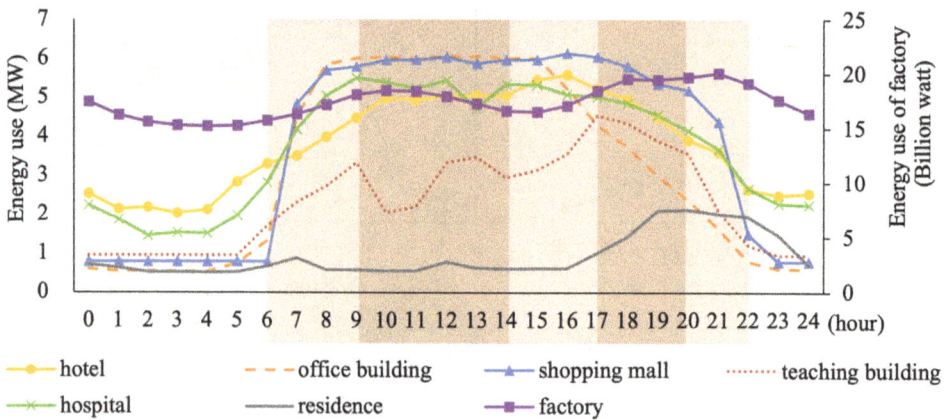

Figure 4.5: Large users' electricity consumption (typical day)
Source: Ye (2018).

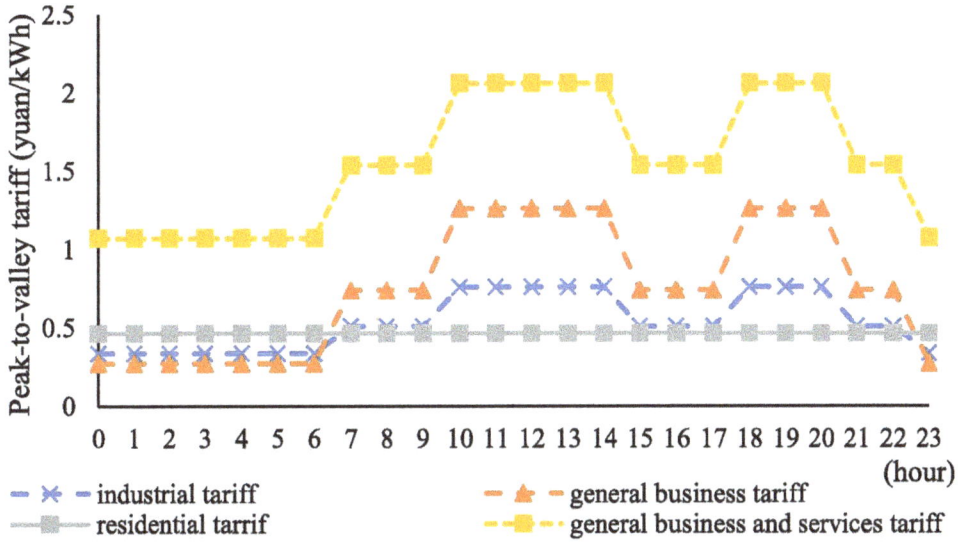

Figure 4.6: Peak-to-valley tariff in Beijing

Source: Beijing Municipal Commission of Development and Reform (2020b).

Table 4.2: Abbreviated symbols and values

Symbol	Definition	Value	Data sources
P_{pv}	Cost of distributed PV	RMB3,380/kW	China PV Industry Association (www.chinapv.org.cn)
P_s	Cost of ES	US$300/kWh	Liu et al. (2021)
$P_{evc,c}$	Cost of each charging pile	RMB533,000/pile	Yang et al. (2020)
$P_{evc,c}$	Land cost of charging pile	RMB1,920,000/group	Yang et al. (2020)
$P_{ev,t}$	Charging fee of EV (RMB/kWh)	Figure 4.6	Beijing Municipal Commission of Development and Reform (2020b)
$P_{s,t}$	Service fee for EV charging (RMB/kWh)	Figure 4.6	Beijing Municipal Commission of Development and Reform (2020b)
P_t	Change in peak-to-valley in Beijing (RMB/kWh)	Figure 4.6	Beijing Municipal Commission of Development and Reform (2020b)
C_{pvm}	Annual operation and maintenance costs of distributed PV power generation	RMB54/kW/year	China PV Industry Association (www.chinapv.org.cn)
η_1	Conversion efficiency of distributed PV	22.8	China PV Industry Association (www.chinapv.org.cn)
η_2	Conversion rate of distributed PV installed capacity and the panel area	10 sq m/kW	Field research
γ	Discount rate	8 per cent annually	Wang et al. (2018b)
EF_{grid}	Marginal emission factor	0.9419 tCO$_2$/MWh	Ministry of Ecology and Environment

Source: Authors' compilation.

Economic and environmental impact of PV–ES–CS in different locations and scales

There are obvious differences in the economic and environmental effects of the PV–ES–CS near different types of buildings. The return on investment of the PV–ES–CS calculated under different scales by building type is shown in Table 4.3. The distribution of PV, ES and CS can be identified when the maximum return on investment is achieved. The distributed PV installed capacity, ES capacity and number of charging piles are all non-zero, indicating that the return on investment of the PV–ES–CS is better than that of distributed PV power generation (a ≠ 0, b = c = 0), ES power station (b ≠ 0, a = c = 0), EV charging station (c ≠ 0, a = b = 0), PV ES power station (a, b ≠ 0, c = 0) or ES charging station (b and c ≠ 0, a = 0).

The economic and environmental effects of PV–ES–CS at different locations and scales can be summarised as follows.

Table 4.3: Maximum return on investment for PV–ES–CS

	Teaching building	Hotel	Shopping mall	Hospital	Residence	Office building	Factory
Maximum investment return rate	12.58%	11.23%	12.81%	13.92%	1.42%	9.81%	4.29%
Investment (RMB million)	26	16	39	41.5	8.5	13	13.5
NPV (RMB thousand)	24,396	12,591	37,955.7	44,954.7	–1,094.6	8,363.6	530.1
Distributed PV curtailment rate	21.92%	27.39%	26.50%	18.68%	36.99%	31.70%	29.20%
CO_2 emissions reduction (thousand tonnes)	9.36	5.33	13.66	16.30	0.40	3.67	4.46
a (kW)	2,440.91	1,493.87	3,785.68	4,082.30	128.84	1,095.02	1,282.69
b (kWh)	5,258.67	2,888.29	8,206.03	8,728.05	184.27	2,312.60	2,246.25
c	5	5	5	5	5	5	5

Source: Authors' calculations.

1) Investment return ratio of PV–ES–CS is highest in the hospital scenario and lowest in the residential scenario

Figure 4.7 shows that if there is sufficient investment, the return on investment of the PV–ES–CS falls by building type from high to low as follows: hospitals, shopping malls, teaching buildings, hotels, office buildings, factories and residences. According to the rate of return on investment, the PV–ES–CS in different scenarios can be classified into three gradients. The first gradient has the highest return on investment—namely, hospitals and shopping malls; the second is hotels, teaching buildings and office buildings; and the third gradient is factories and residences, which have the lowest return on investment. Among them, a CS near a residence has no investment value because the rate of return is lower than the current deposit interest rate of the Bank of China.

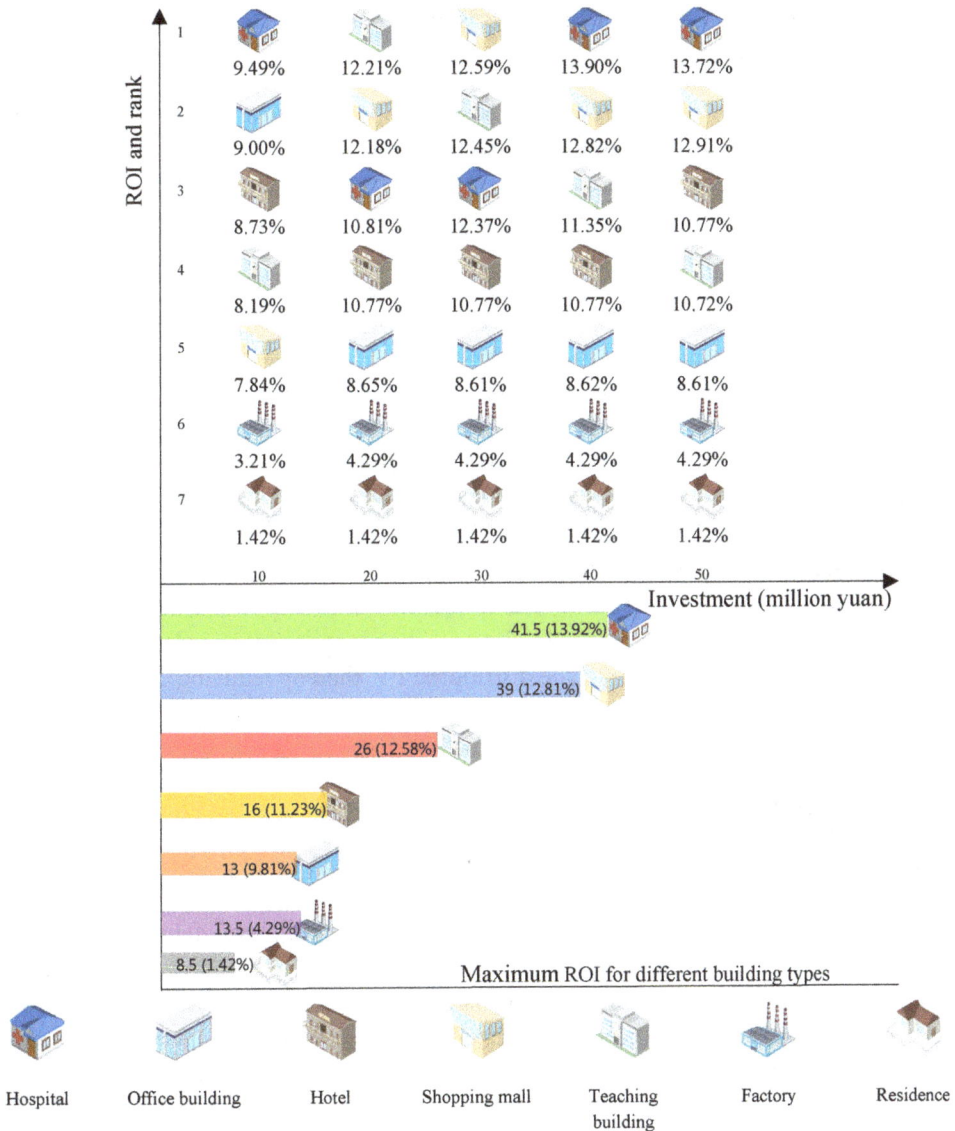

Figure 4.7: Return on investment of various PV–ES–CS with different levels of investment

Notes: On the one hand, a PV–ES–CS system should contain at least 1 kW PV for RMB3,380 (China PV Industry, 1 kWh ES for RMB1,957.47) (Liu et al. 2021) and 1 charging pile for RMB2.45 million (Yang et al. 2020) — that is, the smallest investment should be more than RMB3 million. On the other hand, the largest investment should be lower than RMB50 million; the return on investment of a PV–ES–CS system above this amount is less than zero. Therefore, we set the investment constraint of PV–ES–CS between RMB3 million and RMB50 million for all scenarios.

Source: Authors' calculations.

2) Developing PV–ES–CS near hospitals is more conducive to carbon emissions reduction

The carbon emissions reduction of the PV–ES–CS is related to the curtailment ratio of distributed PV. Under the same investment level, a lower curtailment rate results in a larger reduction of carbon dioxide emissions. With sufficient investment, the PV–ES–CS falls into descending order of carbon dioxide emissions reduction per unit of investment (Figure 4.8): hospitals (3.87 kg/RMB), shopping malls (3.46 kg/RMB), factories (3.30 kg/RMB), hotels (3.24 kg/RMB), teaching buildings (3.18 kg/RMB), office buildings (2.61 kg/RMB) and residences (0.57 kg/RMB), which is basically consistent with the economic ranking. When the diurnal power load of the building is close to the distributed PV power generation curve and nocturnal power consumption is relatively high, carbon dioxide emissions are reduced to a greater extent by the PV–ES–CS. Buildings use a large amount of electricity from the ES at night, leaving enough space for consuming distributed PV during the day and reducing the rate of curtailment. When the building load curve during the day is closer to the distributed PV power generation curve, the distributed PV curtailment rate is relatively low. Hospitals show the largest nocturnal power consumption and diurnal power load curve closest to the distributed PV power generation curve. Therefore, building a PV–ES–CS near a hospital can maximise the use of the distributed PV, thereby reducing carbon dioxide emissions from the large power grid.

Synergistic development of heating system decarbonisation and renewable energy increase

Chinese policymakers are actively exploring the path to heating system decarbonisation. For instance, the *Fourteenth Five-Year Plan for Renewable Energy Development*, jointly issued in 2022 by nine national departments (NDRC 2022), states that the utilisation scale of geothermal heating, biomass heating, biomass fuel and solar thermal for non-power generation will exceed 60 million tonnes of standard coal equivalent by 2025. And the *Implementation Opinions on Further Accelerating the Application of Heat Pump Systems and Promoting Clean Heating* issued by the Beijing Municipal Commission of Development and Reform in 2019 provided that, by 2022, the utilisation area of the new heat pump system in Beijing would reach 20 million square metres, and the cumulative utilisation area 80 million square metres, accounting for about 8 per cent of the total heating area, to substantially enhance the application of the heat pump system in the city. However, the city's heating system decarbonisation has not been achieved because the heating area provided by renewable energy remains limited, and the primary energy supply structure of the heating system is still dominated by fossil fuels (NBS 2005). Given that renewable energy power generation will dramatically grow in the future, it is crucial to step onto the path to achieve the synergistic development of renewable energy generation and heating system decarbonisation.

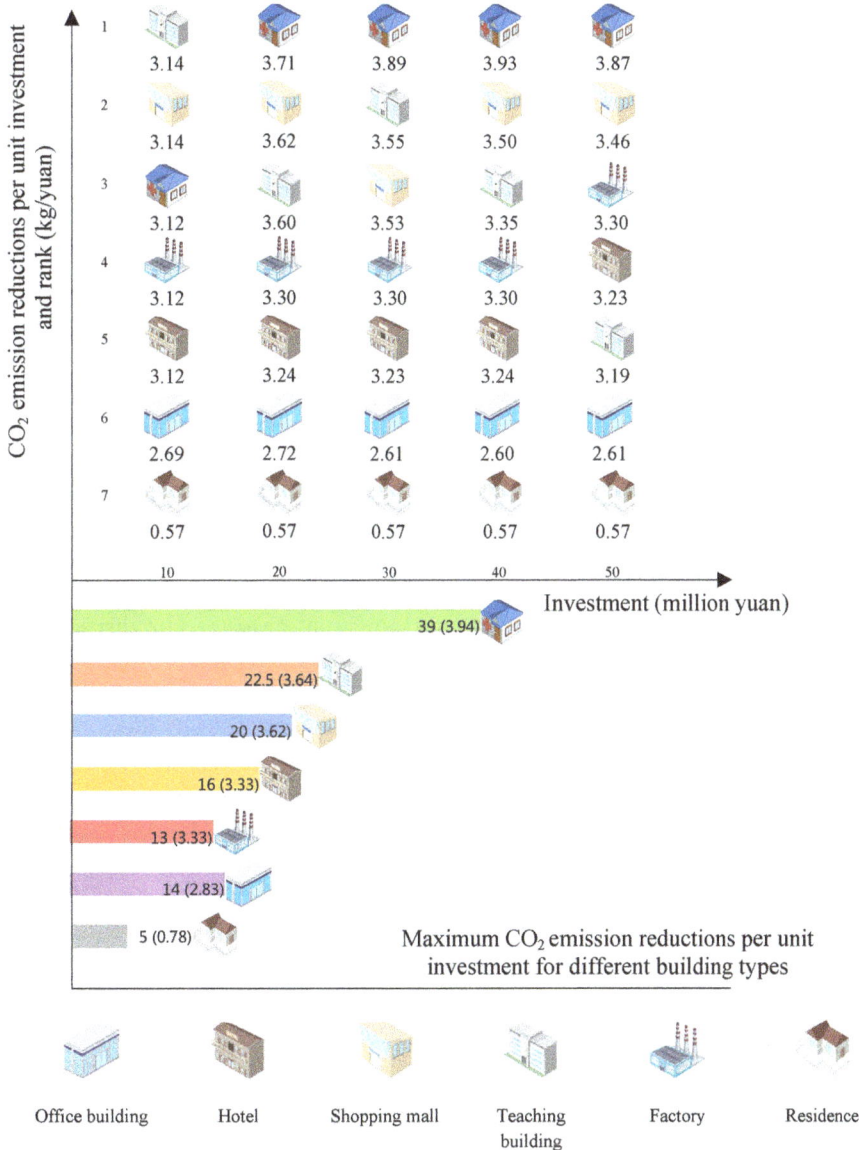

Figure 4.8: Carbon dioxide emissions reduction per unit of investment in PV–ES–CS
Source: Authors' calculations.

Beijing sits on the North China Plain. As China's capital, it plays a leading role in national carbon peaking and carbon neutrality initiatives (Huang et al. 2020; Beijing Municipal People's Government 2022). De-coalification has basically been realised in Beijing, however, the city is still under enormous pressure to cut carbon dioxide emissions from natural gas and oil usage. Beijing has implemented several policies aimed at reducing carbon dioxide emissions, such as its *Fourteen*th Five-Year Plan for Ecological and Environmental Protection, issued in 2021 (Beijing Municipal People's Government 2021). It stated that, by 2025, the total carbon dioxide emissions in the city would be more than 10 per cent lower than the peak in 2020, which was 148 megatonnes (Huang et al. 2020; Zhang et al. 2019; Beijing Municipal People's Government 2017a; Beijing Municipal Bureau of Statistics 2021b), and the proportion of renewable energy consumption would be about 14 per cent.

The current energy system in Beijing mainly comprises the power, heating and transport sectors. Its local power supply system is dominated by natural gas generation, which amounted to 39.7 terawatt hours in 2020, accounting for 87.1 per cent of local power generation, whereas renewable energy, such as wind, PV, hydropower and waste incineration power generation, accounted for only 9.87 per cent of local power generation (CEC 2021), as shown in Figure 4.9. Although Beijing's power system has moved away from coal, supply is still dominated by fossil fuels, with a low proportion of renewable energy. Furthermore, as shown in Figure 4.10, only 40 per cent of the electricity consumed in Beijing in 2020 came from local electricity production, with the remaining demand met by imported electricity. As a result, increasing the amount of local renewable energy is important to decarbonise the power and heating systems in Beijing.

Beijing's heating system, which uses natural gas as the primary energy input, is separated into two segments: district heating and individual heating. The district heating area accounts for 73.67 per cent of the total heating area (MoHURD 2021; Beijing Municipal Commission of Development and Reform 2020a). Gas-fired combined heating and power meets 9 per cent of the district and 6.89 per cent of the individual heating demand. The analysis of Beijing's heating structure shows that the city has established a clean heating system. However, due to the technical structure based on gas-fired boiler heating, additional steps must be taken to replace natural gas with renewable energy to achieve complete decarbonisation of the system.

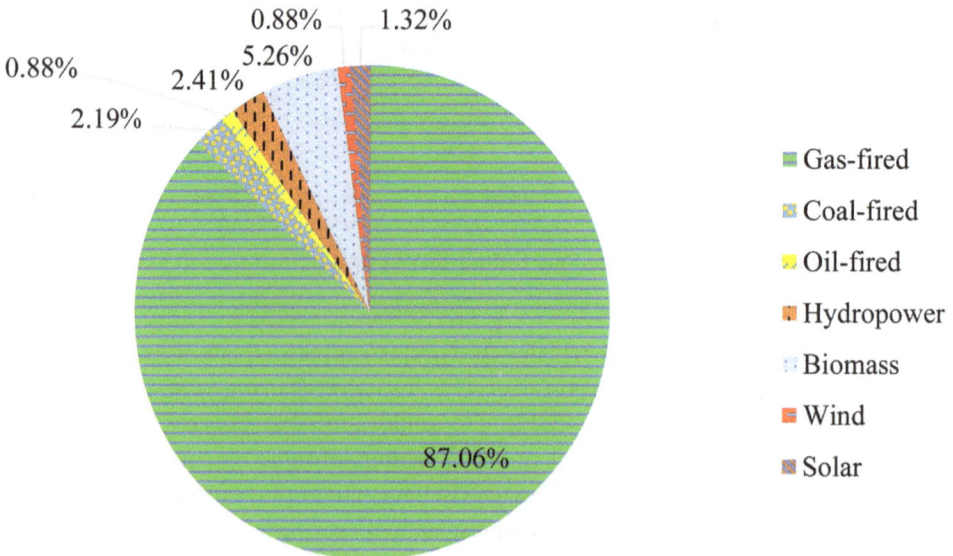

Figure 4.9: Beijing's local power supply structure in 2020
Source: Calculated based on data for power generation by region in CEC (2021).

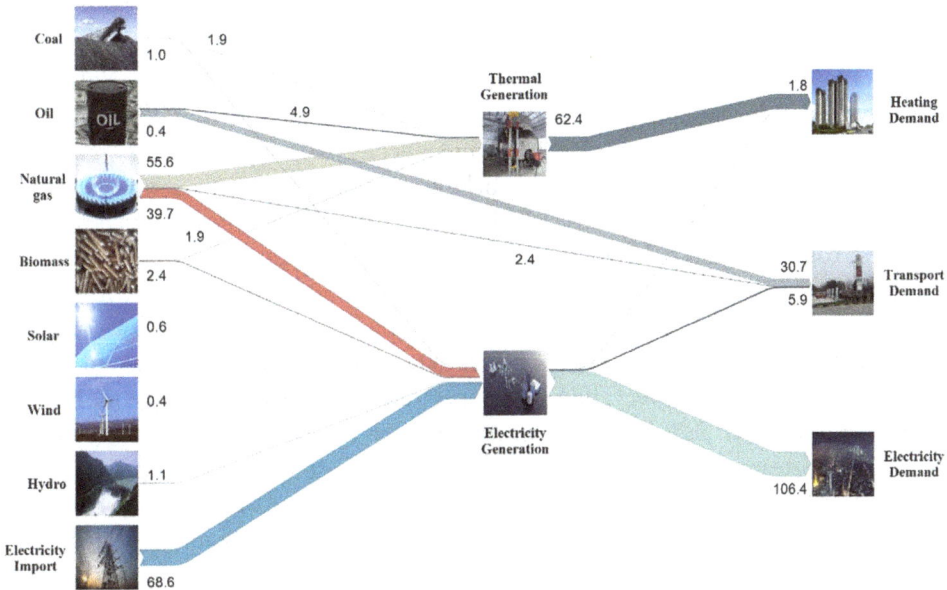

Figure 4.10: Beijing's energy balance in 2020 (terawatt hours)

Sources: Based on data in MoHURD (2021); CEC (2021); Beijing Municipal Bureau of Statistics (2021a).

Future energy system design in Beijing

Key technical paths for heating system decarbonisation

The technical paths to heating system decarbonisation in the existing policy documents mainly include ground-source heat pumps, solar thermal, 'coal-to-electricity' and 'coal-to-natural gas' central heating pilot projects and power plant waste heating (Beijing Municipal People's Government 2017a), which must be fully considered when setting the first technical path scenario in this chapter. This scenario serves as a baseline for other scenarios for heating decarbonisation. Considering Beijing's industrial structure, resource endowment and infrastructure construction, the natural gas demand in the heating system should be further reduced based on the technical paths mentioned in the existing policies during the design of the heating system decarbonisation path (Beijing Municipal People's Government 2021). Therefore, it is necessary to replace natural gas heating with cleaner electricity, which is the second technical path scenario in this chapter—that is, replace traditional gas heating with electric technologies (such as air-source heat pumps). In addition, in the context of China's dual carbon targets, more renewable energy should replace fossil-based energy and cross-sector application of energy technologies should be employed (Luo et al. 2021a). So, it is necessary to analyse how cross-sectoral energy technologies will help achieve synergistic development of the electricity and heating sectors in the low-carbon transition. The third technical scenario for heating system decarbonisation in this chapter is the synergistic development scenario, which is mainly based on improvements in the traditional technology for generating combined heat and power—that is, replacing fossil fuels with renewable energy.

As an efficient production technology, combined heat and power (CHP) can not only achieve the synergistic development of power and heating systems, but also has great potential in reducing energy costs and carbon dioxide emissions (Zhu et al. 2021), and is widely used in the design of the district heating system decarbonisation path (Jimenez-Navarro et al. 2020; Aunedi et al. 2020; Olympios et al. 2020). However, CHP essentially burns fossil fuels for energy production, making it impossible to fully eliminate carbon dioxide emissions. How to replace fossil fuel with renewable energy in CHP affects not only carbon dioxide emissions reduction but also large-scale penetration of renewable energy in the future. As a result, this chapter focuses on the CHP generation of solar thermal power plants for future development. When intermittent renewable electricity is generated, the excess power is transferred to an electric heater to heat salt, which is transported to a high-temperature molten salt tower. The high-temperature steam generated by the molten salt is used to heat water and drive a steam turbine unit to generate electricity. A thermal power plant's steam turbine generator can be employed as a peak-shaving power station for wind or PV power generation (Wu et al. 2017). The new development modes for CHP generation technology can fully utilise the infrastructure resources of a thermal power plant and accomplish the synergistic development of renewable energy and heating system decarbonisation. Finally, it is worth noting that the development of wind and hydropower requires vast natural resources and land; in contrast, the development of distributed photovoltaics is more flexible in terms of location. Therefore, the scenarios in this study focus on the synergistic development of large-scale applications of heat pumps and CHP in combination with distributed photovoltaics.

Forecasting the future energy demand in Beijing

The dynamic trend of Beijing's future energy demand must be analysed to design the city's large-scale renewable energy system. Beijing's power consumption was 114 terawatt hours in 2020 (CEC 2021) and is expected to reach 268.76 terawatt hours by 2050 based on a 2.9 per cent annual growth rate (Beijing Municipal People's Government 2017a). Electricity imported to Beijing from other provinces accounts for 30.3 per cent of future total energy consumption, including that generated by coal, oil and natural gas, and is expected to reach 127.96 terawatt hours in 2050 (Beijing Municipal People's Government 2017a).

The heating demand in Beijing is determined by both heating area and heating supply per unit area. The total heating area in the city was 895 square kilometres in 2020 and 1,000 square kilometres in 2022 (Beijing Municipal Commission of Development and Reform 2019). If the heating area in Beijing continues to grow at its current rate, it will reach 4,725.82 square kilometres by 2050. Thus, the total heating demand will reach 203.03 terawatt hours (TWh) by 2050, considering future improvements in buildings' energy savings (Beijing Municipal People's Government 2022). The dynamic changes in future heating demand in Beijing are listed in Table 4.4.

Table 4.4: Forecast future heating demand in Beijing

Year	Heating area (sq km)	Heating demand per unit area (TWh/sq km)	Total heating demand (TWh)
2020	895	0.0808	72.35
2030	1,558.50	0.0655	102.05
2050	4,725.82	0.0430	203.03

Note: It is assumed that the heat supply per unit area in Beijing will fall by 10 per cent every five years in the future.

Source: Authors' forecasts.

Description of future energy transition scenarios in Beijing

This chapter proposes three development modes for Beijing's future power and heating system based on the city's energy supply structure and important decarbonisation and transition technologies, including the business-as-usual (BAU) scenario, the electricity substitution (ELS) scenario and the synergistic development (SD) scenario. The operation and transition of Beijing's power and heating systems in the three scenarios are simulated year by year from 2020 to 2050. The design and key assumptions of the three scenarios are described in the subsections that follow.

1) Business-as-usual (BAU) scenario

Figure 4.11 depicts the BAU scenario, which is similar to Beijing's energy system in 2020. We analyse the technical path for Beijing's future heating system decarbonisation under existing policy conditions. In the BAU scenario, Beijing's heating system mainly uses fossil fuel and various renewable energy technologies, such as ground-source and air-source heat pumps. Beijing's future electricity demand, heating demand and renewable energy development under the BAU scenario will be the benchmarks for developing future energy system transition pathways and will be determined primarily by policy documents such as Beijing's *Energy Development Plan during the Thirteenth Five-Year Plan Period*, *Thirteenth Five-Year Plan for New and Renewable Energy Development* and *Fourteenth Five-Year Plan for Ecological and Environmental Protection* (Beijing Municipal People's Government 2017a, 2017b, 2021).

In the BAU scenario, the replacement of coal-fired and oil-fired units in the power system will be completed by 2025. The installed capacity of gas turbines will rise at a rate of 3.3 per cent until it reaches the peak in 2025 (Beijing Municipal People's Government 2017a, 2021). The scale of wind and hydropower development remains unchanged, whereas the installed capacity of photovoltaic and biomass power generation is expected to rise at the same rate as in 2020. The future electricity supply structure in the BAU scenario is presented in Table 4.5. The heating system is divided into district heating and individual heating. Table 4.6 lists the future heating supply structure in Beijing. Since the electrification transition pathway for the transport sector is not considered in this chapter, the rate of change in electricity consumption in the transport sector is assumed to be in line with overall energy consumption.

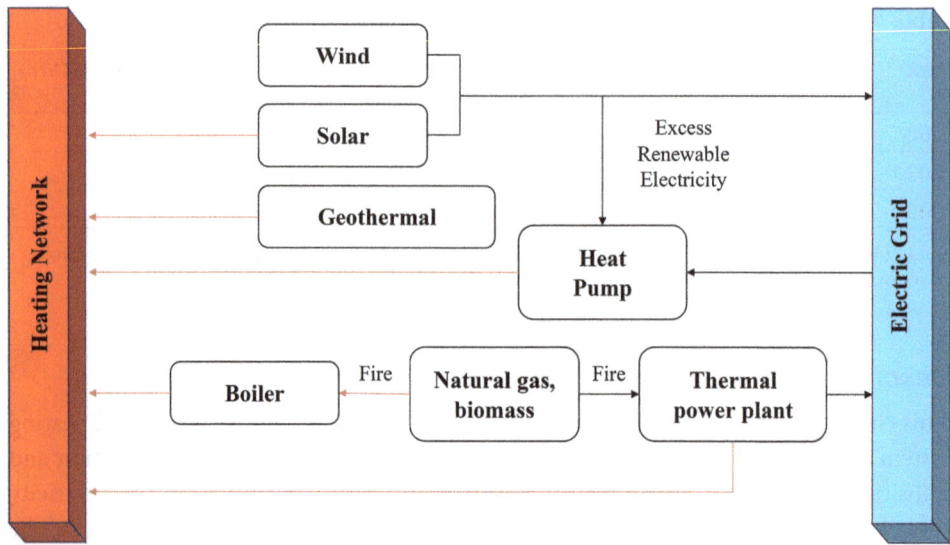

Figure 4.11: Conceptual diagram of Beijing's energy system
Source: Authors' schema.

Table 4.5: Beijing's electricity supply structure under the business-as-usual scenario

Year	Coal (MW)	Oil (MW)	Natural gas (MW)	Wind (MW)	Hydro (MW)	PV (MW)	Biomass (MW)	Import (TWh)
2020	770	200	10,000.00	190	990	560.00	370.00	68.59
2030	0	0	11,762.55	190	990	4,011.84	644.97	84.44
2050	0	0	11,762.55	190	990	167,975.67	1,959.80	127.96

Source: Authors' forecasts.

Table 4.6: Beijing's heating structure under the business-as-usual scenario (square kilometres)

Year	District heating area				Individual heating		
	Gas boiler	Gas CHP	Waste incineration	Ground-source heat pump	Gas boiler	Gas CHP	Air-source heat pump
2020	457.52	59.37	23.32	35.00	154.42	16.23	65.00
2030	538.16	69.83	50.19	124.68	181.64	19.09	574.92
2050	538.16	69.83	232.43	378.07	181.64	19.09	3306.61

Notes: The replacement of coal-fired and oil-fired heating boilers and CHP generation units in the fossil energy heating technology in district heating will be completed by 2025. Beijing's *Fourteenth Five-Year Plan for Ecological and Environmental Protection* (Beijing Municipal People's Government 2021) is the source of the future development trend for gas heating. The heating area of ground-source heat pumps accounts for 8 per cent of the overall heating area and remains unchanged. The heating area of waste incineration technology is computed using the installed capacity, utilisation hours and thermal efficiency of biomass thermal power plants. The future trend for fossil fuel heating in individual heating is consistent with that for district heating, and the remaining individual heating demand is met by air-source heat pumps.

Source: Authors' forecasts.

2) Electricity substitution (ELS) scenario

Unlike the BAU scenario, under which gas dominates the heating structure, the ELS scenario emphasises the use of heat-pump technology to replace gas heating. This substitution mode affects the structure of the heat and power supplies. After electrical heating replaces gas heating, such as gas boilers and gas CHP generation, additional electricity generation technologies—for example, renewable and imported electricity—will be used to also replace gas-generated power. Therefore, the electricity and heating systems must be rebuilt based on the BAU scenario.

Replacing gas heating with air-source heat pumps will increase power consumption. Consequently, Beijing's power consumption will reach 280.34 terawatt hours in 2050. Except for the replacement of fossil energy power production technology, the changes in other power production technologies are consistent with the BAU scenario. Furthermore, the ELS scenario focuses on decarbonising the energy system when gas heating is replaced with electric heating. Thus, it ignores the influence of transport electrification. The future development mode of the transport sector is assumed to be consistent with the BAU scenario.

The assumptions for the future total heating area and the heating demand per unit area are consistent with the BAU scenario. The trend for fossil fuel heating before 2025 was consistent with the BAU scenario. After 2025, the area for gas heating will decline until 2050, when natural gas consumption in the heating sector is completely replaced with electricity. The trends for the other heating technologies are consistent with the BAU scenario. Table 4.7 presents Beijing's future heating structure under the ELS scenario.

Table 4.7: Beijing's future heating structure under the electricity substitution scenario (square kilometres)

Year	District heating area				Individual heating		
	Gas boiler	Gas CHP	Waste incineration	Ground-source heat pump	Gas boiler	Gas CHP	Air-source heat pump
2020	457.52	59.37	23.32	35	154.42	16.23	65
2030	430.53	55.86	50.19	124.68	145.31	15.27	736.66
2050	0	0	232.43	378.07	0	0	4,115.33

Source: Authors' forecasts.

3) Synergistic development (SD) scenario

Although under the ELS scenario fossil fuels are replaced in the heating system, the electricity system will continue to consume a certain amount of fossil fuels to ensure stability. Therefore, not all the electricity consumed by heat pumps comes from renewable sources, preventing the ELS scenario from achieving 100 per cent renewable power and heating systems. We therefore propose the synergistic development (SD) scenario under which the energy consumption of CHP units will gradually shift from the combustion of fossil fuel to renewable energy. It is assumed that the individual heating development model under the SD scenario is consistent with the ELS scenario. The heating area of ground-source heat pumps and waste incineration in district heating systems is comparable with the BAU scenario. In terms of fossil fuel technology, it is expected that thermal power units will be retrofitted, replacing fossil fuel combustion with renewable generation based on the operation of the BAU scenario and the technical principle of CHP generation. Under the SD scenario, zero consumption of fossil fuel in the heating system can be achieved by 2050. Therefore, district heating is the same under the SD and BAU scenarios, while individual heating is comparable with that under the ELS scenario. The key difference is that gas-fired units are powered by renewable energy generation rather than by fossil fuels in the SD scenario. Figure 4.12 depicts the power and heating system operation modes under the SD scenario.

Methods

Energy system simulation tool: EnergyPLAN

The EnergyPLAN model is used to simulate the various sectors of the energy system in Beijing, and the model's accuracy is validated using the current state of Beijing's energy system. The EnergyPLAN model has several advantages as a bottom-up energy system analysis platform: 1) the model is suitable for the simulation of energy systems at various spatial scales; 2) the model can simulate the hourly technical operations of the energy system; 3) the model has advantages over existing institutional frameworks by enabling the adjustment of multiple operational methods and reducing the exogenous impact of the current system on the high proportion of renewable energy in the future; and 4) the model is highly efficient and can quickly compare the technical alternatives of various energy systems and adjust the operating strategy of the energy system in real time.

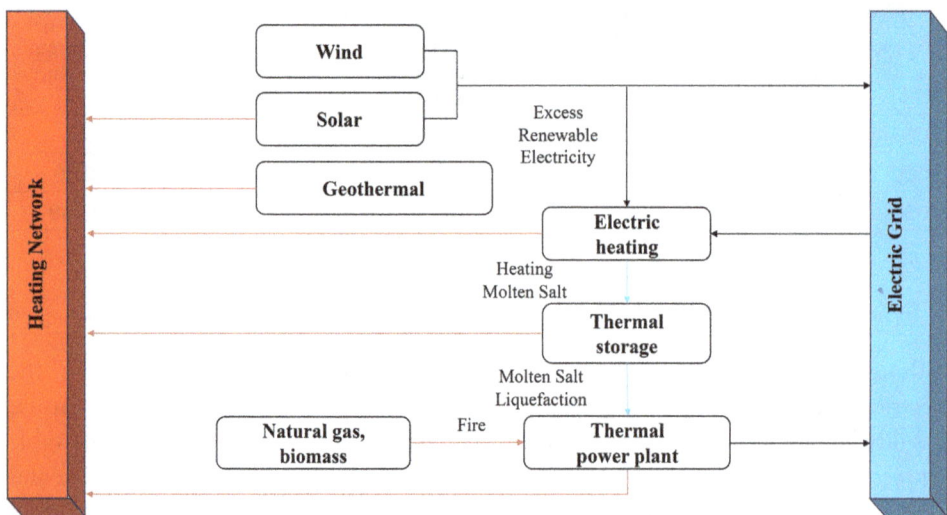

Figure 4.12: Synergistic development of renewable energy and fossil fuels
Source: Authors' schema.

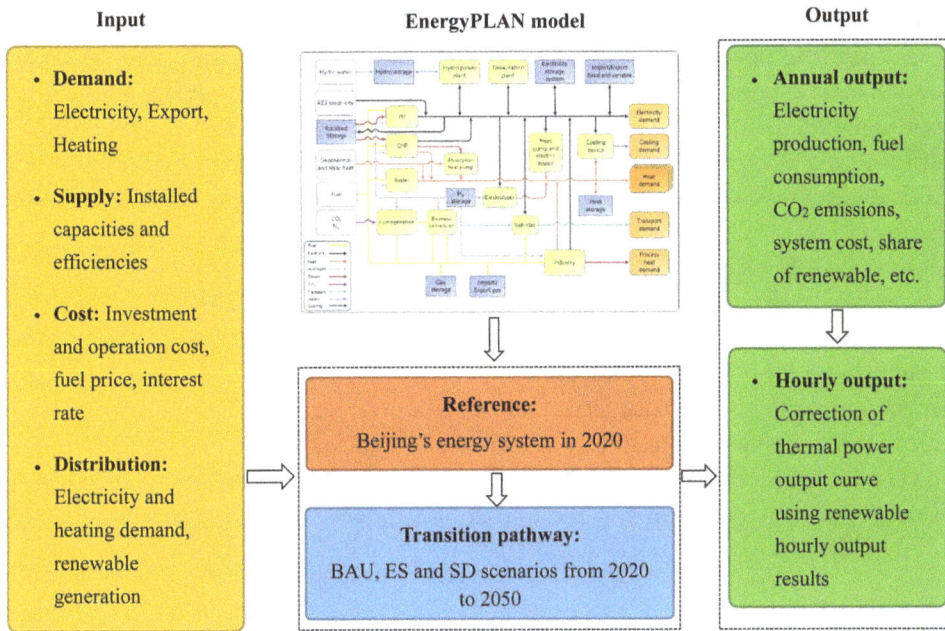

Figure 4.13: Schematic diagram of the EnergyPLAN model and the structure of Beijing's electricity and heating sectors

Source: Authors' schema.

The EnergyPLAN model is primarily utilised in the following areas to accomplish the strategic design of large-scale energy systems through technical and economic analysis of possible technology combinations: 1) determination of the transition pathway for a high-proportion renewable energy system (Hansen et al. 2019; Connolly et al. 2011, 2016; Sáfián 2014; Nielsen et al. 2011; Thellufsen et al. 2020; Kwon and Østergaard 2013); 2) the widespread impacts of certain technology on the overall operation of the energy system (Zhang et al. 2019; Askeland et al. 2019; Groppi et al. 2019; Lund and Salgi 2009; Thellufsen et al. 2019; Hedegaard et al. 2012); 3) evaluation of different energy policies and development strategies (Xiong et al. 2015; Ma et al. 2014); and 4) interconnection of various energy systems (Huang et al. 2020; Askeland et al. 2019; Alves et al. 2020; Drysdale et al. 2019). In addition, the EnergyPLAN model can be used to design energy systems at different levels, including urban (Zhang et al. 2019; Luo et al. 2021a), regional (Yuan et al. 2020; Huang et al. 2020) and national (Hansen et al. 2019; Xiong et al. 2015).

Figure 4.13 depicts the entire structure of the three decarbonisation scenarios for Beijing's power and heating systems using the EnergyPLAN toolbox. The inputs of the model involve all aspects of the energy system, which can be divided into five categories: energy demand, development scale of energy production and storage technology, cost, energy system operating strategy and carbon dioxide emissions. In addition, since renewable energy is intermittent and unstable, the model requires time-series data on energy demand and renewable power production at an hourly resolution, facilitating the matching of energy supply and demand in systems with variable power sources.

In terms of energy system output, the EnergyPLAN model employs various indicators to evaluate the operational status of the energy system under different technology combination modes, including carbon dioxide emissions, excess power production, socioeconomic costs, proportion of renewable energy and primary energy structure. Although the EnergyPLAN model fails to integrate renewable energy generation into CHP under the SD scenario, the operation of thermal power plants before and after the transition is merely the difference in fuel supply rather than a change in power generation technology. As such, EnergyPLAN can be used to indirectly simulate the energy supply after the thermal power plants' transition under the SD scenario. The detailed steps are as follows: 1) operation of the thermal power units is assumed to be dependent on the burning of fossil fuels under the SD scenario; 2) the renewable energy generation required to replace fossil fuel combustion in CHP units can be calculated based on the outputs obtained from the model—the power supply heat of the thermal power unit, the proportion of retrofitted gas-fired units, the power generation efficiency of the retrofitted CHP unit and the utilisation efficiency of renewable energy in the CHP generation unit (Zhang et al. 2019).

Model validation

In this section, the simulated power and heating supply structure of Beijing's energy system in 2020 was compared with the real data from Beijing in 2020 to validate the EnergyPLAN model (Table 4.8). The real data for power production were obtained from the China Electricity Council (CEC 2021), the district heating production data were from the Ministry of Housing and Urban–Rural Development (MoHURD 2021) and the individual heating production data were the difference between the total heating area and the district heating area in 2020. The results show that the relative error between the real data and the EnergyPLAN model simulation results is less than 2 per cent. Evidently, the EnergyPLAN model can accurately simulate Beijing's energy system and can be used to simulate the city's future energy system transition scenarios.

Table 4.8: Validation of the model of Beijing's energy system in 2020 (terawatt hours)

Indicator	Actual data	EnergyPLAN simulation	Difference (%)
Electricity			
Thermal	41.1	41.01	0.22
Hydro	1.1	1.1	0
Wind	0.4	0.4	0
PV	0.6	0.6	0
Waste incineration	2.3	2.31	0.43
Renewables share of electricity	3.95%	3.9%	1.28
Heating			
District boiler	43.01	43.01	0
District CHP	5.58	5.58	0
Individual heating	19.05	19.04	0.05

Source: Authors' validations.

Cost data

It is necessary to consider local resource endowments, the matching of energy supply and demand, system operating stability and the costs of various combinations of technologies —that is, the total annual cost of the energy system—to identify the optimal decarbonising strategy for the future energy system in Beijing. The EnergyPLAN model divides the total annual cost of the energy system into four components: investment cost, operation and maintenance (O&M) cost, fuel cost and external electricity prices. The technology lifecycle and interest rates should also be considered in the total system cost (Yuan et al. 2020). Most of the cost data come from the China model cost dataset (Luo et al. 2021a; Xiong et al. 2015) and the EnergyPLAN model cost database; fuel price information comes from Luo et al. (2021a) and fuel handling and other O&M costs come from Sorknæs (2020). Appendix 4.1 provides detailed cost statistics for various technologies, fuel price data and predictions of future costs for various technologies and gasoline prices.

Results and discussion

The impact of different heating decarbonisation paths on renewable energy integration

The impact of the decarbonisation of different heating sectors on renewable energy integration differs substantially between the BAU, ELS and SD scenarios. Figure 4.14 demonstrates that the local renewable energy share in total electricity and heating demand in Beijing is the lowest under the BAU scenario (72.57 per cent), without considering the imported power supply structure. Under the ELS and SD scenarios, the proportion of local renewable energy is more than 80 per cent and 100 per cent, respectively, of total electricity and heating demand in Beijing. Although fossil energy is fully replaced with renewable energy under the ELS scenario, thermal power units must produce power to match supply and demand in the power system in real time. Therefore, the electricity and heating demands cannot be met by 100 per cent renewable energy integration under the ELS scenario. Under the SD scenario, however, the operating mode of the retrofitted thermal power unit is essentially a fuel supply change rather than technical substitution. As a result, even if many thermal power production technologies are utilised under the SD scenario, renewable energy could meet 100 per cent of the electricity and heat demand.

In addition, although the ELS scenario achieves a high proportion (more than 80 per cent) of locally generated renewable energy, imported electricity still accounts for more than half of the electrical supply in Beijing—greater than under the SD scenario (see Table 4.9). The reason for this is that renewable energy cannot fill the gap resulting from the phasing out of gas-fired units in terms of the hourly balancing of energy supply and demand, so imported power is needed to achieve this balance. However, we are unable to determine the amount of imported decarbonised electricity. Therefore, we assume that the share of imported electricity in future power and heating systems is as low as possible to reduce carbon dioxide emissions. The SD scenario outperforms the ELS scenario in terms of both renewable energy integration and power supply structure.

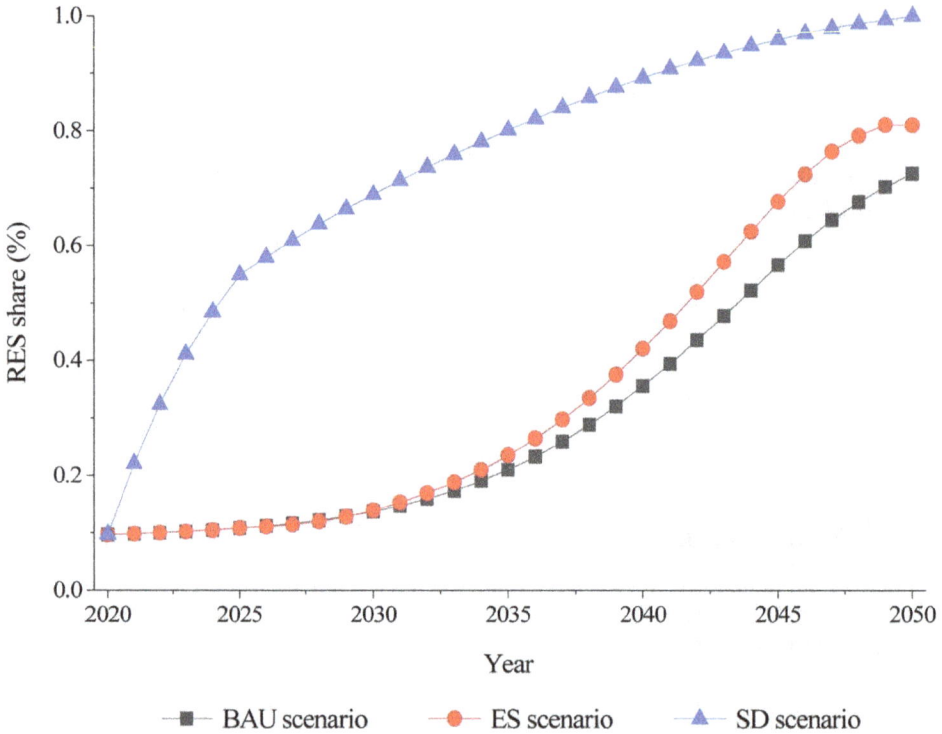

Figure 4.14: Shares of locally generated renewable energy in Beijing's electricity and heating systems

Source: Authors' projections.

Table 4.9: Proportion of imported electricity under the business-as-usual, electricity substitution and synergistic development scenarios (%)

	2025	2030	2035	2040	2045	2050
BAU scenario	57.87	56.12	56.74	56.98	52.99	47.09
ELS scenario	57.87	57.65	62.76	65.04	62.10	55.41
SD scenario	50.58	45.21	43.71	43.77	41.89	38.68

Source: Authors' projections.

Environmental sustainability of different heating decarbonisation paths

The impacts of different heating decarbonisation paths on carbon dioxide emissions are shown in Figure 4.15. The carbon dioxide emissions calculated in the EnergyPLAN model are the sum of carbon dioxide emissions from the use of coal, oil and natural gas in Beijing's energy system. The carbon dioxide emissions from using these fossil fuels are obtained by multiplying the consumption of each fuel by its emission coefficient. The amount of fossil fuel consumption is the direct output of the EnergyPLAN model and the carbon dioxide emission coefficient of each kind of fossil fuel comes from the database of the EnergyPLAN model, which has been widely used in the emissions accounting for China's regional energy system (Yuan et al. 2020; Zhang et al. 2019;

Luo et al. 2021a; Xiong et al. 2015). The results show that the carbon dioxide emissions arising from fossil fuels are lowest under the SD scenario, as its goal is to eliminate carbon dioxide emissions from natural gas combustion in power and heating systems. Under the ELS scenario, in which natural gas is gradually replaced with electricity in the heating system, the carbon dioxide emissions in the energy system cannot be fully eliminated since a small number of gas-fired units are required to maintain the system's stable operation. Thus, the SD scenario is the most environmentally friendly scenario.

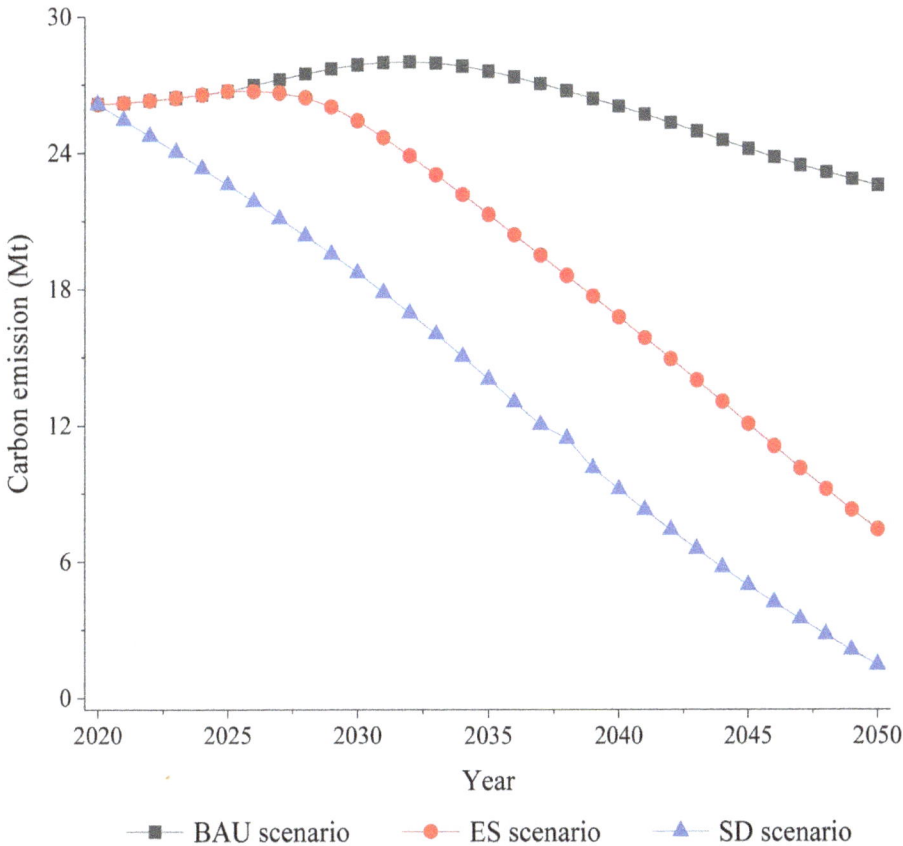

Figure 4.15: Carbon dioxide emissions from fossil fuel combustion in electricity and heating systems
Source: Authors' projections.

The above analysis only considers carbon dioxide emissions resulting from the combustion of fossil fuels. Given that the electricity supply structure in Beijing is dominated by imported electricity, as shown in Table 4.9, it is necessary to further analyse whether imported and exported electricity affect carbon dioxide emissions in the electricity and heating systems. Fortunately, in addition to carbon emissions from fossil fuel combustion, the EnergyPLAN model has an output indicator for carbon dioxide emissions, which represents the system's emissions from imported electricity and exportable excess electricity production (EEEP). This indicator can be used to analyse how imported and exported electricity affect the environmental benefits of the electricity and heating systems. Taking the BAU scenario as an example, when the system requires imported electricity, carbon dioxide emissions in the system would exceed the emissions resulting from fossil fuel combustion in Beijing, as shown in Figure 4.16. This indicates that some of the imported electricity comes from fossil fuel power generation, which will have negative environmental impacts on the electricity and heating systems. However, the growth rate of carbon dioxide emissions (pentagram) gradually slows when the EEEP emerges, and carbon dioxide emissions from imported and exported electricity will peak in 2037 and then decline rapidly. These results indicate that although the export of renewable energy reduces the carbon dioxide emissions of the external electricity system, considering that this electricity is produced locally from renewable energy, it is still viewed as an environmental benefit from the local electricity system. The mechanisms of the impacts of imported and exported electricity on carbon dioxide emissions under the ELS and SD scenarios are the same as under the BAU scenario.

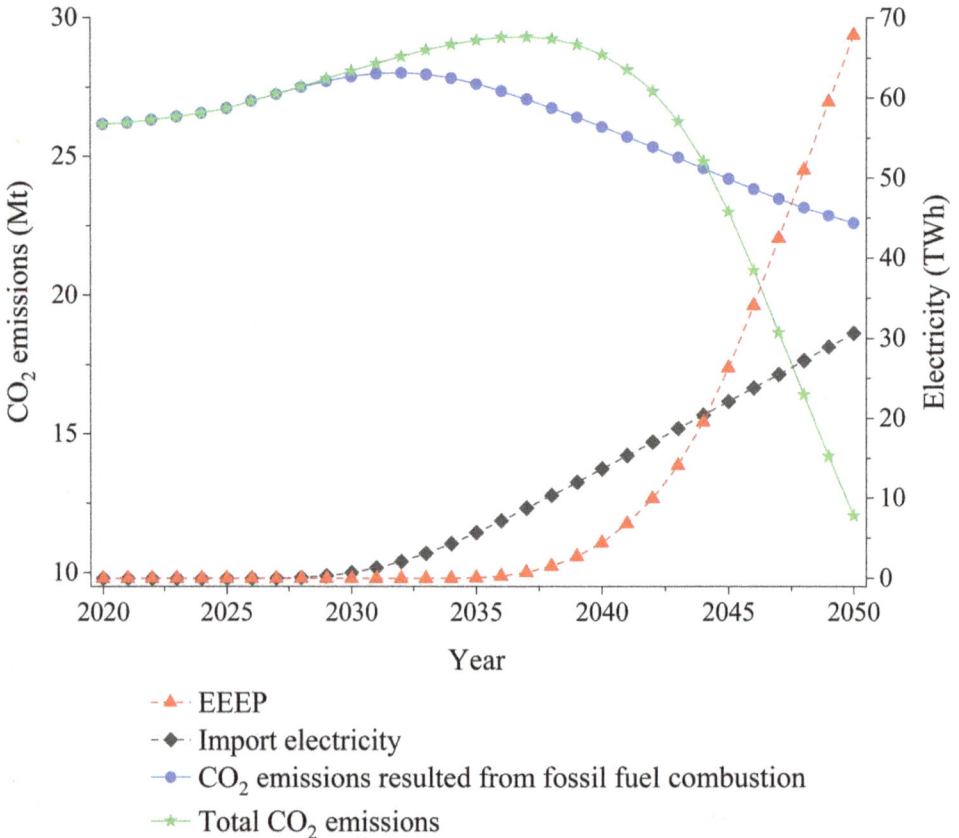

Figure 4.16: Carbon dioxide emissions from electricity and heating systems with and without consideration of imported electricity and exportable excess electricity production under the business-as-usual scenario

Note: The left axis represents carbon dioxide emissions and the right axis represents imported electricity and exportable excess electricity production.

Source: Authors' projections.

Table 4.10 presents the values of carbon dioxide emissions under the three scenarios. Emissions under the SD scenario change from positive to negative in 2047 when imported and exported electricity are considered. This suggests that, when local fossil fuel consumption is reduced, renewable energy can not only eliminate local carbon dioxide emissions, but also reduce emissions from external electricity systems. By 2050, the imported and exported electricity will generate 10.55 megatonnes of carbon dioxide emission reductions in Beijing under the SD scenario, which is equivalent to 30.85 terawatt hours of coal consumption reductions. In contrast, the carbon dioxide emission reductions and coal consumption reductions are only 1.59 megatonnes and 4.67 terawatt hours, respectively, under the ELS scenario. The main reason for this difference is that the ELS scenario requires more imported electricity and produces more negative environmental effects than the SD scenario. A similar conclusion is drawn when comparing the relevant results under the BAU scenario and the SD scenario. In sum, the SD scenario is superior to both the ELS and the BAU in terms of carbon dioxide emissions reduction resulting from imported and exported electricity.

Table 4.10: Carbon dioxide emissions from imported and exported electricity under the BAU, ELS and SD scenarios (megatonnes)

Scenario	Year										
	2020	2021	2022	2023	2024	2025	2026	2027	2028	2029	2030
BAU	26.153	26.202	26.307	26.42	26.555	26.721	26.985	27.247	27.526	27.806	28.082
ELS	26.153	26.202	26.307	26.42	26.555	26.721	26.716	26.726	26.763	26.809	26.861
SD	26.153	25.451	24.756	24.045	23.32	22.601	21.857	21.104	20.359	19.611	18.879
	Year										
	2031	2032	2033	2034	2035	2036	2037	2038	2039	2040	2041
BAU	28.349	28.602	28.837	29.037	29.187	29.291	29.307	29.232	29.025	28.669	28.122
ELS	26.912	26.964	27.004	27.014	26.991	26.922	26.871	26.547	26.197	25.7	25.02
SD	18.17	17.496	16.86	16.249	15.659	15.07	14.439	14.104	12.961	12.046	10.974
	Year										
	2042	2043	2044	2045	2046	2047	2048	2049	2050		
BAU	27.333	26.248	24.813	23	20.894	18.655	16.411	14.196	12.039		
ELS	24.105	22.9	21.351	19.431	17.219	14.752	12.003	9.021	5.832		
SD	9.691	8.138	6.264	4.027	1.503	–1.155	–3.832	–6.488	–9.09		

Source: Authors' projections.

Economic affordability of different heating decarbonisation paths

The cost-effectiveness of various heating decarbonisation paths for electricity and heating systems is critical to determine whether power and heating systems using high proportions of renewable energy are feasible in the future. Table 4.11 demonstrates that the total system cost under the ELS and SD scenarios is 8.88 per cent and 18.08 per cent, respectively, higher than under the BAU scenario. Furthermore, the cost in the SD scenario is 8.44 per cent higher than under the ELS scenario. Although the cost of fossil energy consumption is zero under the SD scenario, the cost of retrofitting thermal power plants is relatively high, as is the power generation cost using renewables. The higher cost under the ELS scenario is the result of the installation of air-source heat pumps. However, the cost increase is much lower than that of the thermal power plant retrofit under the SD scenario. The cost-effectiveness of the SD scenario can be improved in the future by reducing renewable energy generation costs, on the one hand, and improving the efficiency of retrofitted CHP units on the other. Table 4.11 lists the cost structure of future heating systems with high proportions of renewable energy integration. The results show that most of the cost of the heating system under the SD scenario is attributed to the thermal power plant retrofit rather than the fuel cost.

Table 4.11: The total cost and cost components of the energy system under the three scenarios (RMB billion)

		2025	2030	2035	2040	2045	2050
BAU scenario	Total energy system cost	117.88	160.13	204.73	257.87	325.80	402.93
	Investment cost	43.51	70.94	105.14	148.39	207.89	278.72
	Fixed O&M cost	18.81	24.04	30.78	39.49	51.74	66.41
	Variable O&M cost	55.56	65.16	68.81	70.00	66.18	57.80
ELS scenario	Total energy system cost	117.88	172.64	225.32	284.34	355.72	438.72
	Investment cost	43.51	84.35	129.17	180.66	246.41	327.22
	Fixed O&M cost	18.81	26.38	34.96	45.04	58.32	75.23
	Variable O&M cost	55.56	61.91	61.19	58.63	51.00	36.28
SD scenario	Total energy system cost	188.31	228.57	286.92	341.36	404.42	475.76
	Investment cost	106.59	139.94	191.52	240.63	299.85	369.11
	Fixed O&M cost	33.06	39.38	49.85	59.73	71.80	86.03
	Variable O&M cost	48.52	49.01	45.23	40.63	32.37	20.17

Source: Authors' projections.

Conclusions and policy implications

This chapter examined two representative cases of inter-industrial coordination in China's energy transition: photovoltaic–energy storage–charging station (PV–ES–CS) and the synergistic development of renewable energy and heating systems in Beijing. Despite differences in technical architecture and application scenarios, both cases illustrate the crucial role of coordinated planning between renewable energy supply and power demand in improving the economic viability of decarbonisation and reducing carbon emissions.

The study demonstrates that, first, the PV–ES–CS development model is superior to the independent development of distributed PV, ES power stations, EV charging stations, combined distributed PV and ES power stations and combined ES power stations and EV charging stations. The PV–ES–CS also has stronger environmental benefits because distributed PV can improve environmental value while ES is favourable for minimising power fluctuations by cutting peaks and filling valleys, which can improve energy efficiency. Second, PV–ES–CS systems show significant differences in economic and environmental benefits depending on the location in which they are constructed. Considering only the return on investment, the preferred places for PV–ES–CS construction should be hospitals, shopping malls, teaching buildings, hotels, office buildings, factories and residences, in this order; this is because the closer the load curve of the building is to the generation curve of solar PV in the daytime, and the lower is the power consumption at night, the higher will be the return on investment in a PV–ES–CS system. Considering only the carbon dioxide emission reductions per unit of investment, the preferred places for PV–ES–CS construction are hospitals, shopping malls, factories, hotels, teaching buildings, office buildings and residences, in this order; this is because the closer the load curve of the building is to the generation curve of solar PV in the daytime, the higher are the carbon dioxide emission reductions of a PV–ES–CS system. Hence, considering both the return on investment and carbon dioxide emission reductions per unit of investment, the most preferred places for PV–ES–CS construction are hospitals, and the last, residences. Third, the economic and environmental effects of PV–ES–CS systems are basically the same— that is, high economic benefits often accompany high environmental benefits. The economic benefits of PV–ES–CS are closely related to the utilisation rate of distributed PV. A low PV curtailment rate has high economic benefits and carbon dioxide emissions reduction equal to the distributed PV amount multiplied by the grid's carbon dioxide emissions factor.

Bejing's power–heat coupling model reveals that, first, the SD scenario, under which the CHP incorporating renewable energy generation occurs synergistically with heating system decarbonisation, can provide the efficient utilisation of renewable energy and the elimination of carbon dioxide emissions from fossil fuel combustion. Specifically, this scenario achieves 100 per cent renewable energy use to meet electricity and heating demands while reducing emissions from fossil fuel combustion to zero by 2050. Carbon dioxide emissions still occur under the ELS scenario because gas-fired units cannot be completely replaced. In addition, imported and exported electricity will generate 10.55 megatonnes of carbon dioxide emission reductions under the SD scenario, while under the ELS scenario the emission reductions are only 1.59 megatonnes. Furthermore, the overall economic costs under the ELS and SD scenarios are 8.16 per cent and 15.31 per cent higher, respectively, than those under the BAU scenario. Second, as the proportion of renewable energy in the power and heating systems increases, there will be a surplus of renewable electricity produced that exceeds demand. Surplus renewable energy can be used locally or exported to obtain revenue. It is found that the local use of surplus renewable energy can promote decarbonisation of the heating system and the synergistic development of renewable energy. Third, the cost of fossil fuel consumption accounts for less than 14.34 per cent of the total cost in the electricity and heating systems in 2050, suggesting that the cost of fossil fuel consumption is no longer the most important factor impacting the cost–benefit of future electricity and heating systems. Instead, the investment cost of the sophisticated renewable energy infrastructure required to facilitate heating decarbonisation, which accounts for more than 69.17 per cent of the total cost of the electricity and heating systems in 2050, will be the most important factor affecting the cost–benefit.

Based on the two case studies, several policy recommendations are put forward.

1. Promote integrated infrastructure planning and PV–ES–CS deployment: Policymakers should strengthen spatial and investment coordination across power generation, transport and heating sectors to ensure that infrastructure is aligned with renewable generation profiles and demand characteristics. In particular, the construction and development of PV–ES–CS stations should be increasingly promoted.

2. Prioritise flexible demand-side deployment: Incentives should be provided for installing PV–ES–CS systems in locations with favourable load profiles, such as hospitals and commercial buildings, and for electrification of heat demand using smart control technologies.

3. Implement differential incentive policies to unlock system value. For large-scale (investment above RMB41.5 million in this study) PV–ES–CS stations, incentive policies should be implemented to increase the peak and valley electricity price difference; for small-scale systems (investment below RMB13 million in this study), incentives will improve the number of EV charging stations; for all scales of PV–ES–CS systems, the cost reduction of ES plays an important role in promoting the development of PV–ES–CS stations.

4. Facilitate the progressive retrofitting of fossil fuel–based CHP units: To accelerate the decarbonisation of heating systems, it is essential to promote the gradual conversion of coal-fired and gas-fired CHP plants into renewable-powered thermal units. By changing the primary energy source rather than the entire technological system, this approach enables the continued use of existing infrastructure while aligning power and heat production with carbon neutrality goals. Such retrofitting also supports the integrated planning and operation of power and heat systems, improving overall system flexibility and decarbonisation efficiency.

5. The subsidy policies or investment strategies for energy corporations should focus on decreasing capital investment in infrastructure construction to maximise the long-term cost–benefit of high-proportion renewable energy integration and heating system decarbonisation.

Future research

As China's energy transition enters a new phase, marked by high shares of renewables and complex cross-sector interactions, the need for robust mechanisms of inter-industrial coordination becomes increasingly urgent. However, existing academic and policy research falls short in several key areas that are critical for scaling up and institutionalising such coordination. First, existing studies tend to concentrate on bilateral interactions between sectors, lacking a systemic, multisectoral and multilevel coupling perspective. Second, most modelling approaches are based on static optimisation, which fails to capture dynamic features such as load flexibility, market price interactions and dispatch uncertainty. Third, institutional and policy mechanisms supporting coordination are often fragmented—particularly in cross-sectoral investment alignment, carbon price transmission and regionally integrated governance. These gaps point to the urgent need for a more holistic and dynamic framework to understand and guide inter-industrial coordination.

To address these limitations, it is necessary to develop a multidimensional research framework that systematically characterises the mechanisms and features of different types of industrial synergy. This framework is organised around six key dimensions, each targeting a distinct aspect of coordination:

1. Horizontal synergistic development, which focuses on the systemic coupling of renewable energy with sectors such as transportation, heating and energy storage.

2. Vertical synergistic development, which examines the coordination of renewable energy deployment and the electrification of energy-intensive industries.

3. Spatial synergistic development, which addresses the regional mismatch between renewable energy production and consumption, along with mechanisms for cross-regional dispatch and market integration.

4. Temporal synergistic development, which explores the synchronisation of renewable energy expansion and the phase-out of traditional fossil fuels to ensure system stability.

5. Technological synergistic development, which investigates how advances in key low-carbon technologies, pathway selections and cost evolution affect the potential for coordination.

6. Institutional synergistic development, which centres on institutional safeguards, including the design of flexible power systems, market mechanisms and cross-sectoral policy integration.

Together, these six dimensions offer a comprehensive structure for future research and modelling efforts, enabling more robust evaluation of coordination strategies under various energy transition scenarios.

Building on this framework, future research must also overcome critical technical and methodological challenges. Limitations in data granularity, algorithmic integration and spatial resolution constrain the accuracy of coordination modelling. Leveraging advanced tools such as geographic information systems, big-data analytics and machine learning can enhance the identification, simulation and validation of synergistic mechanisms. Furthermore, energy system transformation is not only a technical or institutional endeavour; it also has deep social implications. Issues such as changes in employment structure, local governance capacities and public behavioural responses are vital to consider. Hence, a comprehensive analytical framework should incorporate social coordination, facilitating an integrated approach that links technology, institutions and society to support decarbonisation goals.

Beyond methodological innovations, future research should also aim to elevate the theoretical foundations and practical relevance of coordination mechanisms. At the theoretical level, multisector coordination can be conceptualised as a complex systems coupling process. Relevant theories such as complexity science, institutional embeddedness theory and network evolution theory can help unpack the drivers, interaction pathways and boundary conditions of such coordination. This would enable a shift from empirical identification towards theoretical abstraction and mechanistic interpretation—contributing to a more generalisable theory of synergistic energy transition. At the practical level, research must focus on enhancing the policy relevance of modelling outcomes. This includes linking model outputs with specific policy objectives, improving the evaluability of policies, optimising implementation pathways and establishing a scenario-based comparative framework for coordinated strategy design.

References

Alves, M., R. Segurado, and M. Costa. 2020. 'On the Road to 100% Renewable Energy Systems in Isolated Islands.' *Energy* 198: 117321. doi.org/10.1016/j.energy.2020.117321.

Askeland, Kristine, Kristina N. Bozhkova, and Peter Sorknæs. 2019. 'Balancing Europe: Can District Heating Affect the Flexibility Potential of Norwegian Hydropower Resources?' *Renewable Energy* 141: 646–56. doi.org/10.1016/j.renene.2019.03.137.

Aunedi, Marko, Antonio Marco Pantaleo, Kamal Kuriyan, Goran Strbac, and Nilay Shah. 2020. 'Modelling of National and Local Interactions Between Heat and Electricity Networks in Low-Carbon Energy Systems.' *Applied Energy* 276: 115522. doi.org/10.1016/j.apenergy.2020.115522.

Beijing Municipal Bureau of Statistics. 2021a. *2020 Beijing Energy Balance Sheet*. Beijing: China Statistics Press.

Beijing Municipal Bureau of Statistics. 2021b. 'Gross Regional Product.' *Beijing Statistical Yearbook 2020*. Beijing: China Statistics Press.

Beijing Municipal Commission of Development and Reform. 2019. *Implementation Opinions on Further Accelerating the Application of Heat Pump Systems and Promoting Clean Heating*. Beijing: Beijing Municipal Commission of Development and Reform.

Beijing Municipal Commission of Development and Reform. 2020a. 'In the Winter of 2020, Beijing's Clean Heating Area Exceeded 98%, and Many Projects Were Approved to Weave a Dense Heating Network.' Press release. Beijing: Beijing Commission of Municipal Development and Reform.

Beijing Municipal Commission of Development and Reform. 2020b. *Notice of Beijing Municipal Development and Reform Commission on Adjusting the Sales Electricity Price in Beijing.* [2020], no. 1708 (30 November). Beijing: Beijing Municipal Commission of Development and Reform. www.beijing.gov.cn/zhengce/zhengcefagui/202012/t20201201_2153982.html.

Beijing Municipal People's Government. 2017a. *Beijing Energy Development Plan during the Thirteenth Five-Year Plan Period.* Beijing: Beijing Municipal People's Government.

Beijing Municipal People's Government. 2017b. *Thirteenth Five-Year Plan for New and Renewable Energy Development.* Beijing: Beijing Municipal People's Government.

Beijing Municipal People's Government. 2021. *Fourteenth Five-Year Plan for Ecological and Environmental Protection: Promoting the Large-Scale Application of Local Photovoltaics.* Beijing: Beijing Municipal People's Government.

Beijing Municipal People's Government. 2022. *Beijing Energy Development Plan during the Fourteenth Five-Year Plan Period.* Beijing: Beijing Municipal People's Government.

Beuse, Martin, Bjarne Steffen, Mathias Dirksmeier, and Tobias S. Schmidt. 2021. 'Comparing CO_2 Emissions Impacts of Electricity Storage Across Applications and Energy Systems.' *Joule* 5, no. 6: 1501–20. doi.org/10.1016/j.joule.2021.04.010.

China Electricity Council (CEC). 2021. *2020 Power Industry Statistics Compilation.* Beijing: Statistics and Data Centre of the China Electricity Council.

Connolly, D., H. Lund, and B.V. Mathiesen. 2016. 'Smart Energy Europe: The Technical and Economic Impact of One Potential 100% Renewable Energy Scenario for the European Union.' *Renewable and Sustainable Energy Reviews* 60: 1634–53. doi.org/10.1016/j.rser.2016.02.025.

Connolly, D., H. Lund, B.V. Mathiesen, and M. Leahy. 2011. 'The First Step Towards a 100% Renewable Energy System for Ireland.' *Applied Energy* 88, no. 2: 502–7. doi.org/10.1016/j.apenergy.2010.03.006.

Denholm, Paul, Michael Kuss, and Robert M. Margolis. 2013. 'Co-Benefits of Large Scale Plug-In Hybrid Electric Vehicle and Solar PV Deployment.' *Journal of Power Sources* 236: 350–6. doi.org/10.1016/j.jpowsour.2012.10.007.

Ding, Yiting. 2025. 'Installed Capacity of Wind and Solar Power in China Exceeds 1.4 Billion Kilowatts.' *People's Daily*, 26 January. www.gov.cn/lianbo/bumen/202501/content_70012 10.htm.

Drysdale, David, Brian Vad Mathiesen, and Susana Paardekooper. 2019. 'Transitioning to a 100% Renewable Energy System in Denmark by 2050: Assessing the Impact from Expanding the Building Stock at the Same Time.' *Energy Efficiency* 12: 37–55. doi.org/10.1007/s12053-018-9649-1.

Eldeeb, Hassan H., Samy Faddel, and Osama A. Mohammed. 2018. 'Multi-Objective Optimization Technique for the Operation of Grid Tied PV Powered EV Charging Station.' *Electric Power Systems Research* 164: 201–11. doi.org/10.1016/j.epsr.2018.08.004.

Englberger, Stefan, Archie C. Chapman, Wayes Tushar, Tariq Almomani, Stephen Snow, Rolf Witzmann, Andreas Jossen, and Holger Hesse. 2021. 'Evaluating the Interdependency Between Peer-to-Peer Networks and Energy Storages: A Techno-Economic Proof for Prosumers.' *Advances in Applied Energy* 3: 100059. doi.org/10.1016/j.adapen.2021.100059.

Groppi, D., D. Astiaso Garcia, G. Lo Basso, and L. De Santoli. 2019. 'Synergy Between Smart Energy Systems Simulation Tools for Greening Small Mediterranean Islands.' *Renewable Energy* 135: 515–24. doi.org/10.1016/j.renene.2018.12.043.

Gu, Chaolin, Tan Zongbo, and Liu Wan. 2009. 'Progress in Research on Climate Change, Carbon Emissions and Low-Carbon Urban Planning.' *Urban Planning Forum*, no. 3: 38–45.

Hansen, Kenneth, Brian Vad Mathiesen, and Iva Ridjan Skov. 2019. 'Full Energy System Transition Towards 100% Renewable Energy in Germany in 2050.' *Renewable and Sustainable Energy Reviews* 102: 1–13. doi.org/10.1016/j.rser.2018.11.038.

He, Jijiang. 2019. 'Beijing Dahongmen Jimei DC Photovoltaic Storage and Charging Integrated Project Creates Multiple Firsts in Various Fields.' *Polaris Power News Network*, 28 April. chuneng.bjx.com.cn/news/20190428/977571.shtml.

Hedegaard, Karsten, Brian Vad Mathiesen, Henrik Lund, and Per Heiselberg. 2012. 'Wind Power Integration Using Individual Heat Pumps—Analysis of Different Heat Storage Options.' *Energy* 47, no. 1: 284–93. doi.org/10.1016/j.energy.2012.09.030.

Hochreiter, Sepp, and Jürgen Schmidhuber. 1997. 'Long Short-Term Memory.' *Neural Computation* 9, no. 8: 1735–80. doi.org/10.1162/neco.1997.9.8.1735.

Huang, Pei, Marco Lovati, Xingxing Zhang, Chris Bales, Sven Hallbeck, Anders Becker, Henrik Bergqvist, Jan Hedberg, and Laura Maturi. 2019. 'Transforming a Residential Building Cluster into Electricity Prosumers in Sweden: Optimal Design of A Coupled PV–Heat Pump–Thermal Storage–Electric Vehicle System.' *Applied Energy* 255: 113864. doi.org/10.1016/j.apenergy.2019.113864.

Huang, Xiaodan, Hongyu Zhang, and Xiliang. Zhang. 2020. 'Decarbonising Electricity Systems in Major Cities Through Renewable Cooperation—A Case Study of Beijing and Zhangjiakou.' *Energy* 190: 116444. doi.org/10.1016/j.energy.2019.116444.

Ibrahim, Ibrahim Anwar, Slaiman Sabah, Robert Abbas, M.J. Hossain, and Hani Fahed. 2021. 'A Novel Sizing Method of a Standalone Photovoltaic System for Powering a Mobile Network Base Station Using a Multi-Objective Wind Driven Optimization Algorithm.' *Energy Conversion and Management* 238: 114179. doi.org/10.1016/j.enconman.2021.114179.

Intergovernmental Panel on Climate Change (IPCC). 2014. *Climate Change 2014: Mitigation of Climate Change. Working Group III Contribution to the IPCC Fifth Assessment Report.* Cambridge: Cambridge University Press. www.ipcc.ch/site/assets/uploads/2018/02/ipcc_wg3_ar5_full.pdf. doi.org/10.1017/CBO9781107415416.

Jimenez-Navarro, Juan-Pablo, Konstantinos Kavvadias, Faidra Filippidou, Matija Pavičević, and Sylvain Quoilin. 2020. 'Coupling the Heating and Power Sectors: The Role of Centralised Combined Heat and Power Plants and District Heat in A European Decarbonised Power System.' *Applied Energy* 270: 115134. doi.org/10.1016/j.apenergy.2020.115134.

Kaufmann, Robert K., Derek Newberry, Chen Xin, and Sucharita Gopal. 2021. 'Feedbacks Among Electric Vehicle Adoption, Charging, and the Cost and Installation of Rooftop Solar Photovoltaics.' *Nature Energy* 6, no. 2: 143–9. doi.org/10.1038/s41560-020-00746-w.

Kwon, Pil Seok, and Poul Alberg Østergaard. 2013. 'Priority Order in Using Biomass Resources —Energy Systems Analyses of Future Scenarios for Denmark.' *Energy* 63: 86–94. doi.org/ 10.1016/j.energy.2013.10.005.

Lashof, Daniel A., and Dilip R. Ahuja. 1990. 'Relative Contributions of Greenhouse Gas Emissions to Global Warming.' *Nature* 344, no. 6266: 529–31. doi.org/10.1038/344529a0.

Liu, Xin, Xinran Li, Zhuangxi Tan, Huang Jiyuan, Cui Yiwei, and Zhang Bingyu. 2021. 'Economic Comparison of Optimal Strategies Based on Different Types of Energy Storage Batteries Participating in Primary Frequency Regulation.' *High Voltage Engineering*: 1–9. chn.oversea.cnki.net/kcms/detail/detail.aspx?filename=GDYJ202204018&dbcode=CJFQ &dbname=CJFDLAST2022&uniplatform=NZKPT.

Lund, Henrik, and Georges Salgi. 2009. 'The Role of Compressed Air Energy Storage (CAES) in Future Sustainable Energy Systems.' *Energy Conversion and Management* 50, no. 5: 1172–9. doi.org/10.1016/j.enconman.2009.01.032.

Luo, Shihua, Weihao Hu, Wen Liu, Xiao Xu, Qi Huang, Zhe Chen, and Henrik Lund. 2021a. 'Transition Pathways Towards a Deep Decarbonization Energy System—A Case Study in Sichuan, China.' *Applied Energy* 302, no. 117507: 1–14. doi.org/10.1016/j.apenergy.2021. 117507.

Luo, Xing, Dongxiao Zhang, and Xu Zhu. 2021b. 'Deep Learning Based Forecasting of Photovoltaic Power Generation by Incorporating Domain Knowledge.' *Energy* 225: 120240. doi.org/10.1016/j.energy.2021.120240.

Luo, Z., Zechun Hu, Yonghua Song, Xia Yang, Kaiqiao Zhan, and Junyang Wu. 2011. 'Study on Plug-In EVs Charging Load Calculating.' *Automation of Electric Power Systems* 35, no. 14: 36–42.

Ma, Tao, Poul Alberg Østergaard, Henrik Lund, Hongxing Yang, and Lin Lu. 2014. 'An Energy System Model for Hong Kong in 2020.' *Energy* 68: 301–10. doi.org/10.1016/j.energy.2014. 02.096.

Ministry of Ecology and Environment (MEE). 2022. *China's Policies and Actions for Addressing Climate Change (2022)*. Beijing: Ministry of Ecology and Environment of the People's Republic of China. english.mee.gov.cn/Resources/Reports/reports/202211/P02022111060 5466439270.pdf.

Ministry of Housing and Urban–Rural Development (MoHURD). 2021. *Urban Construction Statistical Yearbook 2020*. Beijing: Ministry of Housing and Urban–Rural Development of the People's Republic of China.

Mohajan, Haradhan Kumar. 2017. 'Greenhouse Gas Emissions, Global Warming and Climate Change.' In *Proceedings of the 15th Chittagong Conference on Mathematical Physics, Jamal Nazrul Islam Research Centre for Mathematical and Physical Sciences (JNIRCMPS), Chittagong, Bangladesh. Volume 16*.

National Bureau of Statistics of China (NBS). 2005. *China Statistics Yearbook 2004*. Beijing: China Statistics Press.

National Development and Reform Commission (NDRC). 2022. *Fourteenth Five-Year Renewable Energy Development Plan*. Beijing: National Development and Reform Commission.

NDRC, National Energy Administration, Ministry of Finance, Ministry of Natural Resources, Ministry of Ecology and Environment, Ministry of Housing and Urban–Rural Development, Ministry of Agriculture and Rural Affairs, China Meteorological Administration, and National Forestry and Grassland Administration. 2022. *Fourteenth Five-Year Plan for Renewable Energy Development*. Beijing: National Development and Reform Commission. www.ndrc.gov.cn/xxgk/zcfb/ghwb/202206/t20220601_1326719.html?code=&state=123.

National Energy Board (NEB). 2020. *NB/T 10394-2020: Specification for Photovoltaic Power Generation System Performance*. National Energy Administration Announcement No. 5 of 2020. Beijing: National Energy Board. www.chinesestandard.net/PDF/English.aspx/NBT 10394-2020.

National Renewable Energy Consumption Monitoring and Early Warning Centre. 2025. *National New Energy Grid Connection and Consumption in 2024*. Beijing: National Renewable Energy Consumption Monitoring and Early Warning Centre. mp.weixin.qq.com/s/2SJA4s 8afIDoe0g3AWSuNw.

Navarra, Antonio. 2023. 'Syukuro Manabe: Recipient of Nobel Prize in Physics 2021.' In *Oxford Research Encyclopedia of Climate Science*. Oxford: Oxford University Press. doi.org/10.1093/ acrefore/9780190228620.013.930.

Nielsen, Steffen, Peter Sorknæs, and Poul Alberg Østergaard. 2011. 'Electricity Market Auction Settings in a Future Danish Electricity System with a High Penetration of Renewable Energy Sources—A Comparison of Marginal Pricing and Pay-as-Bid.' *Energy* 36, no. 7: 4434–44. doi.org/10.1016/j.energy.2011.03.079.

Olympios, Andreas V., Antonio M. Pantaleo, Paul Sapin, and Christos N. Markides. 2020. 'On the Value of Combined Heat and Power (CHP) Systems and Heat Pumps in Centralised and Distributed Heating Systems: Lessons from Multi-Fidelity Modelling Approaches.' *Applied Energy* 274: 115261. doi.org/10.1016/j.apenergy.2020.115261.

Omran, Walid A., M. Kazerani, and M.M.A. Salama. 2011. 'Investigation of Methods for Reduction of Power Fluctuations Generated from Large Grid-Connected Photovoltaic Systems.' *IEEE Transactions on Energy Conversion* 26, no. 1: 318–27. doi.org/10.1109/TEC. 2010.2062515.

Peng, Danni, and Xu Dawei. 2021. 'China's Power Shortage Pain.' *China Newsweek*, 14 October. mp.weixin.qq.com/s/zb6aiSPNG723pFd-HSEQLA.

Pu, Yanru, Peng Wang, Yuyi Wang, Wenhui Qiao, Lu Wang, and Yuhu Zhang. 2021. 'Environmental Effects Evaluation of Photovoltaic Power Industry in China on Life Cycle Assessment.' *Journal of Cleaner Production* 278: 123993. doi.org/10.1016/j.jclepro.2020.123993.

Sabadini, Felipe, and Reinhard Madlener. 2021. 'The Economic Potential of Grid Defection of Energy Prosumer Households in Germany.' *Advances in Applied Energy* 4: 100075. doi.org/ 10.1016/j.adapen.2021.100075.

Sáfián, Fanni. 2014. 'Modelling the Hungarian Energy System—The First Step Towards Sustainable Energy Planning.' *Energy* 69: 58–66. doi.org/10.1016/j.energy.2014.02.067.

Sainato, Michael. 2020. 'The Collapse of Coal: Pandemic Accelerates Appalachia Job Losses.' *The Guardian*, 29 May. www.theguardian.com/us-news/2020/may/29/coal-miners-corona virus-job-losses.

Shang, Wen-Long, Jinyu Chen, Huibo Bi, Yi Sui, Yanyan Chen, and Haitao Yu. 2021. 'Impacts of COVID-19 Pandemic on User Behaviors and Environmental Benefits of Bike Sharing: A Big-Data Analysis.' *Applied Energy* 285: 116429. doi.org/10.1016/j.apenergy.2020.116429.

Thellufsen, J.Z., H. Lund, P. Sorknæs, P.A. Østergaard, M. Chang, D. Drysdale, S. Nielsen, S.R. Djørup, and K. Sperling. 2020. 'Smart Energy Cities in a 100% Renewable Energy Context.' *Renewable and Sustainable Energy Reviews* 129: 109922. doi.org/10.1016/j.rser.2020.109922.

Thellufsen, Jakob Zinck, Steffen Nielsen, and Henrik Lund. 2019. 'Implementing Cleaner Heating Solutions Towards a Future Low-Carbon Scenario in Ireland.' *Journal of Cleaner Production* 214: 377–88. doi.org/10.1016/j.jclepro.2018.12.303.

von Storch, Hans, and Patrick Heimbach. 2022. 'Klaus Hasselmann: Recipient of the Nobel Prize in Physics 2021.' In *Oxford Research Encyclopedia of Climate Science*. Oxford: Oxford University Press. doi.org/10.1093/acrefore/9780190228620.013.931.

Wang, Mudan, Xianqiang Mao, Yubing Gao, and Feng He. 2018b. 'Potential of Carbon Emission Reduction and Financial Feasibility of Urban Rooftop Photovoltaic Power Generation in Beijing.' *Journal of Cleaner Production* 203: 1119–31. doi.org/10.1016/j.jclepro.2018.08.350.

Wu, Y., X. Zhang, H. Wang, and J. Sun. 2017. 'Molten Salt Heat Storage and Supply Technology Based on Heating Using Abandoned Wind Power PV Power or Off-Peak Power.' *Energy* 22: 93–99.

Xiong, Weiming, Yu Wang, Brian Vad Mathiesen, Henrik Lund, and Xiliang Zhang. 2015. 'Heat Roadmap China: New Heat Strategy to Reduce Energy Consumption Towards 2030.' *Energy* 81: 274–85. doi.org/10.1016/j.energy.2014.12.039.

Yang, Meng, Lihui Zhang, and Wenjia Dong. 2020. 'Economic Benefit Analysis of Charging Models Based on Differential EV Charging Infrastructure Subsidy Policy in China.' *Sustainable Cities and Society* 59: 102206. doi.org/10.1016/j.scs.2020.102206.

Ye, X. 2018. 'Economic and Environmental Value Analysis of Replacing Coal by Gas in Buildings.' Thesis, China University of Petroleum, Beijing.

Yuan, Meng, Jakob Zinck Thellufsen, Henrik Lund, and Yongtu Liang. 2020. 'The First Feasible Step Towards Clean Heating Transition in Urban Agglomeration: A Case Study of Beijing–Tianjin–Hebei Region.' *Energy Conversion and Management* 223: 113282. doi.org/10.1016/j.enconman.2020.113282.

Zhang, Hongyu, Li Zhou, Xiaodan Huang, and Xiliang Zhang. 2019. 'Decarbonizing a Large City's Heating System Using Heat Pumps: A Case Study of Beijing.' *Energy* 186: 115820. doi.org/10.1016/j.energy.2019.07.150.

Zhu, Shunmin, Guoyao Yu, Kun Liang, Wei Dai, and Ercang Luo. 2021. 'A Review of Stirling-Engine-Based Combined Heat and Power Technology.' *Applied Energy* 294: 116965. doi.org/10.1016/j.apenergy.2021.116965.

Appendix 4.1 Cost datasets of the future energy system

The cost datasets used for the economic analysis of the future energy system using the EnergyPLAN model were primarily derived from Luo et al. (2021a), although this provided data only for 2016, 2030 and 2050. Therefore, the economic analysis was based on the following assumptions. First, it was assumed that the costs remained constant from 2016 to 2019, changed every five years from 2020 to 2030 and the rate of change remained constant. Second, this study assumed a constant rate of change in fuel prices between 2016 and 2030 and 2030 and 2050. Third, the interest rate remained unchanged at 8 per cent (Yuan et al. 2020). The cost and fuel price data for different technologies in 2020 are listed in Appendix Tables A4.1 and A4.2.

Table A4.1: Costs including investments, fixed operation and maintenance costs and operation lifetime for different technologies in 2020

Technology	Investment cost (RMB million/MW)	Fixed operation and maintenance costs (% of investment cost)	Lifetime (years)
Coal power plant	4.3	4.0	25
Natural gas power plant	3.7	4.0	25
Biomass CHP plant	9.0	4.0	40
Natural gas CHP plant	4.9	4.0	25
Biomass CHP plant	9.9	4.0	40
Pump turbine	8.0	1.5	50
Wind power	9.6	2.0	25
PV power	22.0	2.0	30
Natural gas boiler	0.78	3.0	20
Ground-source heat pump	4.0	1.0	25

Table A4.2: Assumed fuel prices in 2020 (RMB/gigajoule)

Year	Coal	Fuel oil	Diesel	Petrol	Natural gas	LPG	Biomass
2020	23.19	56.33	99.85	106.3	58.34	69.91	40.33

5

The rise of electric vehicles in China: Growth, challenges and future prospects

Shanjun Li, Ha Pham, Yuerong Wang and Lin Yang

This chapter examines the evolution of China's electric vehicle (EV) industry over the past decade, documenting its rapid expansion, policy support and market dynamics. It explores the factors that contribute to China's dominant role in global EV production and sales, including government incentives, the accessibility of charging infrastructure and automaker expansion strategies. The chapter then addresses the ongoing challenges facing the industry, such as intensifying global competition, trade barriers, battery technology constraints and infrastructure disparities. Finally, the chapter discusses prospects for the industry, highlighting technological advancements, supply chain resilience and emerging policy developments to further promote sustainable mobility.

EV market trends

The global EV market has expanded rapidly during the past decade, reaching approximately 17.1 million units by 2024, as shown in Figure 5.1.[1] China, Europe and the United States are the three dominant EV markets. In 2024, China accounted for nearly 65 per cent of global sales, maintaining its position as by far the world's largest EV market. Meanwhile, Europe and the United States contributed 18 per cent and 8 per cent of total EV sales, respectively. Globally, the share of EVs reached nearly 20 per cent of all new passenger vehicle sales in 2024, as depicted in Figure 5.1. Significant disparity in EV penetration exists across markets: the market share of EVs was 48 per cent in China, 24 per cent in Europe as a whole and only 11 per cent in the United States.

---

1   EVs include both full battery electric vehicles and plug-in hybrid electric vehicles.

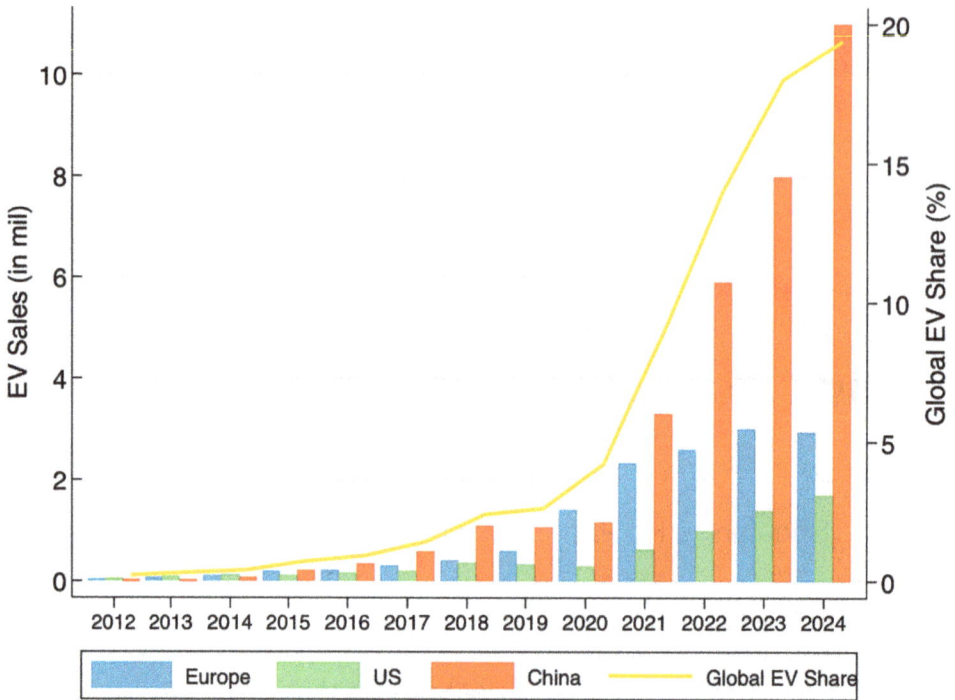

Figure 5.1: Global electric vehicle sales, 2012–24

Notes: The yellow line represents the global EV market share among all new passenger vehicle sales.

Sources: IEA Global EV Outlook (www.iea.org/data-and-statistics/data-tools/global-ev-data-explorer); MarkLines (www.marklines.com/portal_top_en.html).

While the United States was the global leader in EV sales before 2015, sales growth there has since been the slowest among the three markets. From 2016, China consistently outpaced both Europe and the United States in annual EV sales, except in 2020, when Europe temporarily overtook it, largely due to strict Covid-19 shutdowns in China. After 2020, China entered a phase of explosive growth, positioning it as the principal engine of global EV market expansion. In 2024, total EV production reached 12.89 million units, making China the first country to achieve an annual EV production volume of more than 10 million units—a remarkable increase from just 18,000 units in 2013. China's emergence as a global EV leader stems from a decade-long strategic commitment to industry development. The foundation was laid in 2009 with the 'Ten Cities, Thousand Vehicles' pilot program, which initiated state-supported market cultivation. Since then, a combination of generous subsidies, environmental regulations and targeted industrial policies accelerated EV adoption.

The growth of China's EV market has implications that extend beyond its borders. In 2024, China's EV exports surpassed 2 million units for the first time, with Europe and South-East Asia emerging as key destinations. This outward expansion reflects the increasing global competitiveness of Chinese manufacturers, driven by cost advantages, technological innovation and, increasingly, brand reputation. As China continues to lead the global EV transition, it is important to understand the key factors behind its success, such as business strategies and policy frameworks. In the following sections, we explore these key factors, offering insight into how China has emerged as a dominant force in the global EV industry, the challenges and future prospects.

Industry development

This section examines the key drivers behind the rise of China's EV industry by first exploring how automakers have pursued market leadership, competitive pricing and supply chain integration. It then shifts focus to charging infrastructure, which plays a key role in promoting EV adoption and industry growth.

Market leadership and expansion strategies

The rapid rise of Chinese EV manufacturers is evident in Figure 5.2a, which depicts the number of EV brands by country of origin from 2013 to 2023. Chinese brands (represented in blue) make up the largest share of EV manufacturers. While European, Japanese, South Korean and US automakers have also expanded their presence, their growth has been much slower. As of 2024, China had about 50 active domestic EV manufacturers, with firms such as BYD, Geely, NIO and XPENG playing crucial roles in shaping the industry's development. These automakers have adopted different strategies to expand both domestically and internationally while driving innovations in battery technology and production efficiency.

Among these manufacturers, BYD has emerged as the market leader, surpassing both domestic and international competitors. In 2024, BYD sold more than 4.25 million EVs, making it the largest EV producer globally. BYD has also accelerated its global expansion by exporting more than 417,000 EVs to international markets, including Europe, South-East Asia and Latin America. A key factor behind BYD's success is its vertically integrated business model, which allows for inhouse production of key components such as batteries and semiconductor chips. This structure enhances cost control and ensures supply chain reliability, providing competitive advantages over its competitors. Importantly, BYD is one of the few automakers that manufactures battery cells directly from raw materials. This capability is particularly valuable, given that batteries typically account for 30–40 per cent of an EV's total cost.

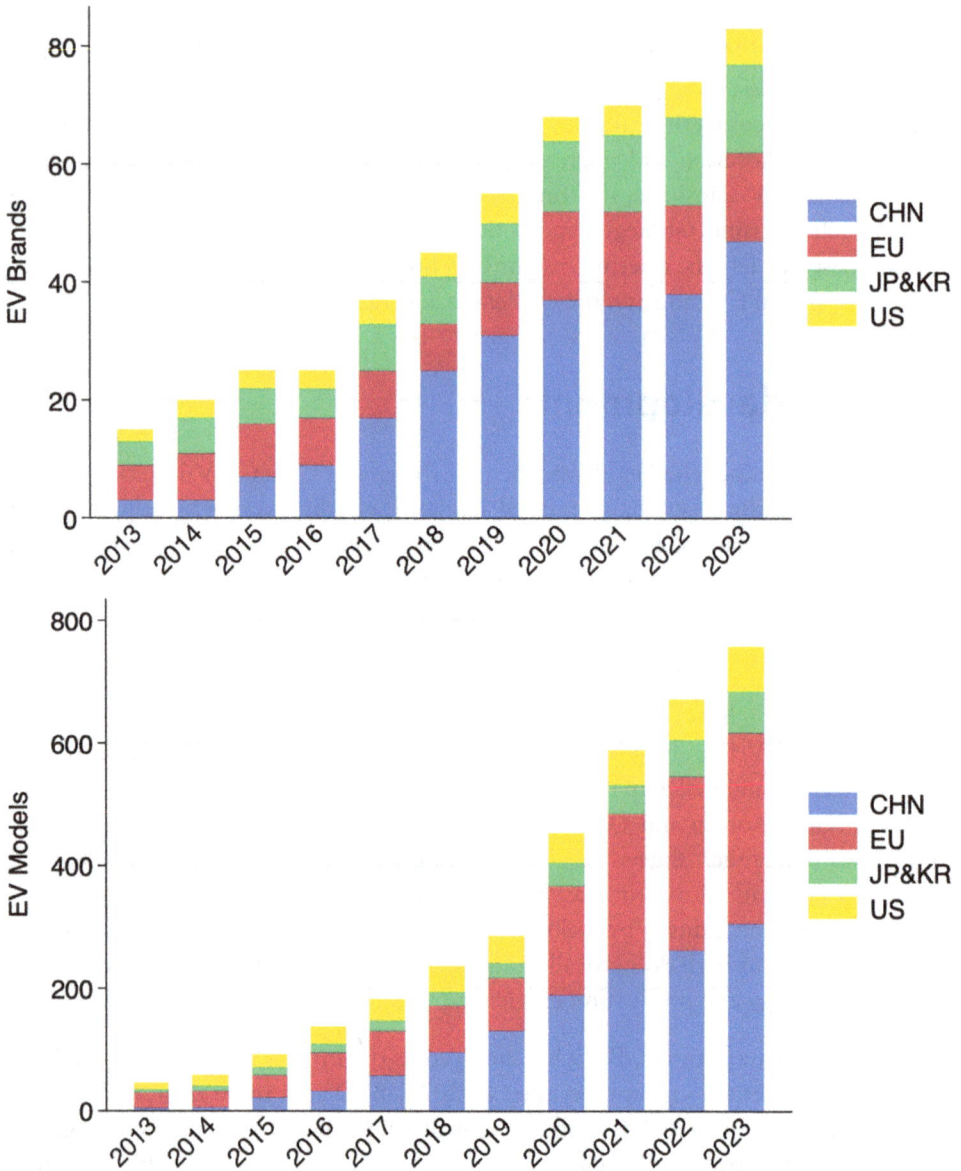

Figure 5.2: Comparison of electric vehicle brands and models by country

Notes: (Top) the number of EV brands by country of origin; (Bottom) the growth in the number of EV models.

Sources: IEA Global EV Outlook (www.iea.org/data-and-statistics/data-tools/global-ev-data-explorer); MarkLines Co. (www.marklines.com/portal_top_en.html).

While BYD is the clear market leader, other Chinese automakers are pursuing different strategies to strengthen their positions. Geely plans for EVs to constitute 50 per cent of its total sales by 2025 and utilises multiple sub-brands to target diverse consumers. Meanwhile, China's leading new-generation EV startups such as NIO, XPENG and Li Auto emphasise technological integration, autonomous driving and digital user experience, setting themselves apart from traditional automotive brands. NIO has gained attention for its battery-swap technology, which allows rapid battery replacement and minimises charging downtime. By the end of 2024, NIO operated nearly 2,800 battery-swap stations across China, with about one-third of those strategically located along highways to facilitate long-distance travel. In contrast, XPENG and Li Auto emphasise autonomous driving technologies, positioning themselves as frontrunners in smart mobility. These differing strategies reflect the broader strategic diversity among Chinese EV manufacturers, who are leveraging innovation and market segmentation to compete for global leadership.

Price competition and model availability

Building on their expansion strategies, Chinese EV manufacturers have further strengthened their market position through competitive pricing and offering a wide range of vehicle models. Supported by substantial government subsidies, economies of scale and dominance in battery manufacturing, Chinese firms enjoy lower production costs than many international competitors. In response to intensifying domestic market competition, automakers in China have engaged in aggressive price competition to attract a greater share of price-sensitive consumers. Throughout 2023 and 2024, the Chinese EV market experienced an intense price war, with virtually all major domestic EV producers, such as BYD, Geely, NIO and Tesla, engaged in price reductions. Since beginning local production of its Model 3 in China in late 2019, Tesla has implemented multiple price cuts in the Chinese market. As of 2024, the price of a Tesla Model 3 in China was approximately 35 per cent lower than its 2019 launch price and about 20 per cent lower than the corresponding price in the United States. On the one hand, these aggressive pricing strategies have squeezed profit margins and directly contributed to the market exit of several brands. On the other hand, they have benefited consumers through improved affordability, thereby expanding domestic EV market share and helping Chinese automakers gain a foothold in overseas markets.

Complementing these pricing strategies is China's leadership in EV model diversity. As shown in Figure 5.2b, China offers more EV models than any other major market, including Europe, Japan, South Korea and the United States. This extensive range caters to a broad spectrum of consumer preferences, from compact urban vehicles to high-performance luxury models. The continued introduction of new designs and configurations has helped Chinese automakers maintain a strong domestic presence while improving their competitiveness in international markets. By combining affordability with product variety, Chinese EV firms are well positioned to compete in the international market, which may or may not be welcoming such competition.

Supply chain integration and technological advancements

Supply chain integration and technological advancements further reinforce China's leadership in the EV sector. China is home to CATL and BYD, the world's two largest EV battery producers, which held 38 per cent and 17 per cent, respectively, of the global market share in 2024, supplying both domestic and international automakers. Overall, China accounted for more than 75 per cent of global EV battery production, underscoring its dominant role in the supply chain—not only in manufacturing, but also in the refining of key materials such as lithium, cobalt and graphite.

Cost reduction through economies of scale has also been critical. China has the world's largest battery manufacturing capacity, with total production and sales of lithium-ion batteries (primarily for EVs) reaching 1.097 terawatt hours (TWh) and 1.040 TWh, respectively. This production scale allows automakers to secure batteries at lower prices, reducing vehicle costs and increasing competitiveness. Additionally, vertical integration across the value chain, exemplified by firms like BYD, enables inhouse production of batteries, chips and other key components. This strategic control over the supply chain enhances operational resilience and offers a significant advantage in a market increasingly shaped by global supply uncertainties.

Due to its complex nature and sensitivity to material purity, battery manufacturing exhibits significant learning-by-doing effects, where unit production costs decline with accumulated production experience. Barwick et al. (2025) estimate that the unit cost of production for a battery manufacturer decreases by about 7.5 per cent with every doubling of production experience in that firm. Learning by doing in upstream battery production creates a snowball effect in response to EV demand shocks, such as consumer subsidies. Increases in EV sales and consequently battery production triggered by a demand shock lead to lower battery and EV costs, which in turn further stimulate EV adoption. The rapid expansion of EV sales in China and the resulting increase in battery production have given manufacturers operating in the Chinese market, primarily domestic producers, a substantial cost advantage.

The widespread adoption of lithium iron phosphate (LFP) batteries, led by BYD, has proven to be another key driver of cost reductions. BYD not only championed the use of LFP batteries in electric vehicles but also pioneered innovations in battery design and manufacturing, such as its proprietary Blade Battery that enhanced energy density, safety and scalability. These advancements significantly lowered production costs and facilitated broader market adoption. LFP batteries offer enhanced safety and a longer lifespan at a lower cost compared with conventional nickel manganese cobalt (NMC) lithium-ion batteries.

Charging infrastructure development

The rapid expansion of China's EV market has been strongly supported by the concurrent development of charging infrastructure, which addresses consumer 'range anxiety' and supports the broader transition to electrified mobility. By the end of 2023, China had deployed approximately 2.7 million public charging stations—up from just 30,000 in 2013. This expansion included 1.52 million alternating current (AC) chargers and 1.2 million direct current (DC) fast chargers, establishing by far the world's largest charging network. Government support has been instrumental in this growth. Between 2015 and 2018, public investment in charging infrastructure totalled RMB13.2 billion, accelerating EV adoption in China (Li et al. 2022). More recently, policy efforts have shifted towards enhancing infrastructure quality and accessibility to better accommodate rising demand and advance national electrification goals.

The spatial distribution of charging infrastructure offers further insights into regional dynamics. Figure 5.3a illustrates the density of charging stations and the scale of EV adoption across provinces, measured as the number of charging stations and EVs per 10,000 people. The figure also reveals that EV adoption is clustered in regions with stronger infrastructure development. Figure 5.3b displays the number of EVs per public charger by province. This metric highlights regional disparities in infrastructure coverage, with notable variation in the EV-to-charger ratio across provinces. Such differences indicate that charging infrastructure is not evenly distributed relative to EV ownership. These patterns underscore the need for continued investment to address regional imbalances and ensure equitable access to charging facilities as EV adoption accelerates.

To further integrate EV infrastructure into urban planning, China has introduced policies mandating new residential and commercial developments to include EV charging capabilities. This ensures that infrastructure expansion keeps pace with urbanisation and vehicle electrification. Additionally, the country is investing in ultra-fast charging stations to support long-distance travel, addressing a key limitation of current EV technology. By simultaneously prioritising urban and highway charging networks, China is developing a comprehensive infrastructure ecosystem that not only supports domestic demand but also establishes a global benchmark for large-scale EV deployment.

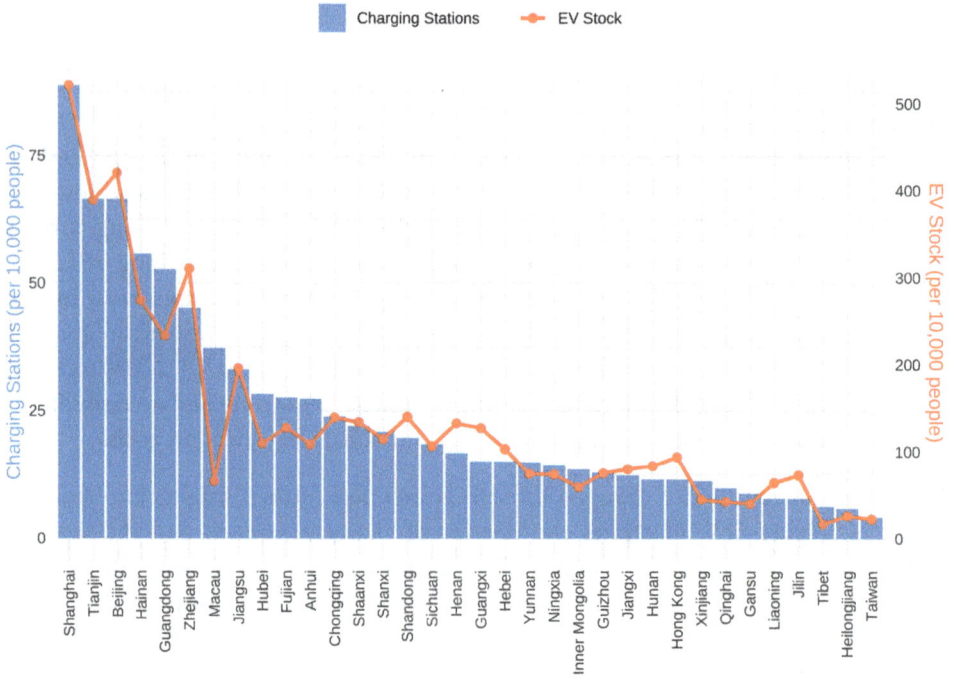

Data source: EVCIPA

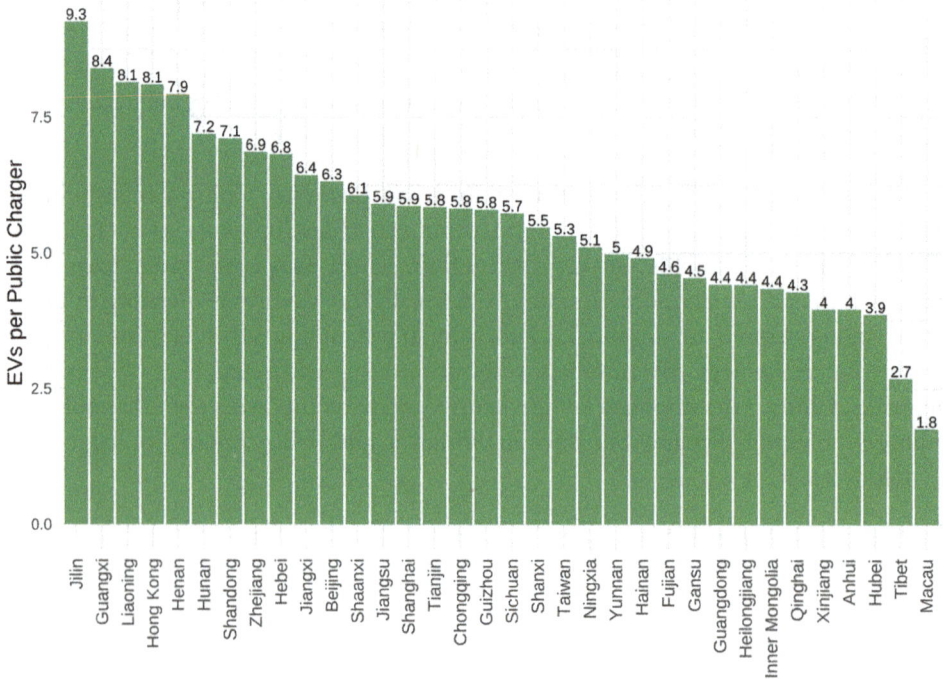

Data source: EVCIPA

Figure 5.3: Distribution of electric vehicle stock and charging infrastructure in China

Notes: (Top) the number of charging stations and EV stock per across provinces, standardised to population; (Bottom) the provincial distribution of EVs per public charger.

Source: Electric Vehicle Charging Infrastructure Promotion Alliance (EVCIPA).

Policy landscape

China's rapid adoption of EVs is largely attributable to its comprehensive policy framework. This section discusses the key components of China's EV policy landscape, including mandates and regulations, financial and non-financial incentives and industrial policies aimed at promoting the EV industry.

Mandates and regulations

In response to global climate change pressures, many countries have set targets for zero-emission vehicle adoption as part of broader efforts to reduce carbon emissions. In 2020, China set a goal for EVs to account for 40 per cent of new car sales by 2030—a target that was easily surpassed in 2024. Given the market's rapid growth, this goal has since been substantially revised upward, to 60 per cent. By setting clear objectives and reducing policy uncertainty, the government has created strong incentives for automakers to invest in EV development, fostering competition and accelerating cost reductions and technological innovation.

A range of policies has been developed to support the achievement of EV adoption targets. One key measure is the tightening of fuel economy standards, aimed at improving vehicle efficiency and encouraging the shift to EVs. Central to China's approach is the Dual Credit Policy, which integrates fuel economy credits with new-energy vehicle credits. This policy functions as a market-based mechanism, allowing automakers to trade credits, thereby providing flexibility while ensuring compliance with fuel efficiency targets. The policy includes progressively stricter fuel consumption targets: 5 litres per 100 kilometres by 2020, 4 litres by 2025 and 3.2 litres by 2030. The fuel economy regulations could encourage innovation in clean vehicle technologies to reduce compliance costs (Barwick et al. 2024b; Rozendaal and Vollebergh 2025).

Financial and non-financial incentives

Beyond regulatory mandates, financial and non-financial incentives have also played a crucial role in promoting EV adoption, particularly during the market's early development. Direct financial subsidies from both the central and local governments were a key component of China's EV policy in its initial expansion phase. Between 2010 and 2016, these subsidies—equivalent to 30–60 per cent of EV purchase prices—were expanded from 10 pilot cities to nationwide, substantially reducing the upfront cost of EVs (Li et al. 2022; Barwick et al. 2024a). As the market matured, these subsidies were gradually phased out, between 2019 and 2022, with the policy focus shifting towards infrastructure investment and long-term market sustainability. Tax incentives have complemented direct subsidies. Full purchase tax exemptions for EV buyers have been extended through to 2027, while reduced registration fees offer additional savings in upfront costs.

While financial incentives have been essential in driving initial adoption, non-financial policies have also played a significant role by enhancing convenience and influencing consumer behaviour. The green licence plate policy, introduced in 2016, allows for easy identification of EVs and grants them preferential treatment in parking and traffic regulations. In cities with strict vehicle registration limits, such as Beijing and Shanghai, EV buyers are either exempt from licence plate lotteries or given priority access. Similarly, in cities such as Shenzhen and Guangzhou, where traffic restrictions apply to gasoline-powered vehicles, EVs are exempt from driving bans. The measures adopted in urban areas could be complementary to regulatory constraints on traditional fuel vehicles in promoting EV adoption.

Industrial policies

In parallel with consumer-focused policies, China has adopted strategies to localise and strengthen its EV supply chain. A key measure was the local content requirement. Between 2016 and 2019, China's whitelist policy mandated that only EVs using batteries from domestically certified suppliers could qualify for government subsidies. By incentivising Chinese EV firms to source batteries from domestic producers, it helped these producers move down the cost curve faster than otherwise, as the policy accelerated learning by doing among them. The policy therefore contributed to the rise of Chinese battery manufacturers such as CATL and BYD, allowing them to be more cost competitive in the global market even after the policy was removed in 2019 (Barwick et al. 2025).

In addition to national policies, a wide range of industrial policies have been implemented by provincial and city governments in China to develop their own local EV industries, often in direct competition with other jurisdictions. These policies include low-cost financing, preferential land access, investment and production subsidies and government procurement support (Fang et al. 2025). These local government efforts have contributed, on the one hand, to the development of a complete EV supply chain—from mineral refining to final vehicle assembly—and, on the other, to the emergence of a highly competitive industry with production capacity exceeding domestic demand. The large number of domestic EV firms and the recent price war among automakers discussed above are manifestations of these policies.

Empirical evidence on policy impacts

Empirical studies broadly confirm the effectiveness of China's EV policies but also highlight their limitations. Large-scale consumer subsidies, which amounted to between 40 per cent and 60 per cent of an EV's sale price at their peak, significantly expanded the market and indirectly supported the development of charging infrastructure. These measures reinforced a positive cycle of adoption and investment (Li et al. 2017, 2022, 2024). Building on this foundation, Hu et al. (2025) adopt a dynamic structural model to assess the long-term impact of subsidy design. They find that a gradual phase-out scheme can induce earlier purchases and strengthen market momentum through peer effects and learning by doing, achieving higher sales at lower fiscal cost. In contrast to financial incentives, non-financial policy instruments generally involve lower fiscal costs but demonstrate varying degrees of effectiveness depending on the policy context and consumer behaviour. Li et al. (2022) show that the introduction of green licence plates in China led to a 37 per cent increase in EV sales for a minimal implementation cost. In cities such as Beijing, Shanghai and Guangzhou, EVs benefit from licence plate and driving exemptions, but their impact is limited as consumers often adapt their behaviour without switching to EVs.

Infrastructure investment complements financial and non-financial incentives by directly addressing range anxiety, thereby strengthening consumer confidence and supporting sustained market growth. Empirical evidence emphasises the importance of indirect network effects, noting that EV demand is influenced by the availability of charging infrastructure, while infrastructure investment depends on growing EV adoption (Li et al. 2017; Zhou and Li 2018; Springel 2021). In the Chinese context, the contrast is even more pronounced, as many urban households lack access to private charging facilities—highlighting the critical role of public infrastructure. Li et al. (2022) estimate that every additional 1,000 public charging stations result in a 20 per cent increase in EV sales. A key insight from the literature is that infrastructure investment delivers greater cost effectiveness than purchase subsidies.

Recent studies have also drawn attention to the unintended consequences of subsidies. Wang and Xing (2023) find that although subsidies stimulate market growth and firm entry, they may inadvertently attract low-quality 'lemon' firms, leading to reputation losses and reducing the overall efficiency of the program. Guo and Xiao (2023) show that the environmental impact of subsidies could be negative due to the increase in vehicle sales and usage. In addition, subsidies often disproportionately benefit high-income households, raising concerns about equity and environmental justice (Jacqz and Johnston 2024; Wang 2024; Xiao and Liang 2024). These findings highlight the need for more targeted and balanced subsidy policies that promote adoption while ensuring quality control and equitable outcomes.

Challenges and future outlook

Despite remarkable growth over the past decade, China's EV industry now faces significant challenges, including intensifying international competition, shifting policy environments and rising trade barriers. At the same time, the industry has substantial opportunities as it continues to consolidate to harness economies of scale and expand both production and sales in global markets.

Global competition and trade barriers

Intensifying global competition and the rise of trade barriers pose significant challenges to China's EV industry. As Chinese EV manufacturers expand into international markets, they face strong competition from established global automakers that have been ramping up investments in EV technology and strengthening their market presence. These legacy manufacturers often benefit from longer histories and stronger brand recognition and hence customer base.

At the same time, rising trade barriers like high import tariffs imposed by key markets, such as the United States and the European Union, create significant obstacles for Chinese EV exports. As of 27 September 2024, the United States increased the tariff on electric vehicles imported from China from 25 per cent to 100 per cent. Similarly, starting on 4 October 2024, the European Union imposed additional tariffs on Chinese EV manufacturers, ranging from 17 per cent to 45 per cent, depending on the company. These tariffs aim to counter unfair subsidies to Chinese EV manufacturers and protect domestic industries from low-cost imports.

These tariffs are also part of broader efforts to strengthen local supply chains and manufacturing capabilities and reduce dependence on China. For example, the US *Inflation Reduction Act*, signed into law in 2022, includes eligibility requirements for EV tax credits that restrict manufacturers from sourcing key minerals and battery components from China and mandate that final vehicle assembly occurs in North America. This trend of 'supply chain decoupling' compels Chinese EV and battery manufacturers to seek alternative markets.

Regulatory developments further exacerbate these challenges. The European Union's Carbon Border Adjustment Mechanism (CBAM) imposes carbon tariffs on high-emission imports, creating risks for Chinese EV exports. CBAM primarily targets energy-intensive industries such as steel, aluminium and chemicals—key materials in EV manufacturing (Clausing and Wolfram 2023). Since China's power generation depends heavily on fossil fuels, its EV production has a relatively high carbon footprint, making it more vulnerable to such tariffs. Although EVs themselves are not yet directly subject to CBAM, rising costs for these materials could weaken the competitiveness of Chinese automakers in the European market.

Technological challenges

Despite significant advancements in EV technology, key challenges—such as limited energy density, slow charging speeds and concerns about battery lifespan—continue to hinder widespread adoption. Low energy density remains a key obstacle, as increasing the number of lithium-ion cells to extend driving range adds both cost and weight to the vehicle.

Beyond performance limitations, the rapid expansion of the EV market has intensified challenges in battery recycling and disposal. China's recycling industry remains complex and largely unregulated, resulting in a low standardised recycling rate. According to a 2023 report by the Development Research Centre of China's State Council, less than 25 per cent of EV batteries in China were properly recycled. Addressing these challenges requires coordinated efforts in policy development, technological innovation and the establishment of a comprehensive recycling system.

The high costs of battery production and raw material sourcing present further difficulties. China has limited domestic reserves of critical minerals such as lithium, cobalt and nickel, making it heavily dependent on imports. As global demand for EV batteries rises, competition for these materials is intensifying, increasing the risks of supply chain disruptions and rising production costs. These constraints could ultimately affect EV affordability, creating challenges for both manufacturers and consumers.

Charging infrastructure disparities

While China has made impressive strides in building its EV charging network, significant regional disparities remain. Urban centres are relatively well served, but rural and less-developed areas still lack adequate infrastructure. This gap limits EV adoption beyond the major cities. In response, both the central and local governments have launched initiatives to expand infrastructure coverage. The National Development and Reform Commission and the National Energy Administration have called for charging stations in every county and charging piles in eligible townships. Beyond expanding infrastructure, the government has been promoting smart charging technologies, integrating them into EVs and charging networks for better energy management.

Automakers are playing a crucial role in infrastructure expansion. NIO has launched the Power County Plan, aiming to establish battery-swap stations in every county across mainland China by 30 June 2025. In 2024, Huawei partnered with automakers and charging operators to deploy more than 100,000 ultra-fast charging piles, expanding national coverage. XPENG has also expanded its charging network, operating more than 1,880 self-operated charging stations across 420 cities. These combined efforts from both the public and the private sectors are gradually reducing regional disparities in charging infrastructure, promoting EV adoption nationwide.

In addition to access, charging speed remains a key concern among consumers. While sales of high-voltage (800 volt) EV models have increased quickly in the past few years, the deployment of compatible ultra-fast charging stations has not kept pace. These stations cost significantly more to build and often require grid upgrades, posing significant financial and technical barriers. At this early stage of ultra-fast charging infrastructure development, government support plays a crucial role. For example, the Shenzhen Government launched the Ultra-Fast Charging City in June 2023 and, by the end of 2024, more than 1,000 ultra-fast charging stations had been built in the city. As demand for faster charging continues to grow and the costs of using ultra-fast chargers decrease with wider adoption, ultra-fast charging stations are expected to become a key trend in the future development of the EV industry.

Future prospects

China's EV industry is entering a new phase following its rapid transformation from a market follower to a global leader. As early drivers such as government subsidies and pilot programs have waned in effectiveness, the industry is shifting towards structural reform and innovation driven by market forces. A key area of progress will be the international harmonisation of technical standards. As Chinese EV makers expand into South-East Asia, Latin America and Europe, aligning with diverse market requirements is crucial. Common standards for charging, battery safety and data protection can lower trade barriers, improve compatibility and strengthen global supply chains, while also easing geopolitical tensions and supporting a more integrated EV ecosystem.

In addition to advancements within the vehicles, the future of EVs will be shaped by their integration into broader energy and digital systems. Beyond transportation, EVs are increasingly seen as mobile energy assets that can support grid stability and renewable energy integration through technologies such as vehicle-to-grid and bidirectional charging. Unlocking these benefits will require coordinated investment in grid modernisation, digital infrastructure and supportive policy frameworks. Policymakers and industry leaders must collaborate to ensure EVs serve not only as modes of transport but also as dynamic components of the energy ecosystem.

As EV adoption continues to accelerate in China, ensuring equitable access to the benefits of electrified mobility is becoming increasingly important. Without deliberate policy action, rural areas and low-income communities risk being left behind due to limited charging infrastructure and higher upfront costs. Future policy efforts may need to shift from broad consumer subsidies to more targeted support that addresses these disparities such as public charging investment in underserved areas, incentives for affordable EV models and programs for secondhand EV markets. Promoting a just and inclusive transition could help ensure that the social, economic and environmental benefits of transportation electrification are shared more broadly across society.

References

Barwick, Panle Jia, Hyuk-Soo Kwon, and Shanjun Li. 2024a. *Attribute-Based Subsidies and Market Power: An Application to Electric Vehicles*. NBER Working Paper No. 32264. Cambridge: National Bureau of Economic Research. doi.org/10.3386/w32264.

Barwick, Panle Jia, Hyuk-Soo Kwon, Shanjun Li, Yucheng Wang, and Nahim B Zahur. 2024b. *Industrial Policies and Innovation: Evidence from the Global Automobile Industry*. NBER Working Paper No. 33138. Cambridge: National Bureau of Economic Research. doi.org/10.3386/w33138.

Barwick, Panle Jia, Hyuk-Soo Kwon, Shanjun Li, and Nahim B. Zahur. 2025. *Drive Down the Cost: Learning by Doing and Government Policies in the Global EV Battery Industry.* NBER Working Paper No. 33378. Cambridge: National Bureau of Economic Research. doi.org/10.3386/w33378.

Clausing, Kimberly A., and Catherine Wolfram. 2023. 'Carbon Border Adjustments, Climate Clubs, and Subsidy Races When Climate Policies Vary.' *Journal of Economic Perspectives* 37, no. 3: 137–62. doi.org/10.1257/jep.37.3.137.

Fang, Hanming, Ming Li, and Guangli Lu. 2025. *Decoding China's Industrial Policies.* NBER Working Paper No. 33814. Cambridge: National Bureau of Economic Research. doi.org/10.3386/w33814.

Guo, Xiaodan, and Junji Xiao. 2023. 'Welfare Analysis of the Subsidies in the Chinese Electric Vehicle Industry.' *Journal of Industrial Economics* 71, no. 3: 675–727. doi.org/10.1111/joie.12337.

Hu, Yunyi, Haitao Yin, and Li Zhao. 2025. 'Subsidy Phase-Out and Consumer Demand Dynamics: Evidence from the Battery Electric Vehicle Market in China.' *The Review of Economics and Statistics* 107, no. 2: 458–75. doi.org/10.1162/rest_a_01295.

Jacqz, Irene, and Sarah Johnston. 2024. 'Electric Vehicle Subsidies and Urban Air Pollution Disparities.' *Journal of the Association of Environmental and Resource Economists* 11, no. S1. doi.org/10.1086/732516.

Li, Shanjun, Lang Tong, Jianwei Xing, and Yiyi Zhou. 2017. 'The Market for Electric Vehicles: Indirect Network Effects and Policy Design.' *Journal of the Association of Environmental and Resources Economists* 4: 89–133. doi.org/10.1086/689702.

Li, Shanjun, Binglin Wang, and Hui Zhou. 2024. 'Decarbonizing Passenger Transportation in Developing Countries: Lessons and Perspectives.' *Regional Science and Urban Economics* 107: 103977. doi.org/10.1016/j.regsciurbeco.2024.103977.

Li, Shanjun, Xianglei Zhu, Yiding Ma, Fan Zhang, and Hui Zhou. 2022. 'The Role of Government in the Market for Electric Vehicles: Evidence from China.' *Journal of Policy Analysis and Management* 41, no. 2: 450–85. doi.org/10.1002/pam.22362.

Rozendaal, Rik, and Herman Vollebergh. 2025. 'Policy-Induced Innovation in Clean Technologies: Evidence from the Car Market.' *Journal of the Association of Environmental and Resource Economists* 12, no. 3. doi.org/10.1086/731834.

Springel, Katalin. 2021. 'Network Externality and Subsidy Structure in Two-Sided Markets: Evidence from Electric Vehicle Incentives.' *American Economic Journal: Economic Policy* 13, no. 4: 393–432. doi.org/10.1257/pol.20190131.

Wang, Jingyuan, and Jianwei Xing. 2023. 'Subsidizing Industry Growth in a Market with Lemons: Evidence from the Chinese Electric Vehicle Market.' 22 November. Available at SSRN: doi.org/10.2139/ssrn.4636568.

Wang, Yucheng. 2024. 'Who Receives Environmental Benefits from Driving Electric Vehicles?' November. Available at The Chinese Economist Society: www.china-ces.org/Files/3055 abstract/202401190211040421.pdf.

Xiao, Junji, and Jing Liang. 2024. 'The Incidence and Distributional Effects of Electric Vehicle Subsidies in China.' 17 September. Available at SSRN: doi.org/10.2139/ssrn.4959489.

Zhou, Yiyi, and Shanjun Li. 2018. 'Technology Adoption and Critical Mass: The Case of the U.S. Electric Vehicle Market.' *The Journal of Industrial Economics* 66, no. 2: 423–80. doi.org/10.1111/joie.12176.

6

The European Union's countervailing duties on electric vehicles from China: Disputes and prospects

ZhongXiang Zhang

China's advantage in the global industrial chain is now not only reflected in the mid to low-end manufacturing industry, but also concentrated in the technology track represented by 'the new three': electric vehicles (EVs), lithium-ion batteries and photovoltaic (PV) products. From 2019 to 2023, the total export value of China's 'new three' increased from US$33.6 billion to US$146.5 billion, with an annual average growth rate of 44.5 per cent (Yang et al. 2024). Given that in 2023 the Chinese yuan depreciated by 4.5 per cent against the US dollar, the growth rate is even higher in terms of the yuan: in 2023, total exports of the 'new three' reached RMB1.06 trillion, breaking through the one-trillion-yuan mark for the first time and reaching 18.3 per cent of China's trade surplus. In 2023, 60 per cent of global EV production was in China and one in every four cars exported by China was an EV, totalling 1.2 million vehicles.

The rapid increase of Chinese-made EVs raises a variety of concerns outside China, particularly in the European Union and the United States. The European Union and the United States as well as a growing number of other developed countries accuse China of heavily subsidising its locally made EVs, and argue that imports of these subsidised vehicles distort their EV markets and hurt their EV industry. A growing number of countries have enacted regulations and trade measures to restrict imports of Chinese-made EVs. The European Union even initiated an anti-subsidy investigation into imports of EVs from China, and its member states have agreed to impose definitive countervailing duties on imports of battery electric vehicles (BEVs) from China. As would be expected, China has claimed that the European Union's decision violates World Trade Organization (WTO) rules and undermines global cooperation on tackling climate change. China has filed an appeal to the WTO dispute-settlement mechanism over the European Union's countervailing measures; requested WTO consultation over tariffs placed on EVs by Canada and Türkiye; and has criticised the United States for abusing the review process of Section 301 tariffs to further raise tariffs on Chinese-made EVs.

This chapter analyses why the European Union began imposing high countervailing duties on Chinese-made EVs when they were not yet common on the streets of Europe, and the differences between and logic of the European Union and the United States. It also examines the level of countervailing duties that the European Union must impose on Chinese EVs to diminish their competitiveness in European markets, and the impact of additional countervailing duties on the European Union itself and Chinese EV exports to Europe. It looks at why the European Union decided to negotiate with China on a minimum price after imposing countervailing duties but was particularly cautious about taking the lowest-price approach. It explores the significance and limitations of the China Chamber of Commerce for Import and Export of Machinery and Electronic Products' authorisation of representatives to negotiate on behalf of Chinese EV manufacturers, the possibility of implementing a hybrid scheme of countervailing duties and minimum prices, and the impact of the re-election of Donald Trump as President of the United States on the direction of the China–EU EV tariff dispute. The chapter also discusses overseas investment by Chinese EV companies and ends with a discussion of the ways to move this dispute forward.

Why do Chinese-made EVs dominate the market?

Why do Chinese-made EVs dominate the EV market? The United States and the European Union accuse China of providing heavy subsidies (see Figure 6.1). Several studies undertaken by US think tanks provide analyses to support this accusation, but China denies the claims.

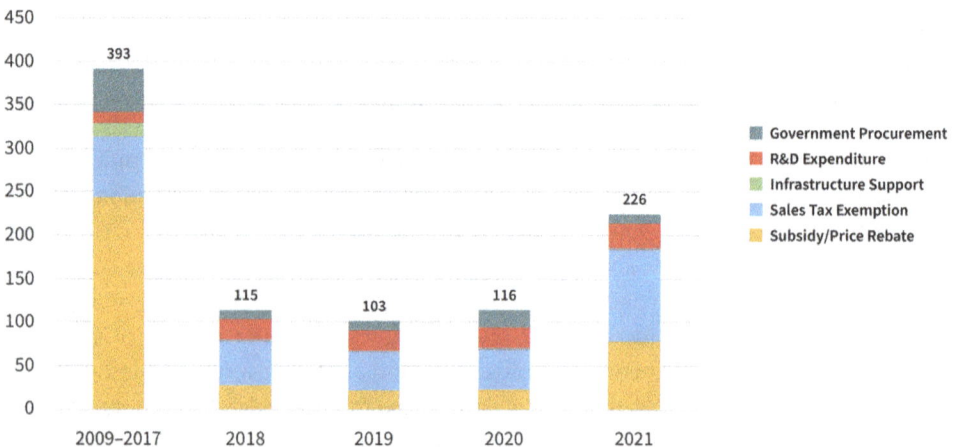

Figure 6.1: Total Chinese Government spending on the new-energy vehicle sector (RMB billion)

Source: DiPippo et al. (2022).

The European Commission (Draghi 2024) estimates that Chinese subsidies for clean tech manufacturing have long been twice as high as those in the European Union as a share of GDP, while China has protected its home market for solar PV, wind power generation equipment and EV batteries. These policies have left the European Union with a significant cost disadvantage—for example, solar PV manufacturing costs in China are 35–65 per cent lower than in Europe and costs for manufacturing battery cells are 20–35 per cent lower (IEA 2024a). According to estimates by the Center for Strategic and International Studies (DiPippo et al. 2022; White et al. 2023), cumulative government support to the EV sector in China[1] totalled RMB393 billion (US$58 billion) between 2009 and 2017. This support jumped further, to RMB560 billion, from 2018 to 2021. The Organisation for Economic Co-operation and Development (OECD 2021, 2023) estimates that Chinese industrial firms received government support equivalent to about 4.5 per cent of their revenue. The largest part of this support comes in the form of below market rate lending. This support, combined with tax concessions and government grants, means Chinese firms may receive almost nine times more government support (relative to sales) than comparable firms in the OECD. According to the information in Chinese automaker BYD's annual reports, direct government subsidies to the company increased from about €200 million in 2020 to €600 million in 2021 and €2.1 billion in 2022. Relative to business revenue, this corresponds to an increase in direct subsidies from 1.1 per cent in 2020 to 3.5 per cent in 2022 (Bickenbach et al. 2024).

China disputes these accusations, and it is fair to say that neither side provides the full story. Subsidies have played a crucial role, particularly in the early phase of EV development, and are widely used in China, but the market advantage of Chinese EVs is not due solely to subsidies, but rather is the result of multiple factors working together: early moves to establish well-planned industrial layouts, long-term industrial policy guidance, long-term subsidies (activating huge domestic demand), continuous increases in research and development investment (upgrading batteries and intelligent technology), an especially large domestic market (the scale effect strengthened by internal competition, providing huge pressure to lower production costs and selling price), industrial chain advantages and a rich pool of engineers and skilled workers.

China's EV production accounts for more than 60 per cent of the global total. However, most of this is consumed domestically. In 2023, almost 9.5 million Chinese-made EVs were sold in China and only 1.2 million were exported—accounting for less than 15 per cent of annual output. Moreover, exports of Chinese EVs are mainly by purely foreign brands, such as Tesla; Sino-foreign brands, such as BMW; or acquired foreign brands, such as MG. In 2023, Tesla's Shanghai factory produced 947,000 EVs, of which 344,000 were exported. Tesla exports alone accounted for 29 per cent of China's total

---

1   The government support provided to industry includes direct subsidies and tax incentives for research and development, other tax incentives, below-market credit to SOEs, support through state investment funds (Government Guidance Funds) and 'China-specific factors', which include, most notably, below market-price land sales (DiPippo et al. 2022).

EV exports. According to the European Federation for Transport and Environment (T&E 2024), Europe's leading advocates for clean transport and energy, in 2023, 19.5 per cent of the approximately 300,000 EVs sold in Europe were made in China (see Figure 6.2). About 60 per cent of newly registered cars from China were foreign brands, such as Tesla (28 per cent), Dacia (20 per cent) and BMW (6 per cent), while SAIC MG accounted for 25 per cent. BYD, Polestar and other Chinese brands account for only 3 per cent of all EVs exported to Europe. This indicates that the European EV market is not only dominated by non-Chinese car companies, but more importantly, overseas capital also profits greatly from such exports.

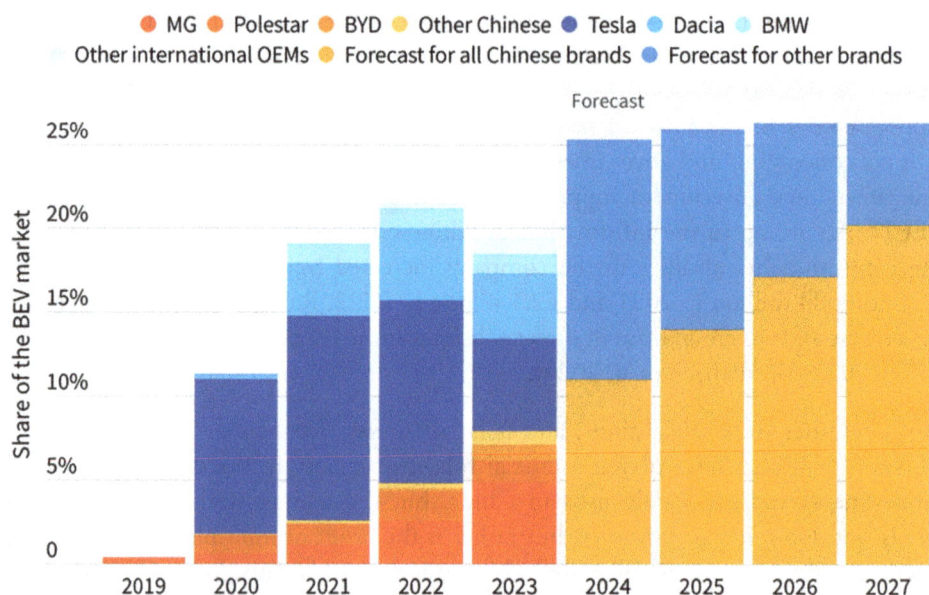

Figure 6.2: Chinese-made electric vehicles set to reach one-quarter of EU sales

Note: OEM = original equipment manufacturer.

Source: T&E (2024).

China as the world's largest exporter of automobiles

According to statistics from the China Association of Automobile Manufacturers, China exported 4.91 million vehicles in 2023, becoming the world's largest exporter of automobiles. The volume of exported vehicles further increased, to 5.86 million units, in 2024, consolidating China's position. The reason China has become the largest exporter is because multinational car companies have long achieved 'localised production'. In 2022, Japanese carmakers sold approximately 23.5 million units globally, only 3.8 million units of which (16 per cent) were sold in the Japanese market; and most of those sold outside the domestic market were sold in the countries in which they were made. Toyota has been ranked first in global sales for four consecutive years.

In its fiscal year ending March 2024, Toyota sold 10.30 million units globally, with 1.53 million sold inside Japan. Clearly, the number of exported Chinese-made vehicles was about one-quarter of the number of Japanese cars sold outside Japan.

While China exported almost 1.3 million EVs in 2024, most of its car exports are conventional fuel vehicles. Logically, if Europe and the United States believe China has overcapacity in the automotive industry, it should be reasonable to demand that China limits its production of traditional fuel vehicles. However, the European Union and the United States are demanding only that China restrict its production capacity of EVs—an important issue that then US treasury secretary Janet Yellen and then German chancellor Olaf Scholz discussed with top Chinese leaders during their visits to China in April 2024.

The reason is that Chinese traditional fuel vehicles cannot compete with foreign manufacturers, particularly high-end brands, but multinational companies lack competitiveness and pricing power for EVs.

In general, multinational enterprises in developed countries enjoy a considerable period of excess profit dividends brought by technological leadership, until later players can become competitive or even surpass them. However, EVs have broken the norm and multinational car companies have not had this opportunity, or at least it has been greatly weakened. Even worse, they are currently at a competitive disadvantage against Chinese EVs, so it is not difficult to understand their accusations of overcapacity in China's EV production.

Europe and the United States are, however, catching up with battery technology for EVs. California requires that by 2035, 50 per cent of all heavy-duty trucks sold will be pure electric, and the state is committed to achieving 100 per cent renewable energy by 2045. Other states in the United States are launching incentive packages to attract investors to build EV factories, sparking a wave of EV subsidies. The US Environmental Protection Agency introduced the strictest climate change regulations in its history, ensuring that two-thirds of new cars sold in the United States by 2032 are EVs. On 2 March 2023, 'Tesla Super Factory Investment Day', its CEO, Elon Musk, stated that Tesla would achieve its annual sales target of 20 million by 2030; Tesla had a delivery volume of just 1.31 million vehicles in 2022. In the European Union, the *Net-Zero Industry Act* sets the basic goal of strategic net-zero technology by 2030. Battery manufacturing capacity must be expanded to 550 gigawatt hours (GWh)—7.3 times that of the existing EU capacity—and meet 90 per cent of the union's annual expected demand by 2030 (European Parliament 2025; Zhang 2024).

Lithium, cobalt and nickel are used in the production of EV batteries. The European *Critical Raw Materials Act* identifies a list of materials crucial for technologies needed for the green and digital transition and sets benchmarks for domestic capacity along the supply chain to be reached by 2030: 10 per cent of the European Union's annual needs for extraction, 40 per cent for processing and 25 per cent for recycling. Moreover, to ensure EU independence over imports from single-country suppliers, the Act specifies that no more than 65 per cent of the union's annual needs of each strategic raw material at any relevant stage of processing should come from a single third country (EC 2024a).

Trade measures to restrict imports of Chinese-made EVs

Given that existing production capacity differs significantly from expected targets, the European Union has set policies to speed up the development of their own battery and EV industries and make up insufficient battery production. To stop or at least slow Chinese EV imports, Europe and the United States have elevated batteries and EVs to the status of strategic technology/industry and introduced subsidies and related support policies, as well as further regulations and trade measures.

In August 2022, the US Congress approved the *Inflation Reduction Act* (*IRA*), with US$369 billion disbursed for measures dedicated to improving energy security and accelerating the clean energy transition. The *IRA* imposes restrictions on imported batteries, battery critical materials and vehicle assembly areas required to produce EVs. Specifically, the *IRA* provides a maximum credit of US$7,500 for each new EV that is assembled in North America: US$3,750 if the specified percentage of battery components was manufactured or assembled in North America and US$3,750 if a percentage of critical minerals in the battery were extracted or processed in the United States or countries with which it has free-trade agreements. Moreover, beginning in 2024, to be eligible, a clean-energy vehicle could not contain any battery components manufactured or assembled by a 'foreign entity of concern' (FEOC) and, beginning in 2025, an eligible clean-energy vehicle may not contain any critical minerals extracted, processed or recycled by a FEOC (US Treasury 2023).

Classifying China as a FEOC means that for the manufacturer's suggested retail price that does not exceed US$80,000 for sports utility vehicles (SUVs), vans, pickups and electric sedans weighing less than 6.9 tonnes and US$55,000 for all other electric sedans, Chinese car companies will not receive a full credit of US$7,500, which will greatly affect their competitiveness. The United States claims these measures have nothing to do with politics or ideology. This is the dilemma that China's EV makers continue to face in exporting to and investing in factories in places such as the United States. The US decision to upgrade batteries to a strategic technology will restrict its ability to achieve energy conservation and carbon emissions reduction faster and at a lower cost.

On 14 May 2024, then president Joe Biden announced that the United States would apply Section 301 tariffs on an estimated US$18 billion worth of goods from China. The new measures included quadrupling tariffs on all EVs from China, from 25 per cent to 100 per cent. The existing tariffs were introduced during the previous Trump administration. The increase was made as part of a planned review of policy. Before he was elected for the second time, President Donald Trump said he would consider a 60 per cent tariff on all Chinese imports. In September 2024, the US Commerce Department (BIS 2024) proposed prohibiting the import and sale of vehicles containing certain hardware and software emanating from China on national security grounds. The proposal focused on hardware and software integrated into the vehicle connectivity system and software integrated into the automated driving system. The move is a significant escalation in the United States' ongoing restrictions on Chinese vehicles, software and components, and would effectively bar nearly all Chinese cars from entering the US market.

The European Union is also very concerned about the rapid rise in low-priced BEVs coming from China, distorting the EU market. On 4 October 2023, the European Commission initiated an investigation into EV imports from China, focusing on whether these were being subsidised by the Chinese Government, whether such subsidies were hurting the European Union's EV industry and whether imposing countervailing measures would be in the European Union's interests. The investigation was to conclude within 13 months of initiation.

The investigation targeted a sample of three Chinese producers: BYD, Geely and SAIC. Tesla (Shanghai) was not included despite its large volume of exports to the EU market. The Chinese Government and the China Chamber of Commerce for Import and Export of Machinery and Electronic Products (CCCME) argued that by not including Tesla (Shanghai) in the sample, the European Commission distorted the subsequent findings by artificially increasing the weighted average duty applicable to the cooperating non-sampled Chinese exporting producers. The European Commission disagreed with these claims. As explained in Recital 30 of the final regulation, the commission accepted the individual examination request of Tesla (Shanghai) given its simple corporate structure, which gave the commission sufficient time and resources to examine the company. No other individual examination requests were received. Furthermore, in Recital 756 of the final regulation, the commission argued that imports of Teslas from China were not expected to increase significantly as the company's spare production capacity was very low (EC 2024b).

On 4 July 2024, nine months after the initiation of the investigation, the European Commission had imposed provisional countervailing duties on imports of BEVs from China. Based on the investigation, the commission concluded that the BEV value chain in China benefits from unfair subsidisation, which is threatening economic injury to EU BEV producers. The individual duties applying to the three sampled Chinese producers were: BYD, 17.4 per cent; Geely, 19.9 per cent; and SAIC, 37.6 per cent. Other BEV producers in China that cooperated in the investigation but were not sampled were subject to the 20.8 per cent weighted average duty. The duty for all other non-cooperating companies was 37.6 per cent.

Following a request, Tesla may receive an individually calculated duty rate (EC 2024c), but otherwise provisional duties were adjusted slightly downwards based on comments on the accuracy of the calculations submitted by interested parties. These provisional duties applied as of 5 July 2024, for a maximum of four months. Within that time frame, a final decision was to be taken on definitive duties, through a vote by EU member states. When adopted, this decision would make the duties definitive for five years (EC 2024d).

On 20 August 2024, based on comments received from interested parties on the provisional measures, as well as the finalisation of the investigation, the European Commission disclosed slightly adjusted countervailing duties on imports of BEVs from China: BYD, 17.0 per cent; Geely, 19.3 per cent; SAIC, 36.3 per cent; other cooperating companies, 21.3 per cent; and all other non-cooperating companies, 36.3 per cent (see Table 6.1). Teslas from China were granted an individual duty rate of 9 per cent. The commission also decided not to retroactively collect countervailing duties (EC 2024e).

On 4 October 2024, the EC proposal to impose definitive countervailing duties on imports of BEVs from China obtained the necessary support from member states for the adoption of the tariffs (EC 2024f). On 29 October, the commission concluded its investigation by imposing definitive countervailing duties on BEV imports from China for five years. Chinese exporting producers were subject to the following duties: BYD, 17.0 per cent; Geely, 18.8 per cent; SAIC, 35.3 per cent; other cooperating companies, 20.7 per cent; and all other non-cooperating companies, 35.3 per cent. Teslas from China were granted an individual duty rate of 7.8 per cent. The commission decided not to retroactively collect the duties.

Table 6.1: EU countervailing duties on battery electric vehicle imports from China (%)

BEV producers from China	Provisional, 12 June 2024	4 July 2024	20 August 2024	Definitive, 29 Oct 2024
BYD	17.4	17.4	17.0	17.0
Geely	20.0	19.9	19.3	18.8
SAIC	38.1	37.6	36.3	35.3
Other cooperating companies	21.0	20.8	21.3	20.7
All other non-cooperating companies	38.1	37.6	36.3	35.3
Tesla	Individually calculated		9	7.8

Sources: European Commission (2024b–e).

The European Commission noted that, between October 2023 and January 2024, the European Union imported 177,839 Chinese EVs—an increase from the period covered by the EC investigation (October 2022 to September 2023) of 14 per cent, and an increase in the average monthly import volume of 11 per cent (EC 2024g). This surge is understandable because car manufacturers were racing to ship EVs produced in China to the EU market before the new tariff measures came into effect, betting that the duties would not be collected retrospectively.

With the European Union and the United States imposing tariffs on Chinese-made EVs, other countries soon followed suit. In August 2024, Canada announced that it would impose 100 per cent tariffs on Chinese-made EVs from 1 October 2024, over concerns about unfair subsidies.

Türkiye had introduced tariffs on EVs imported from China in March 2023 and implemented regulations on EV maintenance and services. In June 2024, Türkiye announced it would impose an additional 40 per cent tariff on all vehicles imported from China, effective from 7 July 2024, and extended to include internal combustion engines and hybrid vehicles from China. The additional tariff was set at a minimum of US$7,000 per vehicle, basically barring budget Chinese vehicles from entering the Turkish market. In September 2024, Türkiye implemented import licence restrictions, stipulating that importers must have 20 authorised service stations in seven different regions of the country before importing plug-in hybrid vehicles not produced by the European Union or countries with a free-trade agreement with Türkiye.

The European Union remains open to a negotiated solution

The share of Chinese-made EVs in the EU market is very small. Based on statistics from the European Commission (EC 2024h), imports of Chinese brands increased to 14.1 per cent in the second quarter of 2024, from 1.9 per cent in 2020. Even fewer Chinese-made EVs are imported into the United States. However, both the European Union and the United States decided to act well before Chinese-made EVs could become widespread in their markets.

Differing from the usual practice, the investigation into Chinese EV subsidies was initiated by the European Commission itself and was not based on an industry complaint. As then US commerce secretary Gina Raimondo said, 'We're not going to wait until our roads are filled with Chinese-made EVs and the risk is extremely significant before we act' (Reuters 2024).

It has been suggested that the European Union should refrain from imposing tariffs as it needs low-priced Chinese EVs to help it achieve its green transition and meet its climate goals. Valdis Dombrovskis, the then European Commissioner for Trade, rebutted such suggestions, contending that the 'argument can be made on any trade-distorting subsidies. But the point is that we have a major EU car industry, and this car industry is at risk if we allow this kind of distortion of the level playing field.' He said the tariffs, which vary by brand but will average 20.8 per cent, on top of an existing 10 per cent, would not close the market to Chinese imports, just level the playing field (Bounds 2024). The European car industry provides 13 million jobs and contributes 7 per cent of the European Union's GDP (EC 2025b). Its precautionary approach suggests the union is very concerned about the rapid increase in imports of low-priced EVs from China distorting the EU market and potentially hurting the auto industry, which is important to the economy and for employment.

The European Commission's proposal has obtained the necessary support from EU member states to impose definitive countervailing duties on imports of BEVs from China (EC 2024f); however, according to EU procedures, if 15 member states and 65 per cent of the EU population vote against the proposal to impose tariffs, it will be put on hold. However, despite the vote in favour of special tariffs for EVs from China, the European Commission remains open to continue negotiations with China to explore an alternative to the tariffs. Both sides are keen to keep negotiating, indicating there is room for compromise.

Intense competition in a saturated Chinese market has led to a price war and forced EV manufacturers to pursue ever-lower production costs. Chinese EV manufacturers have made substantial investments in new production facilities, with additional capacity continuing to come onto the market. Alongside the rapid growth of EV production has been a slowdown in the Chinese economy. This has not only resulted in profit margins in the auto sector plummeting from 8.7 per cent in 2015 to 6.2 per cent in 2020, and to 4.3 per cent in 2024, according to the China Automobile Dealers Association (Wu 2024; Liu 2025), but also raised the need to find new markets to absorb these EVs. However, in response to perceptions of unfair competition, an increasing number of countries are raising tariff and non-tariff barriers against Chinese-made EVs. The US restrictions on Chinese EVs, software and components effectively bar nearly all Chinese EVs from entering the US market. Exports to other markets will be challenging for a variety of reasons. Thus, the European Union, the world's second-biggest EV market, is highly attractive to China-based EV producers. An analysis by Rhodium (2024) indicates that even with the imposition of a 30 per cent duty, many Chinese EV models would still enjoy comfortable profit margins because of the substantial cost advantages they enjoy.

On the EU side, former WTO director-general Pascal Lamy (2024) said in an interview about the EU vote on Chinese EV tariffs that Europe could not remain the only big market open to subsidised Chinese EVs. However, opinion about imposing tariffs on Chinese-made EVs is widely divided. As shown in Table 6.2, while France supported the measure after previously pushing the European Union to start negotiations, 12 member countries abstained from the vote and its largest economy, Germany, together with four other member countries advocated against the step. It is crucial to emphasise Germany's position on this matter. Whether it is for the benefit of its own automotive industry, the overall interests of the European Union or its own leadership, Germany must actively negotiate with and mediate on behalf of the European Union and member states and reach an agreement that is acceptable to both China and Europe. Germany must not stand in opposition to the majority of the European Union. Abstention or opposition—something smaller EU members may consider—is not an option for Germany. It cannot stand alone without undermining the credibility of its leadership and the independence and credibility of the European Union. Germany had spoken out against the proposal until the very end and had been one of those pushing for further negotiations with China.

Table 6.2: How EU member states voted on tariffs for Chinese-made battery electric vehicles

For (45.99% of EU population)	Against (22.65% of EU population)	Abstained (31.36% of EU population)
Bulgaria	Germany	Austria
Denmark	Hungary	Belgium
Estonia	Malta	Croatia
France	Slovakia	Cyprus
Ireland	Slovenia	Czechia
Italy		Finland
Latvia		Greece
Lithuania		Luxembourg
Netherlands		Portugal
Poland		Romania
		Spain
		Sweden

Source: Author's compilation.

The European Union's tariffs are not as severe as those implemented by the United States and Canada; it has not emulated the US approach to systematically shut out Chinese technology. Neither has it set tariffs so high as to block Chinese EVs from entering the EU market. As Rhodium (2024) suggests, to make the European market unattractive for Chinese EV exporters, duties would have to be as high as 40–50 per cent. They might have to be even higher for vertically integrated manufacturers such as BYD. The definitive tariff for BYD is 17 per cent, on top of the existing 10 per cent, which is far lower than the 45 per cent suggested. Even if duties were set high enough to erase the EU premium, BYD might decide that exporting to Europe still makes sense, given slowing demand and competitive pressures in the Chinese market.

The anti-subsidy tariffs imposed on Chinese-made EVs are comparable with previous tariffs. For example, the rates of definitive countervailing duties imposed by the European Commission on imports of certain organic coated steel (OCS) products originating in China differ between manufacturers. While OCS products from Union Steel China are levied at the lowest level of 13.7 per cent, OCS products from key manufacturers such as Ansteel, Baosteel and Tangshan Iron and Steel Group are levied at 26.8 per cent. All other manufacturers are targeted at the highest countervailing duty rate of 44.7 per cent (EC 2019). Imports of certain Chinese-made new or re-treaded pneumatic rubber tyres, used for buses or trucks with a load index exceeding 121, are levied at €27.69

per item for Chinese producers cooperating with both subsidy and anti-dumping investigations, but €57.28 per item for companies cooperating in the original anti-dumping investigation but not in the original subsidy investigation. Against the average Chinese import prices per item into the EU market of €136 in 2020, €156 in 2021 and €208 in 2022, these tariffs are translated into the rate of definitive countervailing duty up to 20.4 per cent and 42.1 per cent, respectively (EC 2025a). Compared with the highest countervailing duty of 44.7 per cent in the OCS case and 42.1 per cent in the pneumatic tyre case for all other non-cooperating companies, the highest definitive countervailing duty of 35.3 per cent in the EV case is low.

The EC decision makes sense because it avoids imposing higher costs on the EU economy. In 2023, the European Union essentially instituted a ban on internal combustion engines by 2035, when all new passenger cars and light vehicles registered in Europe must have zero tailpipe carbon emissions, compelling manufacturers to elevate the proportion of EVs they produce or incur penalties. But EU manufacturers alone will not be able to meet the demand for clean technologies (Zhang 2024). Unless Europe decides to delay its energy transition and scale back its EV ambitions, Chinese EVs may offer the lowest-cost and most efficient route to achieving its decarbonisation targets.

On the other hand, the European Commission cannot afford to take a laissez-faire approach, given that China's state-sponsored competitors represent a threat to EU employment, productivity and economic security in general and its clean tech and automotive industries in particular. According to the simulations of the European Central Bank (ECB 2024), if subsidies to the Chinese EV industry were to follow a similar trajectory to those applied to the solar PV industry, EU domestic production of EVs would decline by 70 per cent and EU producers' global market share would fall by 30 per cent. The application of additional tariffs reflects EU concerns about the influx of low-cost EVs into its market. By not setting tariffs at the high end of the suggested range and implementing differential rates according to the subsidy level and degree of cooperation of Chinese companies, the European Union is signalling that the purpose of the countervailing duties is not to stop imports of EVs from China, but to level the playing field so that EU manufacturers can compete on fair terms, stimulating innovation in the process (EC 2024b).

The European Union's additional countervailing duties, however, hurt China's EV exports badly. Using simulations from a trade model encompassing many countries and sectors, the Kiel Institute for the World Economy, the Austrian Institute of Economic Research and the Supply Chain Intelligence Institute Austria projected that the provisional countervailing duties of 21 per cent, effective on 4 July 2024, would reduce vehicle imports from China by 42 per cent. If the European Union were to reduce duties of 10 per cent imposed on BEVs to zero for WTO member countries with which it has no free-trade agreements, imports from China would only decrease by about 20 per cent. This scenario would enhance EU welfare more than the pure countervailing duty scenario, and the green transition would benefit from fairer trade (Felbermayr et al. 2024).

A 20 per cent reduction is not trivial. It is inevitable that China will appeal to the WTO to safeguard its interests. Indeed, China has filed a Multi-Party Interim Appeal Arbitration Arrangement within the WTO, requested WTO consultation over Canadian, EU and Turkish tariffs on EVs and has criticised the United States for abusing the review process of Section 301 tariffs and imposing further tariffs on Chinese-made EVs.

On 9 August 2024, China appealed to the WTO dispute-settlement mechanism over the European Union's provisional countervailing measures on EVs. China's Ministry of Commerce said that EU actions seriously violated WTO rules and undermined global cooperation on climate change. The ministry urged the European Union to change course and work with China to safeguard economic and trade cooperation and the stability of EV industrial and supply chains.

China's Foreign Ministry spokesperson Lin Jian reacted to the Biden administration's tariff hike in May 2024, saying:

> Tariffs imposed by the first Trump administration on China have severely disrupted normal trade and economic exchanges between China and the US. The WTO has already ruled those tariffs [are] against WTO rules. Instead of ending those wrong practices, the US continues to politicise trade issues, abuse the so-called review process of Section 301 tariffs and plan tariff hikes. (Feng and Cheng 2024)

Based on the same arguments, China has filed a request for WTO consultation over Türkiye's EV tariffs and a ruling on the tariffs Canada imposed on Chinese EVs and metal products.

It is not only the Chinese Government reacting to these trade measures to restrict imports; Chinese EV manufacturers are also taking legal action to challenge the European Union's decision. In January 2025, Chinese EV makers BYD, Geely and SAIC teamed up with German maker BMW to file complaints in the Court of Justice of the European Union (Chilton 2025), hoping to overturn EU tariffs on Chinese EV imports into the EU market.

However, given the increasing number of countries raising tariffs and non-tariff barriers against Chinese-made EVs, whether China's appeal can be settled within the WTO in the short term, and to China's advantage, is an open question. Therefore, in seeking a long-term solution within the WTO, China will have to hope for an alternative settlement, such as a price undertaking, for its benefit.

Minimum pricing as an alternative

While the European Commission's proposal received enough votes from EU member states to impose tariffs on Chinese-made EV imports for the next five years, 15 member countries abstained and its largest economy, Germany, spoke out against the plans until the very end and pushed for further negotiations with China. At the same time, China has pushed for a minimum import price that would avert the need for some of the extra tariffs. It is in the interests of both sides to come to a mutually beneficial agreement, meaning Europe can worry less about the influx of ultra-affordable subsidised EVs from China and Chinese automakers can fret less about tariffs and still profit overseas by honouring minimum pricing. Both sides are still open to negotiation.

Indeed, in August 2024, Chinese EV manufacturers SAIC, BYD and Geely offered to commit to certain prices when selling their vehicles in the European Union to avoid additional tariffs, and the CCCME submitted proposals on behalf of other Chinese automobile manufacturers. Tesla, which receives a separate tariff rate, did not participate in price commitments. The European Commission rejected the proposals.

Since the vote, technical teams from the European Commission and China have been negotiating on a flexible price commitment to avoid tariffs. The European Commission is reported to have rejected China's latest proposal to sell imported EVs in Europe for no less than €30,000.

Pascal Lamy (2024) summed up the difficulty in striking an EU–China deal on EVs at Politico Competitiveness Week in October 2024: 'Things like voluntary export restrictions or price undertakings are extremely difficult for cars.' He said it would be challenging to reach an agreement due to the complexity of minimum price stipulations in the negotiations. Europe and China had agreed on a minimum price stipulation to replace tariffs on Chinese solar panels 10 years ago, but Europe was not happy about the outcome. The European Union lost its PV manufacturing capacities, with China now dominating the entire production chain: China's share in all manufacturing stages of solar panels (such as polysilicon, ingots, wafers, cells and modules) exceeded 80 per cent in 2021 (IEA 2022) and 85 per cent or more thereafter. China's share in the EU photovoltaic market now exceeds 90 per cent. This level of concentration in any global supply chain represents a considerable vulnerability.

One failure of the PV case was attributed to the adoption of extensive avoidance measures. The European Union does not want what happened with solar panels to be repeated with EVs.[2] While minimum pricing has been used for uniform commodities such as PVs, an EV is a highly sophisticated product, with multiple models and configurations, resulting in substantial price differences between each one. The high number of product types entails a high risk of cross-compensation. Moreover, the EU tariffs do not apply to plug-in hybrid electric vehicles and conventional fuel vehicles

---

2   In a speech on technology and politics at the Institute for Advanced Study in Princeton, New Jersey, on 9 April 2024, the European Commission's then executive vice-president Margrethe Vestager contended:
We saw the playbook for how China came to dominate the solar panel industry. First, attracting foreign investment into its large domestic market, usually requiring joint ventures. Second, acquiring the technology, and not always above board. Third, granting massive subsidies for domestic suppliers, while simultaneously and progressively closing the domestic market to foreign businesses. And fourth, exporting excess capacity to the rest of the world at low prices. The result is that nowadays, less than 3% of the solar panels installed in the EU are produced in Europe. We can't afford to see what happened on solar panels, happening again on electric vehicles, wind or essential chips (EC 2024g).

produced overseas, the manufacturers of which may also produce and sell other products and services through the same sales channels and to the same customers. Combined with complicated corporate structures entailing hundreds of companies within each manufacturing group and complex global trade channels, the sheer number of different product models would render price undertakings unenforceable and impractical.

The European Commission is determined to go ahead with individual solutions for the auto industry. It has reportedly received several minimum price proposals from the CCCME to cover a number of EV makers. At the same time, it has received offers directly from multiple EV manufacturers and separately conducted price commitment negotiations with them, which the CCCME has opposed. The CCCME believes allowing parallel negotiations with individual manufacturers will weaken its negotiation capacity.

Issues with the variety of different vehicles suggest that focusing on a single price commitment will not work, so the European Commission has conducted separate price commitment negotiations with EV manufacturers and designated official representatives. In China, generally local governments rather than the central government provide incentives such as tax breaks, cheap land and direct procurement. This supportive role can enhance and amplify central government support. Indeed, different companies within a group establish production bases in different locations to produce different models and may receive different levels of subsidy, meaning setting a single tariff for each group is not justified. Regardless of whether the EU approach is consistent with WTO provisions, it should not apply one price to all models and manufacturers within one group. Prices must be differentiated across companies and locations within a group to avoid penalising companies that receive lower levels of subsidies.

The CCCME does not have the legal authority to punish manufacturers that do not stick to commitments made with the European Commission. Given its poor experience with China regarding PVs, and the greater risk of circumventing undertakings for EVs, it is easy to understand why the European Commission is negotiating separately with individual EV manufacturers. Such deals would be viewed as easier to monitor and enforce.

One solution could be an individually calculated minimum price for each vehicle maker and model type, depending on the size of the vehicle and its range. This could be coupled with import quotas. This option would apply to major EV manufacturers selling large numbers of EVs in the European Union. For other EV manufacturers, negotiations could determine whether to apply minimum prices or tariffs. Provided the CCCME can reach a deal with the European Commission for all EV manufacturers, the agreed minimum prices across models would be applied; otherwise, they would be subject to the EC tariffs decided on in October 2024.

Olof Gill, the European Commission's trade spokesperson, in an interview with the *Financial Times*, said that 'it's not up to the Commission to be prescriptive about what that solution looks like' (White et al. 2024). This implies that it is up to China to find a solution to the EV dispute that addresses the risk to EU industry that the European Commission's investigation identified. It will then be up to China to decide how many concessions it is willing to make to obtain recognition from the European Union.

China's car manufacturers going abroad: Issues for consideration

EU tariffs are imposed only on EVs imported from China and do not apply to those produced in other countries, plug-in hybrid EVs and conventional fuel vehicles. Therefore, brands such as BYD, NIO, Great Wall Motors and XPENG are moving to expand their presence in Europe and have begun exploring local production sites to circumvent the tariffs.

There is a demand for Chinese EVs abroad and Chinese car companies need time to absorb production capacity, so going global is one option. However, it should be emphasised that this will not be the same experience as resource companies investing overseas. China lacks endowments of oil and natural gas and must import large quantities of these resources to meet increasing domestic demand. Extraction and production of such natural resources are highly capital intensive and do not generate much employment. Therefore, resource companies investing overseas have little impact on job creation in China. And, by increasing global supply, they help lower prices of these resources—a win-win outcome for China and the globe (Zhang 2012).

There is another difference. When Japanese car companies built factories overseas in the 1980s, Japan was already facing increasing labour costs and labour supply shortages. By going abroad, Japanese car companies could expand their business while keeping production costs low, thus making their cars competitive on the global market. China now is in a very different position: it needs to create huge numbers of jobs at home.

In terms of providing domestic employment and tax revenue, automobile exports are superior to car companies building factories overseas. In an unfavourable geopolitical environment, therefore, it is crucial for China to make every effort to maintain good economic and trade relations with major consumer countries.

There are several issues to consider when Chinese automakers go global. Chinese car companies must recognise that when building factories overseas it will be difficult to replicate the virtuous cycle operating in China. Due to the size of the Chinese market, pushing down costs enables cars to be sold cheaper, thereby promoting further demand, increasing sales and forming a virtuous cycle. But when building factories overseas, policy barriers and regulations on investment, environmental protection and labour will affect the overseas production capacity of Chinese companies making EVs and batteries, meaning it will be difficult to reach the economies of scale that operate in China, and car prices will remain high.

Rules-of-origin or local content requirements are scheduled to be tightened. With the European Union having imposed duties on EVs from China, Chinese companies have begun setting up plants in Europe to avoid the duties. In July 2024, BYD announced it would invest US$1 billion in a plant in Türkiye, with a planned annual production of 150,000 vehicles, which is expected to begin operation in 2026. In April 2024, Chery signed a joint venture deal with Spanish EV Motors to produce cars in Catalonia, Spain. Some, although not all, EU countries have welcomed Chinese car manufacturers to build factories in their territory. However, the European Union has initiated a probe into the BYD manufacturing plant in Szeged, Hungary, arguing it is of minimal economic value to the union as a whole because it was built using Chinese labour and non–EU materials (Evertiq 2025a). It is therefore expected that the European Union will tighten rules-of-origin requirements for a minimum level of value that must be created in the European Union. Such moves would not necessarily exempt Chinese companies from duties. Chinese factories in Europe should learn from the lessons of past localisation, employ local talent to form the core strength of their workforce, give priority to local suppliers and integrate with local supply chains. Similar rules are likely to be tightened when the United States–Mexico–Canada Agreement (USMCA) is reviewed in July 2026.

A related and more crucial issue is technology transfer. EU Commissioner for Trade and Economic Security Maroš Šefčovič suggested ahead of an official visit to Beijing that technology transfers should be a condition of Chinese investment in Europe's EV industry (Evertiq 2025b). The European Commission's Director-General for Trade Sabine Weyand stated that Europe was 'not interested in investments that are simple assembly businesses without added value and without technology transfer' (Bounds et al. 2025; SCMP 2025). The commission has always insisted that the construction of factories must meet specific conditions, including technology transfer. China has used the strategy of technology swaps for market access since the early period of economic reform and opening, attracting much-needed advanced technologies to invest in the country. The West has long accused China of resorting to such a strategy to force Western companies to transfer technology, which it always denies. The European Union has not tried to hide that it is now emulating China's practice. This suggests Chinese EV and battery manufacturers should be prepared for technology transfers if they invest in the European Union. Chinese EV investment in the United States would face the same issue. The Chinese Government may delay or pause approval for Chinese EV manufacturers to build plants in some regions amid fears of technology leaking into the European Union and North America. This issue must therefore be addressed at the company and country levels as well as between countries.

Moreover, decisions about whether to locate overseas should consider multilateral trade agreements. Some of China's EV original equipment manufacturers are establishing a manufacturing presence in Mexico as a means of reaching Latin American markets and serving North American markets indirectly. However, this move could lead to more action on USMCA trade policies. When building a factory in Mexico, the focus should first be on the Mexican market, to see whether it is profitable to specialise there, rather than on exporting EVs to the United States. If manufacturers specialising in the Mexican market can make a profit there and find they can still export to the United States, this will be an additional dividend. Investing based on this logic will reduce risk. Chinese car companies investing overseas should also learn from Japanese companies, which united overseas.

Concentration of EV investment in a few European countries could trigger concerns in both China and the European Union. One country of concern is Hungary, which has so far attracted most of the EV-related investments by Chinese firms. This China-friendly country has been in conflict with the European Commission on a variety of issues of strategic importance. While the European Union has remained open to Chinese investments in the EV sector, any concentration of Chinese brands in a member country that frequently challenges the majority of EU member states and the European Commission could trigger a backlash against Chinese manufacturers within the broader European Union. Indeed, as mentioned, the European Union has initiated a probe into the BYD manufacturing plant in Hungary, which is expected to bring in investment of €4 billion and create 10,000 jobs (Evertiq 2025a). The investigation is to determine whether the plant receives government aid. If the BYD plant is found to benefit from unfair Chinese subsidies, under the European Commission's rules, it could be forced to reduce capacity, repay subsidies or sell its assets. Another country of concern is Türkiye. Chinese vehicle manufacturers take advantage of a customs union agreement between Türkiye and the European Union. For example, BYD announced a plan to build a US$1-billion plant in Türkiye in July 2024; once in production, the factory would improve BYD's access to the EU market. However, as discussed earlier, Türkiye has implemented severe tariffs and import licence restrictions on Chinese-made EVs. Given the Turkish Government's unpredictable behaviour and disregard for the rules, it is possible the assets of Chinese investors could be nationalised or confiscated. The extent to which Chinese EV manufacturers benefit from Türkiye's customs union status remains an open question.

If prices in foreign markets are higher than those in the domestic market, there will be no dumping, but it does not mean there are no subsidies. This is why the European Union is undertaking an investigation of Chinese EV subsidies, not an anti-dumping investigation. Moreover, the absence of global overcapacity does not mean that Europe and the United States will have to absorb the cars exported from China. The International Energy Agency (IEA 2024b) estimates that under the Stated Policies Scenario, global demand for EVs will reach 45 million units by 2030 and close to 65 million in 2035—up from about 14 million in 2023 and far above China's production capacity. Insufficient production capacity in other developing countries does not mean there is a market to absorb all China's exports, and these countries cannot sacrifice their future industrial development opportunities by not imposing restrictions on Chinese exports. Therefore, for mega-economies like China's, the issue of overcapacity and its impact on China and other countries should not be underestimated. There are many complex factors involved here, which require objective and rational analysis and reflection.

Conclusions

Europe and the United States have enacted regulations and trade measures to restrict the imports of Chinese EVs. For example, the US *Inflation Reduction Act* imposes restrictions on the batteries and raw materials required for EV production. After one year of its subsidy investigation, the European Union imposed countervailing duties on EVs imported from China for five years. These actions are aimed at protecting local manufacturing industries and buying time for the construction of local supply chains. Although these measures protect the auto industries of Europe and North America in the short term, they could be counterproductive in the longer term.

However, given that Chinese EVs enjoy full industrial chain and cost advantages, the EU tariffs are unlikely to dent the expansion of Chinese EVs into Europe, and may not help promote the development of EVs there. Moreover, the European Union's actions limit its consumers' access to more cost-effective Chinese EV services and the means to achieve energy savings, emission reductions and carbon neutrality faster and at a lower cost.

China and Europe must compromise in their dispute over EV tariffs. Whether to continue tariffs or take on price commitments depends on stakeholders within the European Union. Because tariffs are set based on perceived, not material, injury, it will also depend on the extent to which China's price commitments meet EU expectations and evaluations of achieving similar effects through the imposition of countervailing duties.

Overcapacity in China is related to its current industrial policies, which prohibit private enterprises from entering the high-end service industry. This leads to excess capital flow into manufacturing, including high energy-consuming and highly polluting industries, thus resulting in overcapacity in the manufacturing industry and generating large volumes of avoidable pollution emissions. This issue must be paid due consideration, otherwise other industries will repeat the same mistakes.

China's appeal to the WTO over EU duties on its EVs was inevitable, but, facing new tariff and non-tariff barriers from more and more developed and developing countries, can China's demands be effective or beneficial in the short term? Faced with the severe overcapacity of EVs, the fact that nearly all Chinese EVs are barred from entering the US market and exports to other markets will be challenging for a variety of reasons, only the EU market can help absorb China's production capacity in the short term. So, which option, strong countermeasures or low-key negotiation, is more in line with China's interests? Should some Chinese EV manufacturers be given a green light to separately negotiate with the European Union on price commitments or should all Chinese EV manufacturers suffer from tariffs? These are the issues that China must consider.

As the Draghi (2024) report highlights, policy objectives and the corresponding industrial policies and support must be coordinated. Setting extremely strict reduction targets or complete replacement of fuel vehicles by 2035 without fully considering the impact on competitiveness, industrial policies and other supports has put the European Union in a vulnerable position in terms of technological competition with some emerging countries. Unless the European Union amends its carbon dioxide standards regulation[3] to increase flexibility and recalibrates the 2035 internal combustion engine target, in 2035, the sale of new fossil fuel vehicles will be banned. The development of EV vehicles within Europe is therefore inevitable, and more urgent than in the United States. The EU anti-subsidy tariffs are buying time for existing fuel vehicles. Depending on the speed of development and the competitiveness of EVs made in the European Union, it may have to compromise to accept China's price commitments in the early stages of tariff enforcement.

The re-election of Donald Trump as US President could affect the direction and timeline of the dispute over EV tariffs between China and Europe. It could make conditions more unfavourable to China or it could help solve the problem faster. If it is the latter, it must be the European Union and China weighing the potential impact on each of Trump's presidency, each lowering its expectations of the other and the European Union (China) moderately reducing (increasing) expectations about China's price commitments and reaching a compromise.

If China and Europe cannot reach a compromise on the minimum price applicable to all Chinese EV exports, a hybrid alternative of implementing minimum prices for major EV manufacturers and imposing countervailing duties on others may be possible.

In terms of American interests, the United States must consider whether to continue to adhere to its current aggressive restrictions on Chinese EVs, which harm China and do not benefit the United States, or to implement policies that welcome Chinese car companies to build factories in the United States, hiring American workers and accelerating the development of US EVs and batteries.

If the United States chooses to embrace Chinese car companies investing in and building factories on its shores, it can consider increasing tariffs on or tightening the local content requirements of key goods that enter the United States through Mexico based on the importance of those goods when renewing or revising the USMCA. This will limit or block the use of this loophole, which would benefit US employment,

---

3   The CO2 Standards Regulation requires that carbon dioxide emissions from the sale of cars shall not exceed 93.6 grams per kilometre travelled. Any manufacturer that does not comply with the standards faces fines of up to €95 per excess gram, multiplied by the number of cars sold. The specific limit of 93.6 grams of carbon dioxide per kilometre came into effect in 2025 and will continue until 31 December 2029, dropping to 49.5 grams of carbon dioxide per kilometre between 2030 and 2034. This regulation requires a gradual increase in the registration of zero-emission vehicles. Sales of BEVs in the European Union rose continually until the beginning of 2024. In 2023, the share of new BEV sales reached 14.6 per cent. However, the growth curve required for compliance with the carbon dioxide standards by 2030 is steep. The registration share of electric cars in 2024 dropped by 1 percentage point to 13.6 per cent on the EU market. There was a risk that several EU auto manufacturers would not meet their targets and the 2025 passenger vehicle emission targets could result in significant penalties. To increase flexibility, the European Commission has proposed amending the CO2 Standards Regulation for cars and vans. This would allow manufacturers to meet the targets by averaging their performance over a three-year period (2025–27) and to offset any shortfalls in one or two years with excess achievements in the other year(s), while still aiming for the 2025 targets (EC 2025b).

taxation, cheaper and faster domestic production of EVs and cheaper prices for EVs early on. At the same time, it would help achieve stricter energy saving, climate and environmental protection goals. During his presidential campaign, Trump invited Chinese EV manufacturers to build factories and manufacture cars in the United States. It remains to be seen whether he will fulfil this promise. The *America First Investment Policy* released by the White House (2025) aims to protect US economic security and prevent investment threats from China. It not only further restricts Chinese investment in technology and infrastructure critical to the United States, but also restricts US investment in critical technology in China. A careful reading of all the provisions reveals this policy does not rule out Chinese EV manufacturers investing in the United States if they bring Chinese capital and technology with them. Whether this is the case remains to be seen. Chinese EV and battery manufacturers should be prepared for technology transfers if they invest in the United States and the European Union, although the Chinese Government may delay or pause approvals for such ventures amid fears of technology leaking into other countries.

Acknowledgements

An earlier version of this chapter was presented as a keynote address at the Inaugural Australia–China Energy Transition Forum on Electrifying Tomorrow, Sydney, August 2024; China Update 2024, The Australian National University, Canberra, November 2024; thirty-fourth annual meeting of the Chinese Economic Society of Australia, Adelaide, November 2024; China Spatial Economics 2024 Conference, Guangzhou, December 2024; and the Chinese Society for Electrical Engineering Forum on Smart Energy and Energy-Saving Tech Development, Beijing, December 2024; as well as at seminars at the National Development and Reform Commission, Beijing, January 2024, and Fudan University, Shanghai, December 2024. The author gratefully acknowledges financial support from the National Social Science Fund of China (23&ZD095, 20&ZD109). The author bears sole responsibility for any errors and omissions that remain.

References

Bickenbach, Frank, Dirk Dohse, Rolf J. Langhammer, and Wan-Hsin Liu. 2024. *Foul Play? On the Scale and Scope of Industrial Subsidies in China*. Kiel Policy Brief No. 173, April. Kiel: Kiel Institute for the World Economy.

Bounds, Andy. 2024. 'EU Capitals Set to Back Tariffs on Chinese Electric Cars, Trade Chief Says.' Financial Times, 5 August. www.ft.com/content/e9b32eb9-bee6-49bb-a907-953f3324723a.

Bounds, Andy, Henry Foy, and Marton Dunai. 2025. 'EU Probes BYD Plant in Hungary Over Unfair Chinese Subsidies.' *Financial Times*, 20 March. www.ft.com/content/0ef 28741-6194-4ee6-8f23-945415de7458.

Bureau of Industry and Security (BIS). 2024. 'Commerce Announces Proposed Rule to Secure Connected Vehicle Supply Chains from Foreign Adversary Threats.' Press release, 23 September. Washington, DC: US Department of Commerce. www.bis.gov/press-release/commerce-announces-proposed-rule-secure-connected-vehicle-supply-chains-foreign-adversary-threats.

Chilton, Chris. 2025. 'BMW Teams Up with Chinese EV Makers to Fight EU Tariffs in Court.' *CarScoops*, 24 January. www.carscoops.com/2025/01/bmw-teams-up-with-chinese-ev-makers-to-fight-eu-tariffs-in-court/.

DiPippo, Gerard, Ilaria Mazzocco, Scott Kennedy, and Matthew P. Goodman. 2022. *Red Ink: Estimating Chinese Industrial Policy Spending in Comparative Perspective.* Washington, DC: Center for Strategic and International Studies. www.csis.org/analysis/red-ink-estimating-chinese-industrial-policy-spending-comparative-perspective.

Draghi, Mario. 2024. *The Future of European Competitiveness: Part A. A Competitiveness Strategy for Europe.* 9 September. Luxembourg: Publications Office of the European Union. commission.europa.eu/topics/eu-competitiveness/draghi-report_en#paragraph_47059.

European Central Bank (ECB). 2024. 'The Evolution of China's Growth Model: Challenges and Long-Term Growth Prospects.' *ECB Economic Bulletin*, no. 5/2024. Frankfurt: ECB.

European Commission (EC). 2019. *Commission Implementing Regulation (EU) 2019/688 of 2 May 2019 Imposing a Definitive Countervailing Duty on Imports of Certain Organic Coated Steel Products Originating in the People's Republic of China Following an Expiry Review Pursuant to Article 18 of the Regulation (EU) 2016/1037 of the European Parliament and of the Council.* Document 32019R0688, 2 May. Brussels: European Commission. eur-lex.europa.eu/eli/reg_impl/2019/688/oj/eng.

EC. 2022. *Regulation (EU) 2022/2560 of the European Parliament and of the Council of 14 December 2022 on Foreign Subsidies Distorting the Internal Market.* Document 32022R2560, 23 December. Brussels: European Commission. eur-lex.europa.eu/eli/reg/2022/2560/oj.

EC. 2024a. *Regulation (EU) 2024/1252 of the European Parliament and of the Council of 11 April 2024 Establishing a Framework for Ensuring a Secure and Sustainable Supply of Critical Raw Materials and Amending Regulations (EU) No. 168/2013, (EU) 2018/858, (EU) 2018/1724 and (EU) 2019/1020 (Text with EEA Relevance).* Document 32024R1252, 3 May. Brussels: European Commission. eur-lex.europa.eu/legal-content/EN/TXT/?uri=CELEX%3A32024R1252&qid=1720020986785.

EC. 2024b. 'EU Imposes Duties on Unfairly Subsidised Electric Vehicles from China While Discussions on Price Undertakings Continue.' Press release, 29 October. Brussels: European Commission. ec.europa.eu/commission/presscorner/detail/en/ip_24_5589.

EC. 2024c. 'Commission Investigation Provisionally Concludes That Electric Vehicle Value Chains in China Benefit from Unfair Subsidies.' Press release, 12 June. Brussels: European Commission. ec.europa.eu/commission/presscorner/api/files/document/print/en/ip_24_3231/IP_24_3231_EN.pdf.

EC. 2024d. 'Commission Imposes Provisional Countervailing Duties on Imports of Battery Electric Vehicles from China While Discussions with China Continue.' Press release, 4 July. Brussels: European Commission. ec.europa.eu/commission/presscorner/detail/en/ip_24_3630.

EC. 2024e. 'Commission Discloses to Interested Parties Draft Definitive Findings of Anti-Subsidy Investigation into Imports of Battery Electric Vehicles from China.' Press release, 20 August. Brussels: European Commission. ec.europa.eu/commission/presscorner/detail/en/ip_24_4301.

EC. 2024f. 'Commission Proposal to Impose Tariffs on Imports of Battery Electric Vehicles from China Obtains Necessary Support from EU Member States.' Press release, 4 October. Brussels: European Commission. ec.europa.eu/commission/presscorner/detail/en/statement_24_5041.

EC. 2024g. 'Speech by Executive Vice President Vestager on Technology and Politics.' Institute for Advanced Study, Brussels, 9 April. ec.europa.eu/commission/presscorner/detail/en/SPEECH_24_1927.

EC. 2024h. *Commission Implementing Regulation (EU) 2024/2754 of 29 October 2024 Imposing a Definitive Countervailing Duty on Imports of New Battery Electric Vehicles Designed for the Transport of Persons Originating in the People's Republic of China, C/2024/7490 Final.* Document 32024R2754, 29 October. Brussels: European Commission. eur-lex.europa.eu/eli/reg_impl/2024/2754/oj/eng.

EC. 2025a. 'Commission Implementing Regulation (EU) 2025/61 of 15 January 2025 Imposing a Definitive Countervailing Duty on Imports of Certain Pneumatic Tyres, New or Retreaded, of Rubber, of a Kind Used for Buses or Lorries, with a Load Index Exceeding 121 Originating in the People's Republic of China Following an Expiry Review Pursuant to Article 18 of Regulation (EU) 2016/1037 of the European Parliament and of the Council.' *Official Journal of the European Union* 2025/61 (15 January). Brussels: European Commission. eur-lex.europa.eu/legal-content/EN/TXT/PDF/?uri=OJ:L_202500061.

EC. 2025b. *Communication from the Commission to the European Parliament, the Council, the European Economic and Social Committee and the Committee of the Regions: Industrial Action Plan for the European Automotive Sector, COM/2025/95 Final.* Document 52025DC0095, 5 March. Brussels: European Commission. eur-lex.europa.eu/legal-content/EN/TXT/?uri=celex:52025DC0095.

European Parliament. 2025. 'Implementing the EU's Net-Zero Industry Act.' *Briefing*, February. Brussels. European Parliamentary Research Service. www.europarl.europa.eu/RegData/etudes/BRIE/2025/769489/EPRS_BRI(2025)769489_EN.pdf.

Evertiq. 2025a. 'European Union Launches Investigation into Chinese EV Subsidies.' *Evertiq*, 21 March. evertiq.com/design/2025-03-21-european-union-launches-investigation-into-chinese-ev-subsidies.

Evertiq. 2025b. 'EU Trade Chief Suggests Tech Transfers in Exchange for Chinese EV Investment.' *Evertiq*, 21 March. evertiq.com/design/2025-03-21-european-union-trade-chief-suggests-tech-transfers-in-exchange-for-chinese-ev-investment.

Felbermayr, Gabriel, Klaus Friesenbichler, Julian Hinz, and Hendrik Mahlkow. 2024. *Time to Be Open, Sustainable, and Assertive: Tariffs on Chinese BEVs and Retaliatory Measures.* Kiel Policy Brief No. 177, July. Kiel: Kiel Institute for the World Economy. www.ifw-kiel.de/publications/time-to-be-open-sustainable-and-assertive-tariffs-on-chinese-bevs-and-retaliatory-measures-33083/.

Feng, Difan, and Cheng Cheng. 2024. 'New US Tariffs on Chinese EVs Would Breach WTO Deal, Lawyer Says.' *Yicai*, 14 May. www.yicaiglobal.com/news/anticipated-new-us-tariffs-on-chinese-evs-would-violate-wto-deal-expert-says.

International Energy Agency (IEA). 2022. *Special Report on Solar PV Global Supply Chains*. Rev. vers., August. Paris: International Energy Agency. iea.blob.core.windows.net/assets/d2ee 601d-6b1a-4cd2-a0e8-db02dc64332c/SpecialReportonSolarPVGlobalSupplyChains.pdf.

IEA. 2024a. *Advancing Clean Technology Manufacturing: An Energy Technology Perspectives Special Report*. Paris: International Energy Agency. iea.blob.core.windows.net/assets/7e7f4b17-1bb2-48e4-8a92-fb9355b1d1bd/CleanTechnologyManufacturingRoadmap.pdf.

IEA. 2024b. *Global EV Outlook 2024: Moving Towards Increased Affordability*. April. Paris: International Energy Agency. iea.blob.core.windows.net/assets/a9e3544b-0b12-4e15-b407-65f5c8ce1b5f/GlobalEVOutlook2024.pdf.

Lamy, Pascal. 2024. 'EU Cannot Remain the Only Big Market Open to Subsidize Chinese EVs, Former WTO Director General Says.' *CNBC*, 3 October. www.cnbc.com/video/2024/10/03/eu-cannot-remain-the-only-big-market-open-to-subsidize-chinese-evs-ex-wto-director-general.html.

Liu, Xiaomeng. 2025. 'Cui Dongshu: In 2024, the Automotive Industry's Revenue Will Exceed 10 Trillion Yuan, A Year-on-Year Increase of 4%.' *Beijing Business Daily*, 27 January. news.qq.com/rain/a/20250127A06EQ900.

Organisation for Economic Co-operation and Development (OECD). 2021. *Measuring Distortions in International Markets: Below-Market Finance*. OECD Trade Policy Papers No. 247, 12 May. Paris: OECD Publishing. doi.org/10.1787/a1a5aa8a-en.

OECD. 2023. *Measuring Distortions in International Markets: The Rolling-Stock Value Chain*. OECD Trade Policy Papers No. 267, 15 February. Paris: OECD Publishing. doi.org/10.1787/fa0ad480-en.

Reuters. 2024. 'Biden Proposes Banning Chinese Vehicles from U.S. Roads with Software Crackdown.' *CNBC*, 23 September. www.cnbc.com/2024/09/23/biden-software-crackdown-proposes-banning-chinese-vehicles-from-us.html.

Rhodium. 2024. 'Ain't No Duty High Enough.' *Rhodium*, 29 April.

South China Morning Post (SCMP). 2025. 'Tech Transfer a Must for Chinese EV Investors in Europe, EU Trade Chief Suggests.' *Yahoo*, 20 March. www.yahoo.com/news/tech-transfer-must-chinese-ev-093000523.html.

The White House. 2025. *America First Investment Policy*. 21 February. Washington, DC: The White House. www.whitehouse.gov/presidential-actions/2025/02/america-first-investment-policy/.

Transport & Environment (T&E). 2024. 'How Europe Can Use Tariffs as Part of An Industrial Strategy.' *Briefing*, 27 March. Brussels: European Federation for Transport and Environment. www.transportenvironment.org/articles/how-europe-can-use-tariffs-as-part-of-an-industrial-strategy.

US Department of the Treasury. 2023. 'Treasury Releases Proposed Guidance to Continue U.S. Manufacturing Boom in Batteries and Clean Vehicles, Strengthen Energy Security.' Press release, 1 December. Washington, DC: Department of the Treasury. home.treasury.gov/news/press-releases/jy1939.

White, Edward, Cheng Leng, and Claire Bushey. 2023. 'China's "Battery King" Faces Scrutiny Over EV Market Dominance.' *Financial Times*, 4 April. www.ft.com/content/9f411244-eb72-493f-86d2-e7bf77de757e.

White, Edward, Daria Mosolova, and Andy Bounds. 2024. 'EU Rejects Chinese EV Makers' Bid to Avert Hefty Tariffs.' *Financial Times*, 13 September. www.ft.com/content/e1321972-def6-4980-a52e-b9cc5fab2ae0.

Wu, Ziye. 2024. 'The New Energy Vehicle Industry Chain is Uneven: CATL and BYD Hold Most of the Profits, and Experts Suggest that Car Companies Must Produce Batteries.' *Yicai*, [China], 26 August. www.yicai.com/news/102247076.html.

Yang, Jun, Ma Xinyu, and Wang Yang. 2024. 'Why Did Exports in the First Two Months of "The New Three" Have Negative Year-on-Year Growth? Data Analysis and Countermeasure Suggestions.' *Jiemian News*, 12 April. www.jiemian.com/article/11036139.html.

Zhang, ZhongXiang. 2012. 'The Overseas Acquisitions and Equity Oil Shares of Chinese National Oil Companies: A Threat to the West but a Boost to China's Energy Security?' *Energy Policy* 48: 698–701. doi.org/10.1016/j.enpol.2012.05.077.

Zhang, ZhongXiang. 2024. 'Can the World Build Trust and Consensus on Climate Change? Discussions on the Importance and Prospects of China–US–EU Climate Cooperation.' *Frontiers*, no. 8: 31–48.

7

China's green finance system: Policies, market and international role

Mengdi Yue and Christoph Nedopil

China's understanding of sustainable development and approach to addressing environmental challenges have evolved significantly in the past decade: from tackling air, water and soil pollution resulting from rapid industrialisation in the early 2010s (Schmid and Xiong 2023) and its 2020 pledge to achieve carbon peak by 2030 and carbon neutrality by 2060, to the imperative to fully establish 'a beautiful China' by mid-century, which was announced in 2024. It is estimated that to achieve carbon neutrality by 2060, China will need RMB487 trillion in investment in green and low-carbon industries between now and 2050 (Green Finance Committee 2021). With limited fiscal resources, a green financial system is critical in mobilising ('financing green') and redirecting ('greening finance') private capital to achieve this goal and facilitate China's sustainable development goals.

China's green finance system has evolved in parallel with, and to serve, these sustainable development goals. Since the issuance of the 'Green Credit Guidance' in 2012 to encourage banks to use green credits to manage environmental and social risks and the official establishment of the green financial system in 2016, China's green finance has matured into a complex structure covering most of the financial system, including multiple instruments and several regulators.

China's green bond market, for example, remained the world's largest, with a total issuance of US$131 billion both onshore and offshore in 2023—nearly double that of Germany, the second-place issuer (Climate Bonds Initiative 2024). Similarly, by the end of 2024, China's outstanding green loans reached RMB36.6 trillion (approximately US$5.1 trillion), representing 14 per cent of total outstanding loans. China's national carbon market—the world's largest by emissions coverage—encompassed 5.24 billion metric tonnes of carbon dioxide, accounting for more than 40 per cent of China's carbon emissions (Velev 2025).

Despite these achievements, challenges persisted, including drops in green bond issuance in two consecutive years (Yue and Nedopil 2025), a small percentage of green credits in the total credit market, concerns about oversupply of allowances in the carbon market, limited participation by the private sector and the unclear (understudied) effectiveness of green finance in terms of carbon emissions reduction.

Internationally, China has initiated a wide range of green investments and infrastructure in the Global South through the Belt and Road Initiative (BRI). China has also become an increasingly active actor in advancing global green finance and biodiversity standards and regional taxonomy harmonisation. China's engagement in the green finance field outside its borders is seen as using green finance to build soft power in the BRI countries (Nedopil 2022) or seeking global leadership on the environment and sustainability (Rodenbiker 2025).

With this context, this chapter aims to reflect on the decade-long development of China's green finance system and the status of policies, markets and China's international role. It also provides ideas for future research questions relevant to this topic. The remainder of this chapter is structured as follows. Section two examines the evolution of China's green finance policy from the 1990s to the present, highlighting key stages of development, the institutional landscape and the roles of governmental and nongovernmental actors. Section three analyses the components of China's green finance market, including environmental, social and governance (ESG) reporting standards, financial instruments (green bonds, equity, loans and funds), risk management tools (green insurance) and market-based mechanisms, such as the China Certified Emission Reduction (CCER) program and its emission trading system (ETS). Section four explores China's influence in shaping global green finance standards and promoting green investment through the BRI. Section five concludes by summarising key findings, offering an outlook on China's green finance trajectory and identifying areas for future research.

China's green finance evolution and governance

Understanding China's green finance system requires an awareness of historical developments to appreciate potential path dependencies (Heine and Kerber 2002; Nedopil et al. 2021b) and the role of different actors in China's top-down system.

China's green finance evolution

In a historical context, the development of China's green finance governance can be conceptualised in four main stages (see Figure 7.1). During the 1990s and 2000s, awareness of integrating environmental considerations into bank lending began to rise. In 1995, the People's Bank of China issued the *Notice on Implementing Credit Policies and Strengthening Environmental Protection* (PBOC 1995), followed by the joint issuance of the *Opinions on Implementing Environmental Protection Policies and Regulations to Prevent Credit Risks* (SEPA 2007).

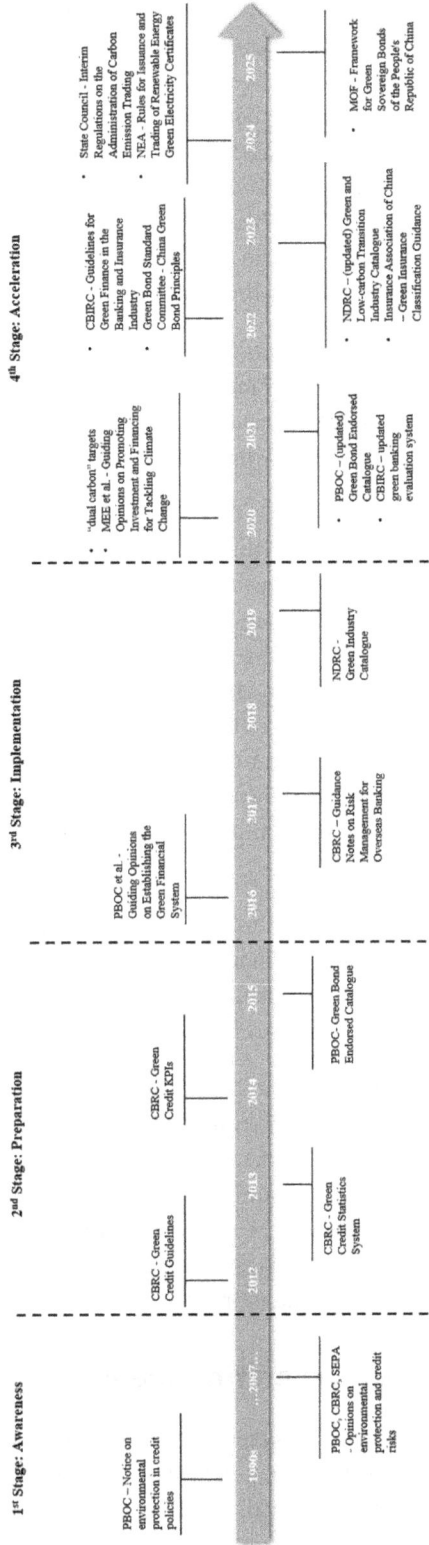

Figure 7.1: Stages of green finance policy development in China
Source: Authors.

From 2012 to 2015, policies on green credit continued to grow, with the China Banking Regulatory Commission (CBRC) establishing its green credit guidance and statistics system. The *Green Credit Guidelines* (CBRC 2012) further strengthened this initiative by requiring banks to integrate environmental, social and governance (ESG) factors into their risk assessments and lending decisions. At the end of 2015, the *Green Bond Endorsed Project Guidelines* were released, listing the sectors eligible for green bond issuance (NDRC 2016). Further preparation work for the green financial system was also underway with the establishment of a taskforce by the PBOC and the United Nations Environment Programme.

The 2016 *Guiding Opinions on Establishing a Green Financial System* issued by a PBOC-led coalition of seven ministries laid out the components of the green finance market (green credits, green securities, green funds, green insurance, environmental rights and regional green finance), marking the third stage of green finance implementation (PBOC et al. 2016).

The development of green finance in China has entered a stage of expansion and acceleration since the government announced its 'dual carbon' goals in 2020. A key policy framework serving these goals, introduced in 2021, was the '1+N' policy framework, which outlines a comprehensive strategy to help the country achieve its dual carbon targets: peaking carbon emissions by 2030 and reaching carbon neutrality by 2060. The '1' represents the overarching guiding document that defines the objectives, fundamental pathways and key tasks for achieving these goals, while the 'N' refers to a series of policies and plans formulated by various departments and local governments. The financial sector is considered an integral part of these goals.

This evolution of China's green finance ecosystem highlights two key features elaborated in the next sections. First, reflecting China's underlying model of political economy, the system is predominantly governed by a top-down approach (Nedopil and Larsen 2023). Policies establishing standards and guidance are typically issued before large-scale application occurs, but once policies are in place, the application accelerates.

Second, through the setup of pilot zones, city or provincial-level governments play a crucial role in experimenting with green finance policies. One example is the seven pilot emissions trading markets established in 2011, which preceded the national ETS that was initiated in 2017 and became operational in 2021. More recently, several provinces and cities have begun exploring transition finance taxonomies, even in the absence of a unified national standard. Local pilots of innovative insurance products also started in the early 2010s, while more formal policies and guidelines were not in place until 2022.

Top-down green finance governance

The development of green finance in China is mainly achieved through top-down and central-to-local governance, with a high-level guiding policy issued before any market practice is implemented. This remains true after China's financial governance system underwent a significant transformation in 2023 with the newly established Central

Financial Commission (CFC) and National Financial Regulatory Administration (NFRA) (see Figure 7.2). The CFC replaces the former Financial Stability and Development Committee and has become the primary body for centralised financial decision-making and coordination. The NFRA replaces the China Banking and Insurance Regulatory Commission (CBIRC), which had replaced the CBRC, while taking on broader responsibilities. Accordingly, the responsibilities of the PBOC and the China Securities Regulatory Commission (CSRC) were also slightly adjusted.

The responsibilities of the various bodies are as follows:

- **People's Bank of China:** The PBOC implements monetary policies that encourage green lending; issues guidelines for green bonds and financial instruments; oversees green investments and supports the development of green industries.
- **National Financial Regulatory Administration:** The NFRA ensures financial stability while aligning financial regulations with China's sustainability goals. It promotes green finance by regulating the greening of financial industries (for example, insurance and banks) and preventing financial risks.
- **National Development and Reform Commission:** The NDRC sets national policies and strategies for green finance, integrating environmental objectives into economic planning. It also leads the issuance of the *Green Industry Catalogue*, a taxonomy of green and low-carbon transition industries.

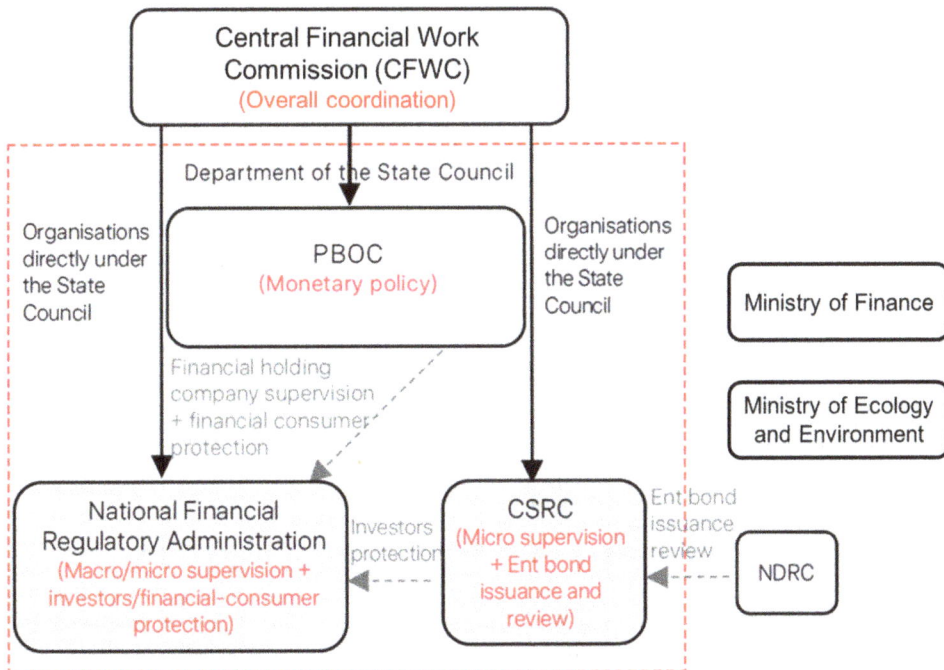

Figure 7.2: China's green finance regulatory environment
Source: Adapted from Zhang et al. (2024).

- **Ministry of Finance:** The MoF supports green finance through fiscal policies, subsidies and tax incentives for green industries. It also mobilises development funds and collaborates with international organisations to secure funding for China's green transition.
- **Ministry of Ecology and Environment:** The MEE advances green finance by setting carbon accounting standards, managing the national carbon market and developing carbon sink methodologies. It ensures environmental compliance of companies and translates ecological value into actionable financial data and tools. It also supports greening overseas investment through the BRI Green Investment Principles.
- **China Securities Regulatory Commission:** The CSRC regulates green securities including green bonds and stocks and promotes ESG disclosures for publicly listed companies.

At the subnational and local levels, provincial and city government offices support and partly implement national green finance policies through local Development and Reform Commissions (the subnational ministries of the NDRC), the Bureaus of Finance (subnational level of the MoF) or the Bureaus of Ecology and Environment (subnational level of the MEE). They also publish local policies based on national high-level policies for more targeted implementation.

Nongovernmental actors in China's green finance ecosystem have limited standard-setting roles and are responsible for implementing government and regulatory policies:

- Market exchanges (such as Shenzhen, Shanghai and Beijing stock exchanges): These provide standards for the ESG information disclosures of listed companies.
- Associations: Associations, such as the National Association of Financial Market Institutional Investors (NAFMII) and the Green Finance Committee (a network of Chinese financial institutions), explore rules and principles for innovative green finance products and provide capacity building as well as standards. These associations are often supported by government regulators (for example, the Green Finance Committee is supported by the PBOC).
- Financial institutions: These are the main bodies for implementing green finance products (such as green insurance, green credits) and policies (stress tests and scenario analyses for climate risks, capacity building).
- Service providers: These offer services such as assessment, certification of green bonds, capacity building, etcetera.

The role of subnational green finance pilot programs

Subnational green finance pilot programs are integral to China's efforts to build a sustainable financial system, serving as testing grounds for innovative policies and mechanisms at the regional level. At the subnational level, two types of pilot zones have been established to explore replicable and scalable green finance practices: the Green Finance Reform and Innovation Pilot Zones, since 2017, and the Climate Investment and Financing Pilots, since 2022. In August 2022, 23 climate finance pilot zones across China were approved. While some cities belong to both categories, the missions of climate finance pilot zones focus more on discouraging high-emission projects and developing carbon finance and carbon accounting.

Figure 7.3 illustrates the first and second batches of green finance pilot zones, outlining their key focus areas. Initially, the program was implemented in eight cities across Zhejiang, Jiangxi, Guangdong, Guizhou and Xinjiang provinces. Later, green finance reforms were extended to Lanzhou in Gansu and Chongqing in Sichuan.

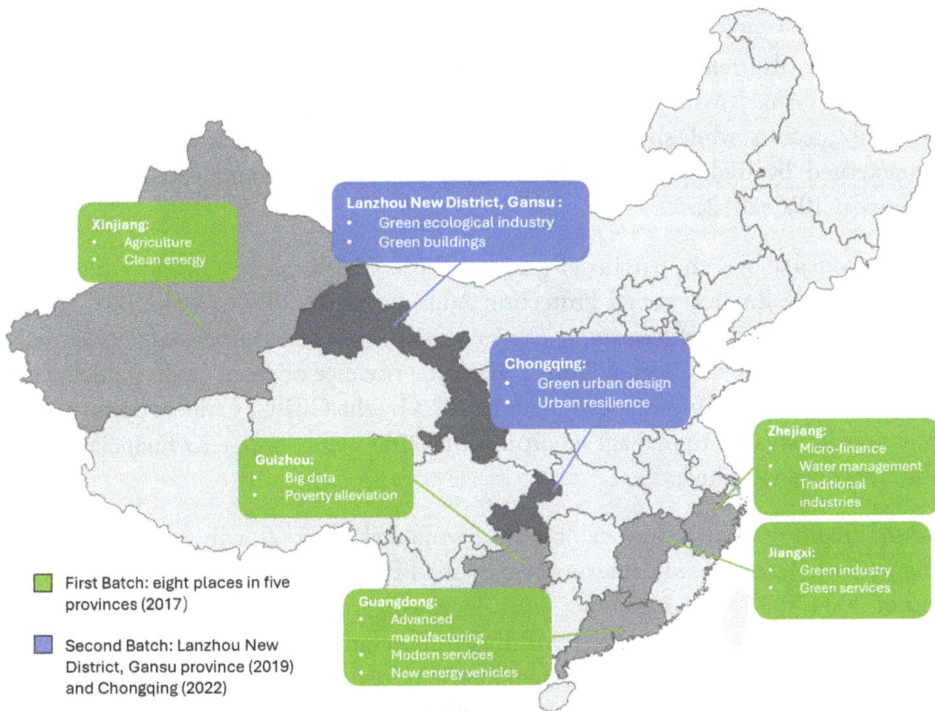

Figure 7.3: Subnational green finance pilot programs in China
Source: Adapted from Zhang et al. (2024).

Green finance application and market development

The development of China's green finance system has led to the development of green financial instruments (for example, green credit, bonds, as well as equity and loans), ESG disclosure standards, environmental risk management through green insurance and carbon market mechanisms. The following subsections examine how these elements have been applied and how they contribute to the broader development of China's green finance market.

Green financial instruments

The application of China's different green financial instruments has evolved significantly, with green credits introduced in 2012 and now the instrument with the largest green finance value, while green bonds, introduced in 2015, have traditionally received the most attention due to their higher disclosure requirements, which facilitate a greater level of research (Wang and Li 2020; Wang et al. 2020; Zhang 2020). Other green securities (such as stocks and mutual funds) have received more attention lately but lag the development of credit and bonds.

Green credit

Green credit, also referred to as green loans, is a key financial instrument employed by Chinese banks to support projects and enterprises operating in environmentally sustainable sectors, while simultaneously imposing strict credit constraints on industries characterised by high pollution, high energy consumption and excess production capacity (CBRC 2012).

As mentioned above, the origins of green credit in China can be traced back to 2007 when the State Environmental Protection Administration, PBOC and CBRC jointly issued the *Opinions on Implementing Environmental Protection Policies and Regulations to Prevent Credit Risks* (SEPA et al. 2007). With the issuance of green credit guidelines in 2012 and a green credit statistics system in 2013 by the CBRC, Chinese banks started to use credit issuance to manage environmental risks before the green financial system was officially established.

Green credit in China has exhibited a sustained high growth rate since 2018 (Figure 7.4). According to data compiled by the PBOC from 21 major banks, as of the end of 2024, the total outstanding green loans—denominated in both renminbi and foreign currencies—amounted to RMB36.6 trillion (approximately US$5.1 trillion). A significant proportion of these green loans are allocated to the ecological modernisation of infrastructure, with primary investment directed to industries related to electricity, heat, gas and water supply (Fu and Sun 2025).

Since 2021, the PBOC has implemented an assessment framework to evaluate banks' green finance performance based on qualitative and quantitative indicators. These metrics include, among others, the volume of outstanding green loans and the holdings of green bonds by banks. This regulatory emphasis may partly explain the strong growth of outstanding green loans since 2018 and serve as an incentive for banks to further expand their green loan issuance in the coming years.

Green bonds

China's green bond market was officially launched in late 2015 (the international green bond market emerged in 2007). China quickly became a dominant global player, accounting for roughly one-quarter of the total global issuances of US$81 billion in 2016, positioning the country as a leader in annual issuance volume (Climate Bonds Initiative and CCDC Research 2017).

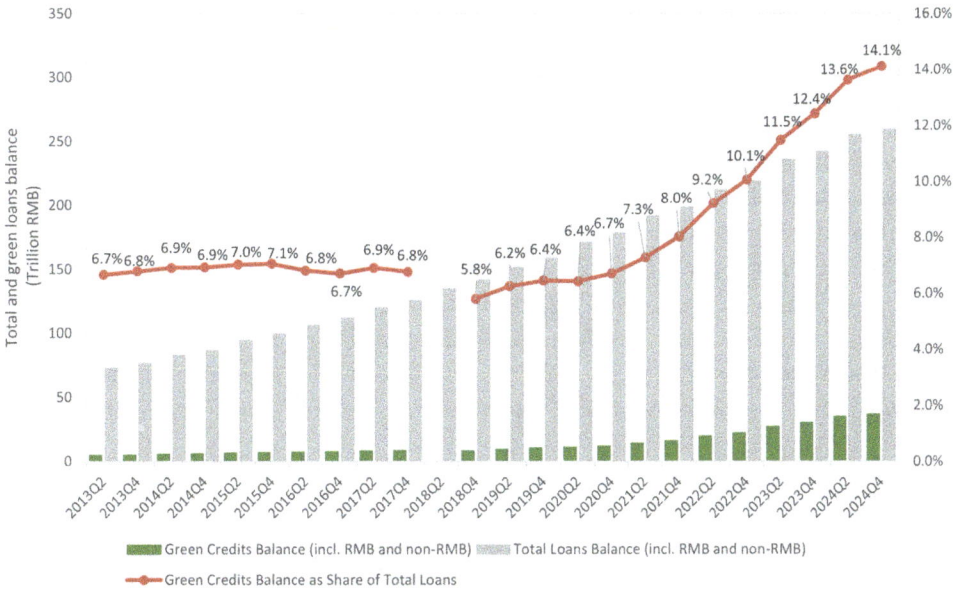

Figure 7.4: Green credits in China, 2013–24

Notes: Data only include 21 major Chinese banks. Data for 2018Q2 are not available, likely due to the transition of green credit accounting responsibilities from the CBRC to the PBOC.

Sources: Compiled by authors based on public data from the PBOC and CBRC.

China's green bond market operates within a fragmented regulatory framework (see Table 7.1), with oversight distributed across multiple entities, including the PBOC, the NDRC, the CSRC, the NAFMII and stock exchanges, depending on the bond type (Lin 2023).

Table 7.1: Green bond types and regulators in China

Types of green bond	Issuer	Regulatory authority	Percentage of funds used for green projects	
			Before CGBP	After CGBP
Green financial bonds*	Financial institutions	PBOC	100	100
Green debt financial instruments	Non-financial corporations	NAFMII	100	100
Green corporate bonds	State-owned and private corporations	CSRC and stock exchanges	70	100
Green enterprise bonds	Mostly state-owned enterprises	NDRC	50	100

Note: * Issued by financial institutions to raise funds for green loans.

Sources: Compiled by authors based on Escalante et al. (2020) and publicly available information.

Before 2022, green financial bonds (regulated by the PBOC) and green debt financial instruments (regulated by NAFMII) required that 100 per cent of proceeds be allocated to green projects. In contrast, green corporate bonds (regulated by the CSRC and stock exchanges) must allocate at least 70 per cent of proceeds to green projects, while green enterprise bonds issued by state-owned enterprises (regulated by the NDRC) had a minimum threshold of 50 per cent. In 2022, the China Green Bond Principles (CGBP) were introduced by the China Green Bond Standards Committee to enhance alignment with internationally recognised standards (especially those of the International Capital Market Association, ICMA) and harmonisation within China. The CGBP mandate a 100 per cent use-of-proceeds, ensuring that all funds raised through green bonds are directed to the construction, operation and acquisition of green projects, as well as refinancing of working capital or repayment of debts associated with such projects (ICMA 2022).

The convergence of Chinese and international green bond standards has been progressing to align Chinese with international taxonomies. As of 2024, the eligibility of green projects for onshore green bond issuance is determined based on the 2021 version of China's *Green Bond Endorsed Project Catalogue*, whereas offshore issuance may reference international taxonomies, such as ICMA standards (as well as the Common Ground Taxonomy, discussed below). Notably, the 2021 revision of the catalogue removed categories such as clean gas production, clean coal utilisation and the retrofitting of coal-fired power plants, which had been part of the 2015 version. This update resulted in an increased share of China's labelled green bonds aligning with the definitions of the Climate Bonds Initiative—from 54 per cent in 2020 to 62 per cent in 2021 (Climate Bonds Initiative and CCDC Research 2022).

China's efforts to align with international standards were further evidenced by the publication of the China–European Union Common Ground Taxonomy (CGT) in 2022, which identifies 72 climate change mitigation activities shared between the EU and Chinese taxonomies regarding the 'substantial contribution' criteria. As of February 2025, 245 green bonds issued in the interbank market comply with the CGT standards, with a combined volume of RMB351.8 billion (US$48.6 billion) and accounting for 20.31 per cent of total outstanding green bonds in the interbank market (Green Finance Committee 2025).

In 2024, despite a slight decline (Figure 7.5), newly issued green bonds onshore and offshore totalled more than RMB850 billion (US$120 billion). The proportion of green bonds in the domestic bond market remains trivial: newly issued green bonds accounted for less than 1 per cent of the total bond market in 2024. Central and provincial state-owned enterprises (SOEs) continued to dominate the issuance landscape, contributing more than 80 per cent of green bonds by volume. Most green bonds have a maturity of less than five years and therefore specialise in providing short-term financing (compared with green loans).

In the offshore market, the Hong Kong Stock Exchange is the largest venue for green bonds issued by Chinese entities, and more than half those issued in 2024 were denominated in US dollars, followed by 37 per cent in renminbi (Fu and Sun 2025).

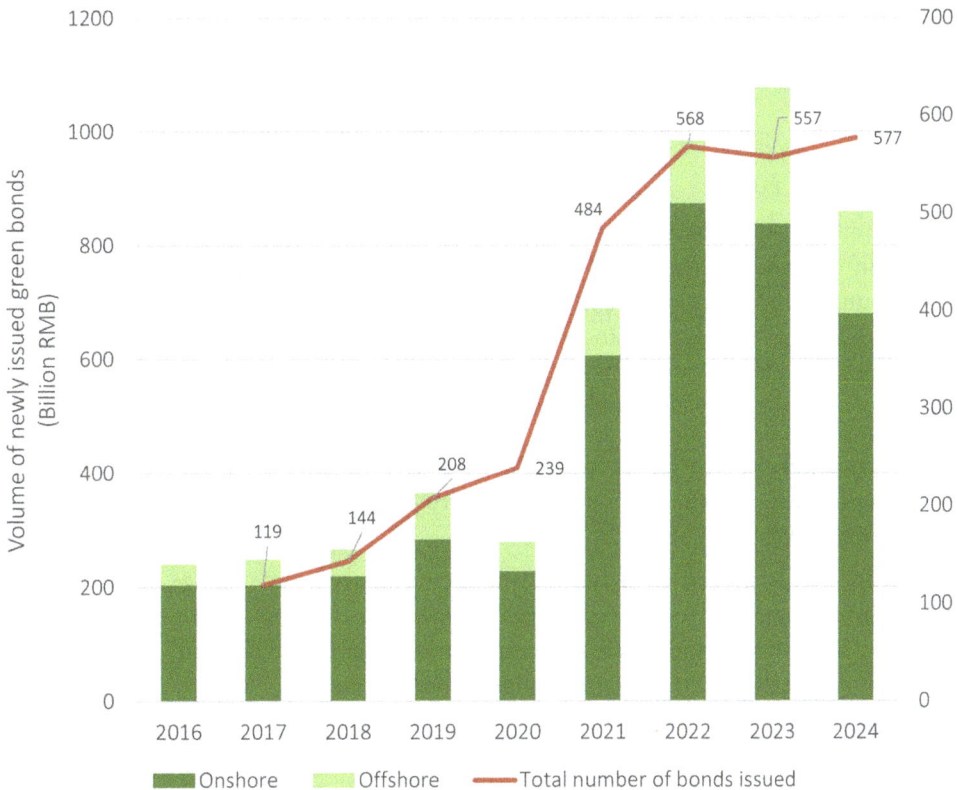

Figure 7.5: China's green bond issuance, 2016–24

Sources: Compiled by authors based on data from the International Institute of Green Finance (iigf-china.com/); Climate Policy Initiative (www.climatepolicyinitiative.org/); Wind (www.wind.com.cn/).

Green stocks

Green stocks are shares in companies that demonstrate they contribute to the green economy (Alnes et al. 2024). The concept of green stocks has been briefly mentioned in Chinese policy documents, such as through encouraging green enterprises in their initial public offerings and refinancing and supporting the compilation of green stock indices. However, so far there are no established standards for or definitions of green stocks. In March 2024, a coalition of seven ministries, led by the PBOC, issued the *Guidelines on Further Strengthening Financial Support for Green and Low-Carbon Development*, emphasising the need to 'develop standards for green stocks and unify the business rules for green stock operations' (PBOC et al. 2024), indicating that this area may witness more rapid progress in the future.

An analysis of 248 A-share listed companies falling under the categories of 'New Energy' and 'Energy Conservation and Environmental Protection' reveals that, as of December 2023, their total market value constituted only 1.81 per cent of the total market value of A-share listed companies in China. Furthermore, their average market value per company was RMB6.4 billion (US$883 million)—significantly lower than the A-share market average of RMB33 billion (US$4.5 billion) (Jin and Jiang 2024). These figures highlight the relatively small scale and limited market presence of green listed companies in China's equity markets.

ESG mutual funds

Environmental, social and governance (ESG) mutual funds and indices have emerged as rapidly expanding categories of green financial products in China (Shen et al. 2023). ESG mutual funds pool money from various (including private) investors to invest in securities such as bonds and stocks. While no official definition or dedicated policy guidelines for 'ESG funds' have been established in China, research institutions and data providers generally classify these funds into two categories:

1. Pure/strongly related ESG funds, which explicitly include keywords such as 'ESG', 'sustainable development', 'carbon neutrality' or 'carbon peak' in their fund names or investment objectives.

2. Broad/weakly related ESG funds, which encompass a wider range of funds with environmental and social responsibility themes in their investment strategies and often include keywords such as 'energy conservation', 'low carbon', 'clean energy' and 'biodiversity' in their names or investment objectives.

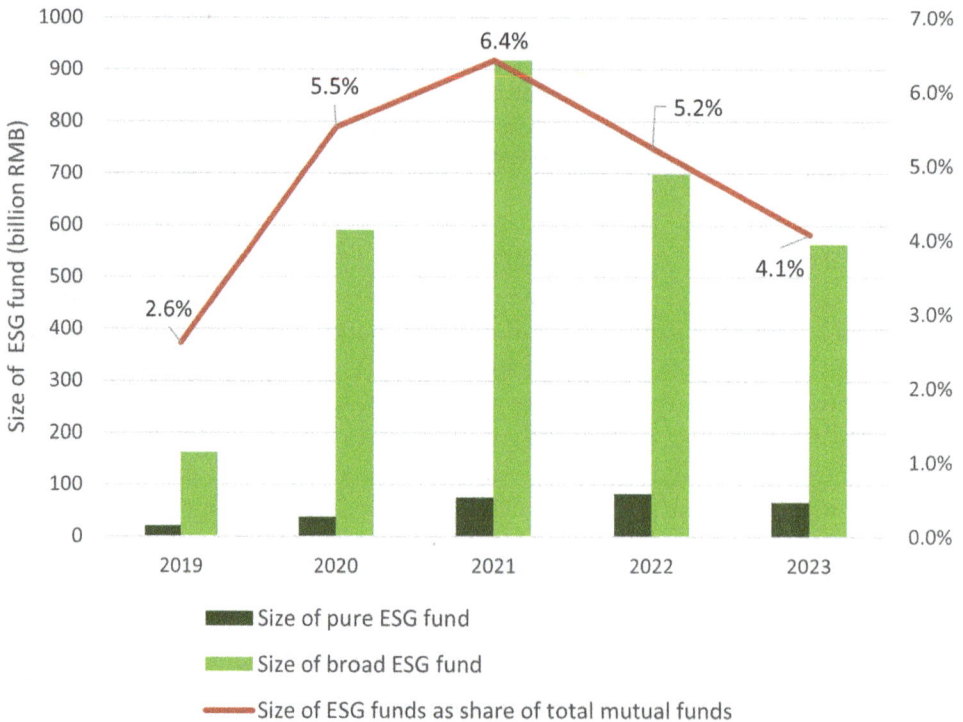

Figure 7.6: Number and share of ESG funds in China, 2019–23
Source: Compiled by authors based on data from IIGF (2024).

A large majority of ESG public mutual funds in China are equity or equity-oriented hybrid funds. As of the end of 2023, there were 101 pure ESG mutual funds (RMB66.4 billion/US$9.15 billion of scale) and 485 broad ESG mutual funds (RMB563.9 billion/ US$77.7 billion of scale) in the Chinese market. The proportion of ESG funds within total mutual funds remains small, declining in both 2022 and 2023, to 5.2 per cent and 4.1 per cent, respectively (see Figure 7.6).

As of November 2024, there were 858 ESG indices in China's domestic market, with bond indices the most common (463 indices, 53.76 per cent), followed by stock indices (392 indices, 49.46 per cent). Indices with environmental protection themes still dominate, but the share of ESG indices covering all three dimensions (environmental, social and governance) has been rising, ranking second only to environmental themes (Li et al. 2025).

ESG disclosure

China's stock markets have been at the forefront in implementing ESG disclosure requirements, as demonstrated by the Shenzhen Stock Exchange SZSE 100 Index and the Shanghai Stock Exchange (SSE), which mandated companies to disclose ESG or corporate social responsibility (CSR) reports in 2008. However, the quality of these ESG and CSR reports remained poor and uninformative due to limited ESG awareness and the absence of clear regulations (Yuan et al. 2022).

By 2025, ESG disclosure requirements were quite disparate, varying between entities (Table 7.2). Most of these requirements have been voluntary, allowing organisations to prepare conceptually and institutionally before mandatory implementation. This approach has facilitated a gradual transition towards more comprehensive ESG reporting practices.

Two significant developments in China's ESG disclosure framework in 2024 are worth noting. First, the *Corporate Sustainability Disclosure Standards* (MoF 2024) were published, aligning closely with the International Sustainability Standards Board's IFRS S1 standard. It is voluntary but represents the first step towards China's goal of establishing a nationally unified ESG disclosure framework by 2030.

Second, China's three major stock exchanges (Shanghai, Shenzhen and Beijing) jointly published consistent guidelines that mandate ESG information disclosure for specific listed companies, including those on the SSE180, KIC 50, SZSE 100 and GEM indexes, as well as companies with dual domestic and international listings. The required disclosures include Scope 1 and 2 emissions, with Scope 3 emissions and scenario analysis being encouraged.

Table 7.2: Evolution of ESG disclosure requirements for different Chinese entities

Entity type		2021	2022	2023	2024
Companies (general)		MEE: Certain companies[a] are required to disclose environmental ('E') information annually	–	Revised Company Law provides a legal explanation for ESG disclosure	MoF et al: *Basic Principles of Corporate ESG Disclosure*; all companies voluntarily disclose ESG information similar to IFRS S1
Listed companies (including listed financial institutions)	Shanghai	–	Certain companies[b] are required to publish environmental and social responsibility reports together with annual reports	–	Certain listed companies[c] are required to disclose ESG information including Scope 1 and 2 emissions (Scope 3 emissions and scenario analysis are encouraged)
	Shenzhen	–		–	
	Beijing	–	–	–	
SOEs		–	SASAC: All listed central SOEs should publish dedicated ESG reports by 2023	SASAC publishes ESG report reference documents for listed central SOEs	SASAC promotes ESG governance and disclosure in overseas investment
Financial institutions[d]		–	PBOC: Voluntary disclosure of environmental and climate information annually	Insurance Association of China: Insurers are encouraged to disclose ESG information based on metrics and framework provided	–
Green bond issuers		–	–	Green Bond Standards Committee publishes *Information Disclosure Guide* to improve the transparency of use of funds	–

Notes: SASAC = State-Owned Assets Supervision and Administration Commission.
[a] Such as key pollution-discharging entities, listed or bond-issuing companies with violations of environmental laws.
[b] SSE: listed companies; SZSE: Shenzhen 100 and GEM-listed companies.
[c] SSE180, KIC 50, SZSE 100, GEM indexes and companies listed both domestically and overseas.
[d] Commercial banks, asset management institutions, trust companies and insurance companies.

As of the end of 2024, about 41 per cent of listed companies in China had published the previous year's ESG report, compared with 27 per cent in 2022 and 34 per cent in 2023 (Lu and Deshangyu 2025); 99.6 per cent of listed companies held by central SOEs had published ESG reports, but only 14 per cent had worked on ESG standards compilation (Wang 2024). These developments show that harmonisation of the now scattered ESG disclosure policies would motivate more stringent ESG disclosure practices.

Green insurance for risk management

One of the earliest categories of green insurance implemented in China was environmental pollution liability insurance (EPLI), which covers a company's legal obligation to compensate third parties for damages resulting from pollution incidents. EPLI could effectively strengthen environmental risk management among companies, mitigate their financial burdens in the event of pollution incidents and ensure timely compensation for affected parties. Between 2007 and 2015, EPLI was piloted with high-polluting companies in selected regions with positive results, which underscored the effectiveness of this market-based approach to environmental governance (Chen et al. 2022). In 2018, EPLI became mandatory for enterprises operating with significant environmental risks or those with a history of major environmental emergencies since 2005.

Since 2018, the definition of green insurance in China has expanded significantly, encompassing a broader range of categories including agriculture, natural disasters, green buildings and carbon finance, while China seeks a more robust green finance framework for managing environmental risks. The development of green insurance gained momentum following the issuance of statistical standards in 2022 and a classification guide in 2023, which delineated 69 categories of insurance products (Table 7.3).

As well as a taxonomy for green insurance products and services innovation, insurance funds in China are also provided with a list of green financial products in which to invest, as well as a dedicated framework for ESG information disclosure. This holistic approach reflects China's commitment to integrating sustainability principles across the insurance sector.

Table 7.3: Green insurance classification guidelines in China

Setting	Category	Insurance product
Cope with extreme weather and climate events	Meteorological disaster	Meteorological disaster Public infrastructure disaster Agricultural weather index
Promote the development of green industries	Clean energy	Solar power Wind power Utilities and electricity Nuclear energy Energy storage Hydrogen energy Power grid
	Industrial optimisation and upgrading	Green manufacturing system Circular economy Optimising and upgrading insurance in other industries
	Green transportation	New-energy automobile industry Rail transit New-energy ships and aircraft Green and efficient transportation system Green transportation infrastructure
	Green building	Green building
	Green low-carbon technology	Green environmental protection equipment Green low-carbon material Green low-carbon technology
Promote low-carbon transformation economic activities	Low-carbon transition	Fossil energy low-carbon transition Low-carbon transition in the industrial sector Low-carbon transformation in the construction sector
Support environmental improvement	Pollution reduction	Environmental pollution liability Ship pollution liability Oil pollution Hazardous materials liability Environmental protection infrastructure
Support biodiversity conservation	Ecological environment	Ecological planting industry Ecological forestry Green livestock Green fishery Ecological functional zone Wildlife Ecological restoration Landscaping
Support the construction of green financial market	Green financing	Green loan
	Carbon market	Carbon trading Carbon sink
Promote green, low-carbon and safe social governance	Green and low-carbon social governance	Production safety liability Public safety liability Critical infrastructure Green service provision

Setting	Category	Insurance product
Promote green and low-carbon exchanges and cooperation	Green and low-carbon trade	Green foreign trade Green 'Belt and Road' Green domestic trade
	Green and low-carbon activity	Green and low-carbon exhibition Green and low-carbon event
Promote green and low-carbon lifestyle	Green life	New-energy automobile Non-motor vehicle Residential decoration quality
Other	Corporate sustainability	Corporate sustainability

Source: Compiled by the authors.

Carbon market mechanisms

China's ETS and the China Certified Emission Reduction (CCER) program operate as distinct but interconnected carbon market mechanisms. The national ETS is a mandatory system covering high-emission sectors. It requires key emitters to manage their carbon emissions within allocated quotas. The CCER program functions as a voluntary carbon market where companies can generate and trade carbon reduction credits from certain projects such as offshore wind and solar thermal power.

The two systems connect through a crucial link: companies in the ETS can use the CCER to offset up to 5 per cent of their annual emissions obligations. This integration creates additional demand for the CCER and provides ETS participants with more flexibility in meeting their compliance requirements.

Emissions Trading System

The development of the ETS in China began in 2011, with the NDRC's approval of seven regional carbon trading pilot schemes, in: Shanghai, Beijing, Guangdong, Shenzhen, Tianjin, Hubei and Chongqing. These pilots started trading in 2013, laying the groundwork for a national system. The national ETS was launched in 2017 and was fully operational by July 2021, initially covering only the power sector. A substantial expansion occurred in March 2025, incorporating 1,500 additional companies from the steel, cement and aluminium smelting sectors. This expansion not only increased the ETS's coverage to more than 60 per cent of China's carbon dioxide emissions but also broadened its scope to include other greenhouse gases (GHGs) such as carbon tetrafluoride and hexafluoroethane.

China has established a comprehensive policy framework to support its ETS (Figure 7.7), the cornerstone of which are the Interim Regulations on the Administration of Carbon Emission Trading, which provide guidelines for the responsibilities of government agencies, key emitting entities and third-party service providers, as well as legal penalties for behaviours such as falsification, failure to pay carbon allowances or obstruction of inspection by supervising government agencies.

```
┌─────────────────────────────────────────────────────────────────────┐
│   Interim Regulations on the Administration of Carbon Emission Trading│
│                        (State Council 2024)                           │
└─────────────────────────────────────────────────────────────────────┘

┌─────────────────────────────────────────────────────────────────────┐
│          Administrative Measures for Carbon Emissions Trading         │
│                             (MEE 2020)                                │
└─────────────────────────────────────────────────────────────────────┘

┌──────────────────┐  ┌──────────────────────┐  ┌──────────────────────┐
│                  │  │   Guidelines for     │  │                      │
│  Allowance/Quota │  │ Measurement, Reporting,│  │ Rules for Registration,│
│  Allocation Plan │  │ and Verification (MRV)│  │ Trading, and Settlement│
│                  │  │                      │  │                      │
└──────────────────┘  └──────────────────────┘  └──────────────────────┘
```

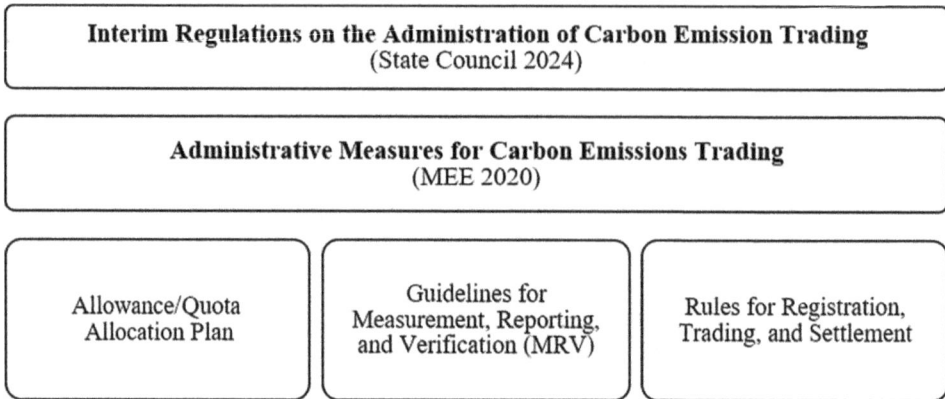

Figure 7.7: Policy framework for China's Emissions Trading System
Source: Modified from MEE (2024).

Of the 36 emissions trading markets globally, China's national ETS is now the largest in terms of carbon dioxide emissions coverage, accounting for about 5 billion tonnes of carbon dioxide (ICAP 2024). A major difference between China's and the European Union's ETSs is the approach to capping emissions: the EU ETS employs an absolute cap, which sets a fixed limit on the total amount of GHG emissions allowed. In contrast, China's ETS employs an intensity-based cap, which limits emissions per unit of production but allows total emissions to fluctuate based on economic output, aligning with its goal of carbon peaking by 2030 (Chen 2024). This makes the Chinese ETS somewhat less predictable as the tradable allowances are not a scarce good a priori, making spot and futures trading of allowances less or not attractive (Liu and Nedopil 2021).

China Certified Emission Reduction program

The CCER program is a voluntary carbon trading mechanism that allows companies to offset their emissions by purchasing credits generated from certified emission reduction projects (Li et al. 2019; Wu 2024). The CCER system was initiated in 2012, entered the trading phase in 2015 and suspended issuance in 2017. In January 2024, China's Ministry of Ecology and Environment officially relaunched and renamed the CCER, the China Carbon Emission Reduction market, after seven years of hiatus. Before this relaunch, several policy documents were in place, providing detailed guidance for all participants (Figure 7.8).

The CCER supplements the mandatory ETS, enabling companies to offset up to 5 per cent of their annual emissions obligations using CCER credits. The initial four sectors supported by the CCER are offshore wind power, solar thermal power, afforestation and mangrove revegetation. Methodology for two more sectors, coalmine gas recycling and energy saving in highway tunnel lighting, were published for consultation in August 2024.

Administrative Measures for Voluntary Greenhouse Gas Emission Reduction Trading (MEE 2023)

| Implementation Rules for Validation and Verification (State Administration for Market Regulation 2023) | Guidelines for Registration, Design and Implementation (Climate Change Strategy and International Cooperation* 2023) | Trading and Settlement Rules (Beijing Green Exchange** 2023) |

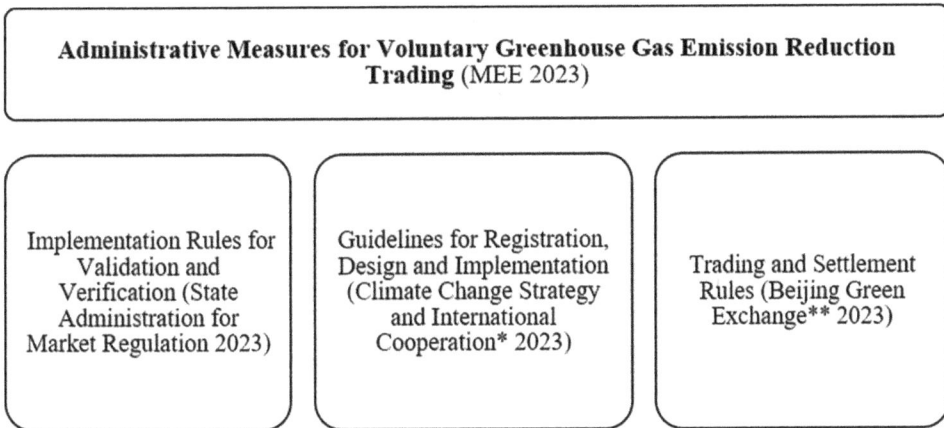

Figure 7.8: Policy framework for the China Carbon Emission Reduction program

Notes: * Serving as the national registration agency for CCER projects. ** Acting as the national trading platform for the CCER.

Source: Compiled by the authors.

China's role in global green finance

China has taken a central role in global green finance, which can be evaluated through international green finance standardisation and application—for example, green investment in its overseas collaboration programs such as the BRI.

International coordination of green finance standards

China's role in coordinating international green finance standards is twofold: aligning its domestic practices with global norms and actively shaping the development of these standards. This dual approach underscores China's commitment to both integrating into and influencing the global green finance landscape.

On the one hand, China has aimed to strengthen the harmonisation of its domestic green finance framework with international standards, including advancements in corporate disclosure requirements, green bond taxonomies and carbon footprint accounting.

- **Cooperation on taxonomy:** In November 2024, the International Platform on Sustainable Finance (IPSF) Multi-Jurisdiction Common Ground Taxonomy (M-CGT) was introduced at the UN Climate Change Conference COP29. Jointly led by the PBOC, the European Commission's Directorate-General for Financial Stability, Financial Services and Capital Markets Union and the Monetary Authority of Singapore, the M-CGT builds on the previous China–EU Common Ground Taxonomy. It aims to harmonise sustainable finance criteria across China, the European Union and Singapore, facilitating cross-border capital flows into green projects.

- **Cooperation on emissions trading schemes:** In August 2024, China's Ministry of Ecology and Environment signed an updated memorandum of understanding (MoU) with the European Commission to enhance cooperation on emissions trading. The MoU outlines collaboration on key areas, including the role of CCER credits, the expansion of China's ETS, the integration of the ETS with green power and green certificates and mutual recognition of carbon accounting and product carbon cost verification. It also emphasises joint efforts in developing carbon market systems, quota allocation methods and monitoring, reporting and verification mechanisms.

Internationally, China has taken a proactive role in global standard-setting platforms and sustainable finance initiatives:

- In the G20 Sustainable Finance Working Group, China has taken a leadership position as co-chair alongside the United States. This working group, elevated from its previous status as a study group in 2021, aims to mobilise sustainable finance to ensure global growth and stability while promoting the transition to greener, more resilient and inclusive economies.

- China is also a founding member of the Network for Greening the Financial System (NGFS), which was established in December 2017. The PBOC has actively participated in NGFS activities, contributing to the development of methodologies and tools for assessing climate-related financial risks. For instance, China conducted its first climate risk stress test in 2021, referencing the network's carbon price scenarios (NGFS 2023).

- China has played a pivotal role in the International Platform on Sustainable Finance (IPSF) since its inception in 2019. Notably, China co-chairs the IPSF Taxonomy Working Group, which led to the development of the EU–China CGT in 2020. In 2024, the CGT was expanded into the M-CGT, incorporating Singapore into the framework to enhance global interoperability in green finance, as discussed previously.

Green investment through the BRI

Since its inception in 2013, the Belt and Road Initiative has evolved into a significant channel for Chinese companies to export green technologies and invest in green projects such as clean energy infrastructure, electric vehicle technology and solar and wind facilities. China's green overseas investment has reached record levels in recent years. In 2021, China announced it would not build new coal-fired power plants abroad. In 2024, China's engagement in green energy (solar, wind and biomass projects) reached an all-time high of US$11.8 billion, representing about 30 per cent of its total energy engagement in BRI countries; this marks a significant increase from 2023 when green energy investments amounted to US$7.9 billion (Nedopil 2025).

Green investment through the BRI has been driven by Chinese regulators, quasi-governmental institutions and industry associations.

Regulators

Before 2017, Chinese companies and financial institutions engaging in overseas investment primarily adhered to a 'host country approach', which entailed complying with local environmental regulations, even when these were less stringent than international standards. Since President Xi Jinping's announcement to build a green BRI in 2017, China's approach to overseas investment has undergone a significant transformation.

In 2022, a key document, *Opinions on Promoting Green Development in the Joint Construction of the Belt and Road Initiative*, was issued by the NDRC, the Ministry of Foreign Affairs (MFA), the MEE and the Ministry of Commerce (NDRC et al. 2022). This policy encouraged enterprises to adopt international or Chinese environmental standards in cases where local standards were absent or inadequate. In its notices to the banking and insurance industry, the former CBIRC also encouraged banks and insurance institutions to ensure that project management was substantively aligned with international best practices (CBIRC 2022).

Quasi-governmental institutions and industry associations

The Belt and Road Initiative Green Development Coalition (BRIGC) is a multi-stakeholder platform co-founded by China's MEE and international partners in 2019, aiming to foster international cooperation on green development and encourage sustainable infrastructure and investment practices. From 2020 to 2023, BRIGC led and published a series of studies of the *Green Development Guidance*, which provides a 'traffic light system' to label projects as 'encouraged', 'neutral' or 'restricted' based on their impact on pollution, climate change and biodiversity (see Figure 7.9; Nedopil et al. 2020, 2021a). This served as a basis for new policy documents issued in 2021 and 2022 that explicitly encouraged the integration of biodiversity, climate and pollution controls throughout all investment phases.

The Green Investment Principles for the BRI were co-initiated by the Green Finance Committee of the China Society for Finance and Banking and the City of London Corporation's Green Finance Initiative in 2018—a set of eight high-level principles for financial institutions to integrate sustainability into BRI-related investments (Green Finance Leadership Program 2018).

Chinese financial institutions are also contributing to a green BRI through industry associations. In November 2024, the Belt and Road Reinsurance Pool held its tenth council meeting, in Kunming, Yunnan Province. The meeting launched 'Green Insurance Principles' for BRI projects, including identifying and assessing environmental risks in projects and developing differentiated underwriting solutions (Xinhua 2024). Similarly, the Belt and Road Bankers Roundtable was established in 2017 with the goal to 'strengthen cooperation in the Belt and Road project financing, transaction bank, and financial market' (ICBC 2019), part of which includes green finance collaboration.

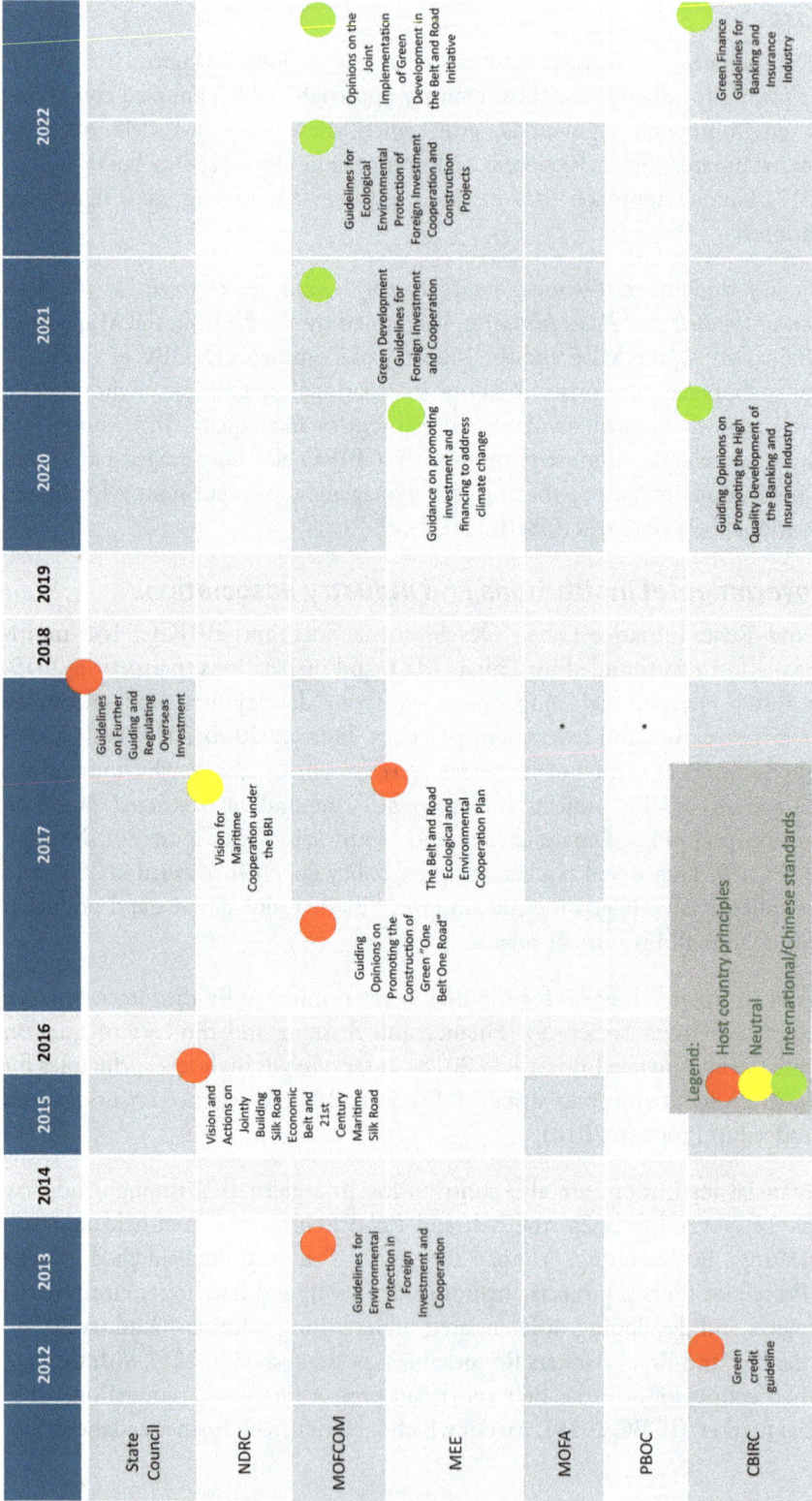

Figure 7.9: Chinese Government-issued guidance and opinions relevant to greening finance in the Belt and Road Initiative

Source: © 2022 Nedopil, Green Finance & Development Center, FISF Fudan University.

Conclusion and outlook

China has made significant progress in developing its green finance sector, positioning itself as a key player in the global sustainable finance landscape. While China is often described as a leader due to the scale of its green bond market and the institutionalisation of green credit policies, this characterisation requires critical examination. The country has implemented a top-down regulatory approach (unlike the European Union, with its more market-driven approach), exemplified by the PBOC's green credit guidelines and the establishment of a national taxonomy for green finance. These efforts have driven substantial capital flows into green projects, particularly in renewable energy and infrastructure. However, challenges remain, including persistent discrepancies in green finance definitions, the prevalence of greenwashing due to less developed disclosure standards, the relatively limited role of the private sector and international investment and the lack of understanding of de facto emission reductions. China persists in its approach of 'establishing the new before abolishing the old' in its green economic transition (Xi 2024)—that is, ensuring energy security by building clean energy alongside traditional energy. The development of its green finance ecosystem has followed a similar approach, taking into consideration the needs of traditional high-emission industries (which also explains the increasing importance of transition finance). As a result, and as some studies have highlighted, there is a 'Panda–Dragon' paradox where China has an aggressive green finance agenda and carbon neutrality goals while expanding the build-out of high-emission fossil fuels (Nedopil 2023). Questions thus remain about the extent to which China's green finance model can serve as a replicable framework for other economies.

Future developments in China's green finance ecosystem will likely focus on enhancing the credibility of green financial products through stricter disclosure requirements and third-party verification mechanisms. The integration of climate risk assessment into macroprudential policies and stress testing will be critical in managing systemic financial risks associated with climate change. Additionally, the digitalisation of finance, including blockchain-based green bond issuance and AI-driven ESG analytics, presents opportunities for improving transparency and efficiency. Further research is required to evaluate the effectiveness of China's green finance policies in mobilising long-term private investment, the role of SOEs in sustainable financing and the implications of China's green finance development for de facto emission reductions and China's global sustainable development leadership. A sectoral analysis of the impact of green finance on industries such as energy, transportation and urban infrastructure would provide deeper insights into the financial mechanisms driving China's low-carbon transition.

References

Alnes, Kristina, Catherine Rothacker, Victor H. Laudisio, and Maria Knudsen. 2024. *Sustainable Finance FAQ: The Rise of Green Equity Designations*. New York: S&P Global. www.spglobal.com/ratings/en/research/articles/241217-sustainable-finance-faq-the-rise-of-green-equity-designations-13345198.

Chen, Qiuping, Bo Ning, Yue Pan, and Jinli Xiao. 2022. 'Green Finance and Outward Foreign Direct Investment: Evidence from a Quasi-Natural Experiment of Green Insurance in China.' *Asia Pacific Journal of Management* 39, no. 3: 899–924. doi.org/10.1007/s10490-020-09750-w.

Chen, Zhibin. 2024. 'China's Carbon Market Model Can Guide Emerging Economies.' *Dialogue Earth*, 3 June. dialogue.earth/en/climate/chinas-carbon-market-model-can-guide-emerging-economies/.

China Banking and Insurance Regulatory Commission (CBIRC). 2022. 'Notice on Issuing Green Finance Guidelines for the Banking and Insurance Industry.' *China Banking and Insurance Regulatory Commission*, no. 15. Beijing: CBIRC. www.gov.cn/zhengce/zhengceku/2022-06/03/content_5693849.htm.

China Banking Regulatory Commission (CBRC). 2012. 'Notice on Issuing Green Credit Guidelines.' *CBRC Notice*, no. 4. *State Council Gazette*, no. 17. www.gov.cn/gongbao/content/2012/content_2163593.htm.

Climate Bonds Initiative. 2024. 'China Solidifies Leadership in Green Finance for 2023.' Press release, 17 May. London: Climate Bonds Initiative. www.climatebonds.net/resources/press-releases/2024/05/china-solidifies-leadership-green-finance-2023.

Climate Bonds Initiative, and China Central Depository & Clearing Co. Ltd (CCDC) Research. 2017. *China Green Bond Market 2016*. London: Climate Bonds Initiative. www.climatebonds.net/resources/reports/china-green-bond-market-2016.

Climate Bonds Initiative, and CCDC Research. 2022. *China Green Bond Market Report 2021*. London: Climate Bonds Initiative. www.climatebonds.net/data-insights/publications/china-green-bond-market-report-2021.

Escalante, Donovan, June Choi, Neil Chin, Ying Cui, and Mathias Lund Larsen. 2020. *The State and Effectiveness of the Green Bond Market in China*. A CPI Report. San Francisco: Climate Policy Initiative. climatepolicyinitiative.org/wp-content/uploads/2020/06/The-State-and-Effectiveness-of-the-Green-Bond-Market-in-China.pdf.

Fu, Yilei, and Ming Sun. 2025. *2024 Green Bond Progress Report*. Beijing: International Institute of Green Finance. iigf.cufe.edu.cn/info/1012/9691.htm.

Green Finance Committee. 2021. *Roadmap for Financing China's Carbon Neutrality*. December. Beijing: Green Finance Committee of China Society for Finance and Banking. www.greenfinance.org.cn/upfile/file/20211204222634_82821_73556.pdf.

Green Finance Committee. 2025. *List of China's Existing Green Bonds Compliant with the China–EU Common Classification Catalogue (February Edition)*. Beijing: Green Finance Committee of China Society for Finance and Banking. www.greenfinance.org.cn/displaynews.php?id=4498.

Green Finance Leadership Program. 2018. *The Green Investment Principle (GIP) for the Belt and Road Initiative*. Shanghai: Green Finance and Development Centre. greenfdc.org/green-investment-principle-gip-belt-and-road-initiative/.

Heine, Klaus, and Wolfgang Kerber. 2002. 'European Corporate Laws, Regulatory Competition and Path Dependence.' *European Journal of Law and Economics* 13, no. 1: 47–71. doi.org/10.1023/A:1013113925093.

Industrial and Commercial Bank of China (ICBC). 2019. 'The Belt and Road Bankers Roundtable Successfully Held in Beijing.' *ICBC News*, 6 May. Beijing: Industrial and Commercial Bank of China. icbc.com.cn/icbc/en/newsupdates/icbc%20news/TheBeltand RoadBankersRoundtableSuccessfullyHeldinBeijing.htm.

International Capital Market Association (ICMA). 2022. *Analysis of China's Green Bond Principles*. Hong Kong: ICMA Asia Pacific. www.icmagroup.org/assets/Analysis-of-Chinas-Green-Bond-Principles.pdf.

International Carbon Action Partnership (ICAP). 2024. *Emissions Trading Worldwide: 2024 ICAP Status Report*. Berlin: ICAP Secretariat. icapcarbonaction.com/en/publications/emissions-trading-worldwide-2024-icap-status-report.

International Institute of Green Finance (IIGF). 2024. *China Green Finance Progress Report (2024)*. Beijing: International Institute of Green Finance. iigf.cufe.edu.cn/info/1014/9340.htm.

Jin, Lei, and Yu Jiang. 2024. *IIGF Annual Report | Green Equity Progress Report 2023*. Beijing: International Institute of Green Finance. iigf.cufe.edu.cn/info/1013/9280.htm.

Li, Lixu, Fei Ye, Yina Li, and Ching-Ter Chang. 2019. 'How Will the Chinese Certified Emission Reduction Scheme Save Cost for the National Carbon Trading System?' *Journal of Environmental Management* 244: 99–109. doi.org/10.1016/j.jenvman.2019.04.100.

Li, Mengjie, Qingyan Zhu, and Chang Gao. 2025. *A Journey of a Thousand Miles Begins with a Single Step, and We Will Continue to Move Forward: A Review and Outlook on the Development of China's ESG Market*. Beijing: China Reform Consulting. www.crhcc.com/esgjm/2025/3/84b89d134f754d568a75c9d7b4cb7bc4.htm.

Lin, Jiun-Da. 2023. 'Explaining the Quality of Green Bonds in China.' *Journal of Cleaner Production* 406: 136893. doi.org/10.1016/j.jclepro.2023.136893.

Liu, Huixin, and Christoph Nedopil. 2021. *Potential Harmonisation of Emission Trading Systems (ETS): China and Southeast Asia*. Regional Programme Energy Security and Climate Change in Asia and the Pacific (RECAP). Berlin: Konrad-Adenauer-Stiftung. www.kas.de/en/web/politikdialog-asien/single-title/-/content/potential-harmonisation-of-emission-trading-systems-ets-china-and-southeast-asia.

Lu, Taoran, and Li Deshangyu. 2025. 'The ESG Disclosure Rate of A-Shares Reached 41% in 2024, with a Significant Increase in Mandatory Disclosures Expected in 2025.' *21st Century Business Herald*, 9 January. www.stcn.com/article/detail/1485956.html.

Ministry of Ecology and Environment (MEE). 2024. *Progress Report of China's National Carbon Market (2024)*. Beijing: Ministry of Ecology and Environment of the People's Republic of China. www.mee.gov.cn/ywdt/xwfb/202407/W020240722528850763859.pdf.

Ministry of Finance (MoF), Ministry of Foreign Affairs, National Development and Reform Commission, Ministry of Industry and Information Technology, Ministry of Ecology and Environment, Ministry of Commerce, People's Bank of China, State-Owned Assets Supervision and Administration Commission of the State Council, and National Administration of Financial Regulation. 2024. 'Notice on the Issuance of the "Corporate Sustainability Disclosure Standards—Basic Standards" (Trial Implementation).' *Finance and Accounting* [2024], no. 17 (20 November). Beijing: State Council of the People's Republic of China. www.gov.cn/zhengce/zhengceku/202412/content_6993358.htm.

National Development and Reform Commission (NDRC). 2016. *Notice on Issuing Green Bond Endorsed Project Guidelines*. NDRCFJ [2015], no. 3504. Beijing: National Development and Reform Commission. www.ndrc.gov.cn/xxgk/zcfb/tz/201601/t20160108_963561.html.

NDCR, Ministry of Foreign Affairs, Ministry of Ecology and Environment, and Ministry of Commerce. 2022. *Opinions of the National Development and Reform Commission and Other Departments on Promoting Green Development in the Joint Construction of the Belt and Road Initiative*. NDRC [2022], no. 408 (16 March). Beijing: State Council of the People's Republic of China. www.gov.cn/zhengce/zhengceku/2022-03/29/content_5682210.htm.

Nedopil, Cristoph. 2022. 'Green Finance for Soft Power: An Analysis of China's Green Policy Signals and Investments in the Belt and Road Initiative.' *Environmental Policy and Governance* 32, no. 2: 85–97. doi.org/10.1002/eet.1965.

Nedopil, Cristoph. 2023. 'Lessons from China's Overseas Coal Exit and Domestic Support.' *Science* 379, no. 6637: 1084–87. doi.org/10.1126/science.adf0126.

Nedopil, Cristoph. 2025. *China Belt and Road Initiative (BRI) Investment Report 2024*. Brisbane and Shanghai: Griffith Asia Institute and Green Finance and Development Centre. doi.org/10.25904/1912/5784.

Nedopil, Christoph, Dimitri De Boer, Danting Fan, and Yingzhi Tang. 2021a. 'What China's New Guidelines on "Green Development" Mean for the Belt and Road.' *Dialogue Earth*, 18 August. dialogue.earth/en/business/what-chinas-new-guidelines-on-green-development-mean-for-the-belt-and-road/.

Nedopil, Christoph, Truzaar Dordi, and Olaf Weber. 2021b. 'The Nature of Global Green Finance Standards—Evolution, Differences, and Three Models.' *Sustainability* 13, no. 7: 3723. doi.org/10.3390/su13073723.

Nedopil, Christoph, and Mathias Larsen. 2023. 'Green Finance in China: System, Practice, and International Role.' In *The Routledge Handbook of Green Finance*, edited by Othmar M. Lehner, Theresia Harrer, Hanna Silvola, and Olaf Weber, 280–99. London: Routledge. doi.org/10.4324/9781003345497-20.

Nedopil, Christoph, Ye Wang, Wenhong Xie, Dimitri De Boer, Shuang Liu, Xiaoting Chen, Yonghong Li, Yuan Zhu, Yan Lan, Panwen Li, and Haishan Zhao. 2020. *Green Development Guidance for BRI Projects Baseline Study Report*. Beijing: BRIGC Secretariat.

Network for Greening the Financial System (NGFS). 2023. 'In Conversation with Mr Xuan Changneng, Deputy Governor, People's Bank of China.' Interview, 8 November. Paris: Network for Greening the Financial System. www.ngfs.net/en/news/conversation-mr-xuan-changneng-deputy-governor-peoples-bank-china.

People's Bank of China (PBOC). 1995. *Notice on Implementing Credit Policies and Strengthening Environmental Protection*. Beijing: People's Bank of China.

PBOC, Ministry of Finance, National Development and Reform Commission, Ministry of Environmental Protection, China Banking Regulatory Commission, China Securities Regulatory Commission, and China Insurance Regulatory Commission. 2016. *Guiding Opinions on Establishing a Green Financial System*. Beijing: Ministry of Ecology and Environment. www.mee.gov.cn/gkml/hbb/gwy/201611/t20161124_368163.htm.

PBOC, National Development and Reform Commission, Ministry of Industry and Information Technology, Ministry of Finance, Ministry of Ecology and Environment, State Administration of Financial Supervision, and China Securities Regulatory Commission. 2024. *Guidelines on Further Strengthening Financial Support for Green and Low-Carbon Development*. Beijing: State Council of the People's Republic of China. www.gov.cn/zhengce/zhengceku/202404/content_6944452.htm.

Rodenbiker, Jesse. 2025. 'Ecological Civilization Goes Global: China's Green Soft Power and South–South Environmental Initiatives.' *Insight & Analysis*. Washington, DC: Wilson Center. www.wilsoncenter.org/publication/ecological-civilization-goes-global-chinas-green-soft-power-and-south-south.

Schmid, Rolf, and Xin Xiong. 2023. 'China's Environmental Solutions.' *Applied Microbiology and Biotechnology* 107, no. 4: 987–1002. doi.org/10.1007/s00253-022-12340-z.

Shen, Hongtao, Honghui Lin, Wenqi Han, and Huiying Wu. 2023. 'ESG in China: A Review of Practice and Research, and Future Research Avenues.' *China Journal of Accounting Research* 16, no. 4: 100325. doi.org/10.1016/j.cjar.2023.100325.

State Environmental Protection Administration (SEPA), People's Bank of China, and China Banking Regulatory Commission. 2007. *Opinions on Implementing Environmental Protection Policies and Regulations to Prevent Credit Risks*. Beijing: Ministry of Ecology and Environment. www.mee.gov.cn/gkml/zj/wj/200910/t20091022_172469.htm.

Velev, Vasil. 2025. 'China's Carbon Market Sees Off Successful 2024 but Challenges Persist.' *Carbon Herald*, 15 January. carbonherald.com/chinas-carbon-market-sees-off-successful-2024-but-challenges-persist/.

Wang, Dong, and Ping Li. 2020. 'The Benefits of Issuing Green Bonds: Evidence from China Green Bonds Market.' November. Available at SSRN: doi.org/10.2139/ssrn.3710646.

Wang, Jiazhen, Xin Chen, Xiaoxia Li, Jing Yu, and Rui Zhong. 2020. 'The Market Reaction to Green Bond Issuance: Evidence from China.' *Pacific-Basin Finance Journal* 60: 101294. doi.org/10.1016/j.pacfin.2020.101294.

Wang, Zilin. 2024. 'SASAC: The ESG Report Disclosure Rate of Central State-Owned Enterprises Listed Companies Has Reached 99.6%.' *Shanghai Securities News*, 9 November. www.cnstock.com/commonDetail/313118.

Wu, Yi. 2024. 'Understanding the Relaunched China Certified Emission Reduction (CCER) Program: Potential Opportunities for Foreign Companies.' *China Briefing*, 3 May. www.china-briefing.com/news/understanding-the-relaunched-china-certified-emission-reduction-ccer-program-potential-opportunities-for-foreign-companies/.

Xi, Jinping. 2024. 'Working Together for a Fair and Equitable Global Governance System: Remarks by H.E. Xi Jinping.' Nineteenth G20 Summit, Rio de Janeiro, Brazil, 18 November. wb.beijing.gov.cn/home/gjjwzx/xxzl/202411/t20241119_3944972.html.

Xinhua. 2024. 'China's Belt and Road Reinsurance Community Releases the "Belt and Road Green Insurance Principles".' *Xinhua*, 26 November. www.news.cn/money/20241126/c581 10f1762b45ee9f52aaa03b52161f/c.html.

Yuan, Xueying, Zhongfei Li, Jinhua Xu, and Lixia Shang. 2022. 'ESG Disclosure and Corporate Financial Irregularities: Evidence from Chinese Listed Firms.' *Journal of Cleaner Production* 332: 129992. doi.org/10.1016/j.jclepro.2021.129992.

Yue, Mengdi, and Christoph Nedopil. 2025. *China Green Finance Status and Trends 2024–2025*. Report. Brisbane and Shanghai: Griffith University and Green Finance and Development Centre, FISF Fudan University. doi.org/10.25904/1912/5786.

Zhang, Hao. 2020. 'Regulating Green Bond in China: Definition Divergence and Implications for Policy Making.' *Journal of Sustainable Finance & Investment* 10, no. 2: 141–56. doi.org/10.1080/20430795.2019.1706310.

Zhang, Jing, Ziying Song, and Christoph Nedopil. 2024. *China Green Finance Status and Trends 2023–2024*. Report. Brisbane and Shanghai: Griffith University and Green Finance and Development Centre, FISF Fudan University. doi.org/10.25904/1912/5205.

8

China's carbon emissions trend after the pandemic

David I. Stern

Is China on a path to peak its greenhouse gas (GHG) emissions soon, as it has pledged under the Paris Agreement? In this chapter, I compare carbon emissions and energy production trends in 2019 and 2023. The former year represents typical conditions before the Covid-19 pandemic, while the latter is the first period after China removed most Covid-19 related restrictions, in December 2022. Carbon emissions increased by 8 per cent over this period, while emissions from the power sector, where decarbonisation efforts might be expected to be first focused, increased by 18 per cent. Though renewable energy production has increased significantly, the production of fossil fuels continues to grow strongly. Coal production in 2023 was 26 per cent higher than in 2019—a compound rate of 6 per cent per annum. Despite heavy rainfall and sluggish economic activity, there were similar trends in the first nine months of 2024. Therefore, there is little sign of 'growing back greener'. Going forward, the changing geopolitical environment could further impede China's emissions peaking.

Climate change remains one of the most pressing global issues. During the pandemic, the decrease in GHG emissions was seen by many as a chance for rapid progress towards a low-carbon economy. For instance, Chen et al. (2022) predicted that Chinese emissions would peak between 2021 and 2026 at a level between 11.7 and 13.1 gigatonnes, with more than 80 per cent probability, and Lui (2022) wrote that China is 'set to significantly overachieve [on] the targets it promised internationally for 2030, with emissions peaking by 2025'. However, as I will show, China's carbon emissions, and especially those from the power sector, continue to grow strongly, raising concerns about China achieving its climate commitments. Furthermore, the sluggish post-pandemic economic recovery and rising energy security concerns amid a changing geopolitical landscape mean that China may struggle to reach its ambitious targets (Ahmed 2023).

China is the world's second-largest economy and both the largest consumer of energy and the largest emitter of carbon dioxide. Its energy consumption amounts to about one-quarter of the global total. Thus, China has a critical role to play in global efforts to combat climate change. Since 2005, China has seen its primary energy demand surge by 47 per cent, and this trajectory is expected to continue with an additional increase of about 30 per cent between 2020 and 2040.[1] Fossil fuels remain dominant among China's primary energy sources, accounting for in excess of 80 per cent of energy use, while also being responsible for most of the country's carbon emissions.

Like many other countries, China is also facing the adverse impacts of climate change (Lu et al. 2022): there has been a rise in the occurrence of extreme weather events such as floods, droughts, heatwaves and sandstorms within the country. The economic cost of such events has significantly increased recently (Dai et al. 2015). Because Chinese emissions form a larger share of the global total than those of any other country and the climate damage it could suffer might also be greater in total value than that of any other country (Tol 2019), China has a greater incentive than any other nation to pursue net-zero emissions.

This chapter aims to understand China's new carbon emissions and energy production trajectories since the pandemic by comparing the trends for both between 2019 and 2023. The former period represents typical conditions before the Covid-19 pandemic, while the latter is the first year after China removed most pandemic restrictions, in December 2022. The research shows that carbon emissions have increased substantially over this period and China is now following a more carbon-intensive path than recent research suggested. It seems less likely that China will peak emissions before 2030. While research has addressed changes in Chinese carbon emissions during the pandemic (for example, Li et al. 2023), the post-pandemic trend had not been investigated before the work of Ahmed and Stern (2023). This chapter updates their research and finds that, adjusting for circumstances, similar trends continued into the first nine months of 2024.

Analysis

I collected data on daily Chinese carbon emissions from Carbon Monitor (Liu et al. 2020). The box plots in Figure 8.1 show the variation in daily emissions over the first eight months of each of the two selected years. The lowest daily emission levels occur during the Chinese New Year period. There is very little variation in daily transport emissions. I test whether there is a statistically significant change in emissions between the two periods by estimating a separate regression equation with just a constant for each of 2019 and 2023 using seemingly unrelated regressions and then computing the difference in regression coefficients between the two equations and its standard error. I estimate this two-equation system for total emissions and for emissions from each sector using Newey–West autocorrelation robust standard errors with 14 lags (Newey and West 1987). Total emissions rose 2.24 million tonnes a day or 8 per cent (standard

---

1   See 'China' at the Climate Action Tracker website: climateactiontracker.org/countries/china/policies-action/.

error = 0.30 million tonnes, p = 0.000) between the two periods. The increase in mean daily emissions from the power sector was also highly statistically significant, totalling 2.21 million tonnes or 18 per cent (0.20, 0.000) between the two periods. There was a statistically significant but small decrease in emissions from the transport sector, of −0.03 million tonnes or −1 per cent (0.013, 0.010). On the other hand, there were no statistically significant changes in emissions from the other two sectors: industry (0.08 million tonnes, 1 per cent, 0.19, 0.67) and residential (−0.02, −1 per cent, 0.05, 0.67). Monthly and annual Chinese energy production data used in the following discussion are from the National Bureau of Statistics of China.[2]

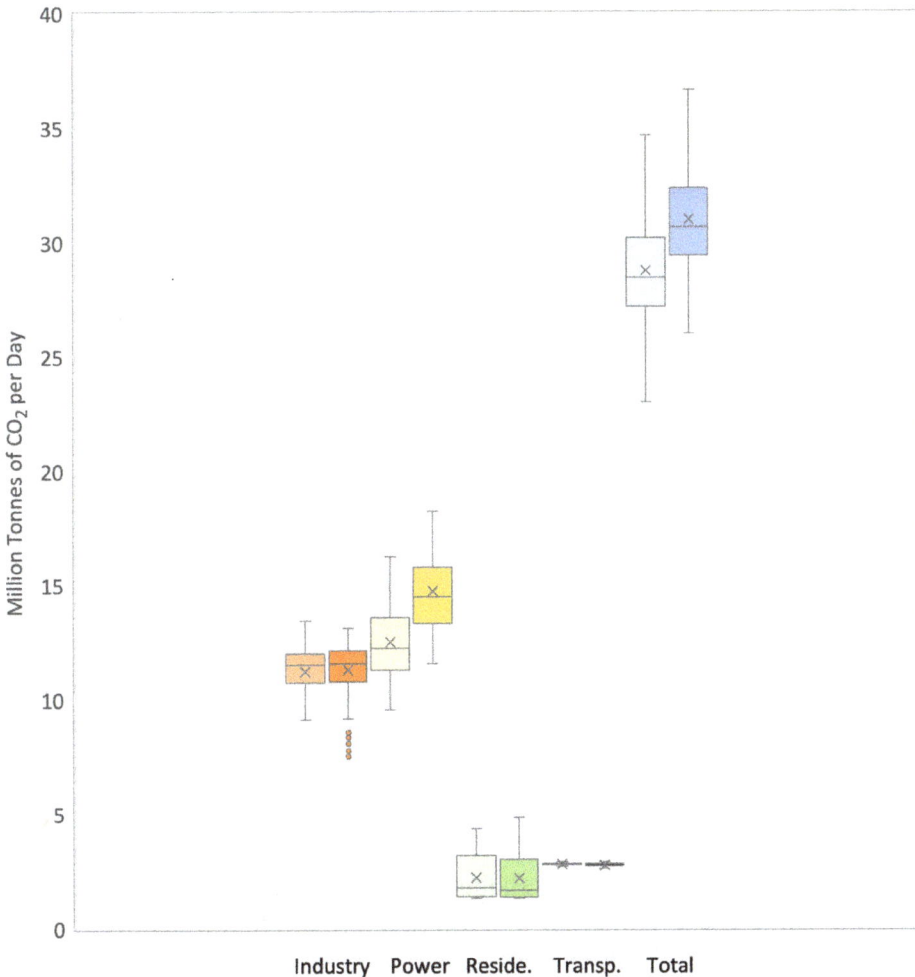

Figure 8.1: Chinese carbon dioxide emissions in 2019 and 2023

Notes: Daily carbon dioxide emissions from fossil fuel and cement production. Each pair of box plots shows daily data for 2019 emissions on the left and 2023 on the right. Reside. = residential sector; Transp. = transportation sector.

Source: Carbon Monitor (carbonmonitor.org/).

---

2   See monthly and annual energy data (data.stats.gov.cn/english/easyquery.htm).

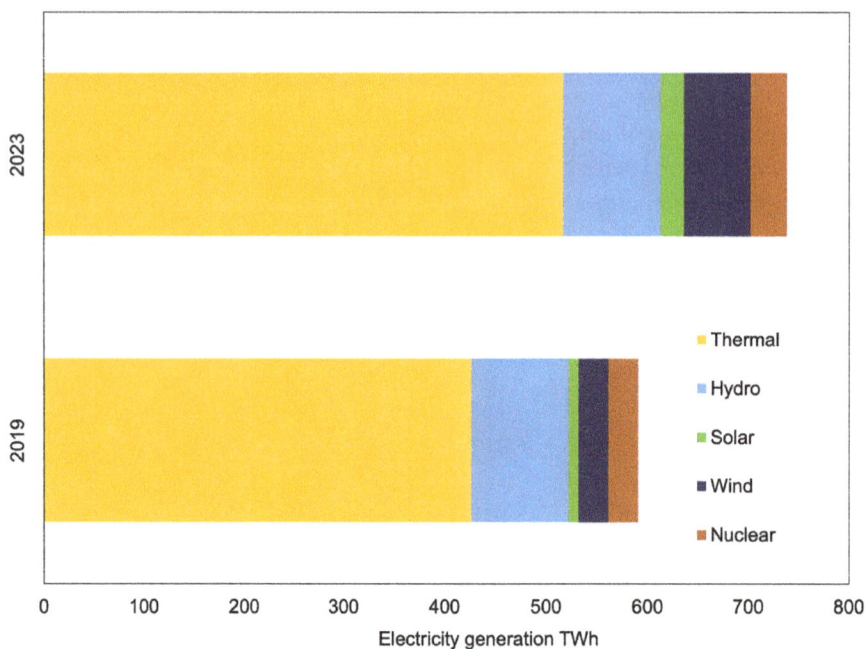

Figure 8.2: Energy sources of Chinese electricity production in 2019 and 2023 (terawatt hours)

Note: The figure shows monthly means for each year.

Source: National Bureau of Statistics of China (data.stats.gov.cn/english/).

The shares of solar and wind in total power generation increased from 1.6 per cent and 5.0 per cent, respectively, in 2019, to 3.2 per cent and 9.0 per cent, respectively, in 2023. Figure 8.2 shows that though electricity production from wind and solar grew substantially over the four years, so did electricity generation from fossil fuels. Average monthly thermal power generation increased 91 terawatt hours or 21 per cent between 2019 and 2023, while generation from new renewables increased 51 terawatt hours or 130 per cent. Nuclear generation increased by 24 per cent, while hydropower fell by 1 per cent. As a result, the share of thermal power in total generation only decreased from 72.2 per cent in 2019 to 70.1 per cent in 2023. So, it is not surprising that, with most of the increase in electricity output coming from fossil fuels, the sector's emissions increased substantially.

Production of energy in general and fossil fuels in particular also continued to increase. Figure 8.3 shows that coal production grew consistently over this period, as shown by the fitted exponential trend. Coal production in 2023 was 26 per cent higher than in 2019—a compound growth rate of 6 per cent per annum. For the first time since 2007, the share of coal in China's total energy consumption increased, from 56 per cent in 2021 to 56.2 per cent in 2022.

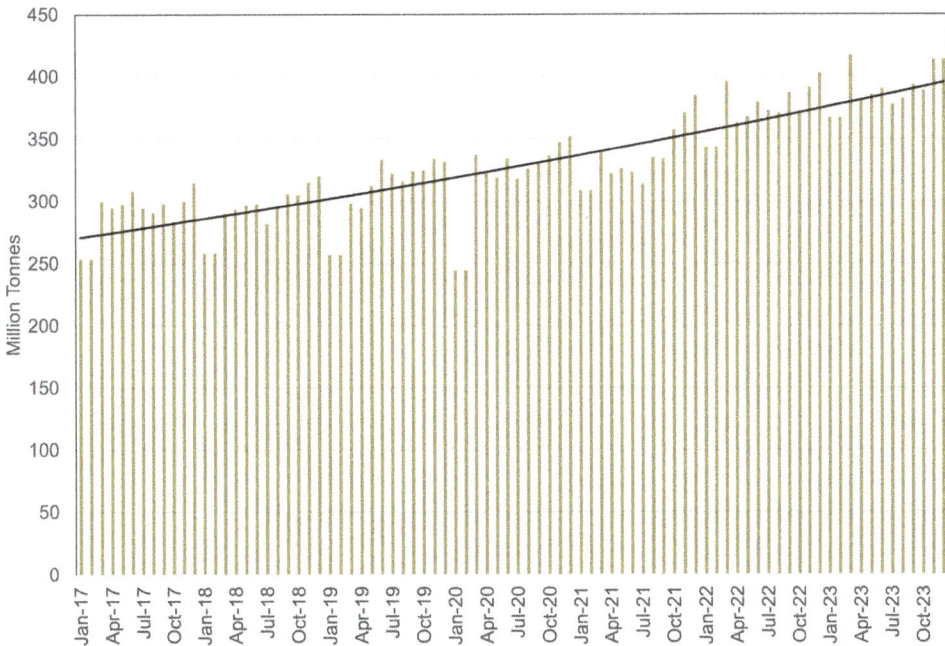

Figure 8.3: Monthly coal production in China, 2017–23

Notes: Values for January and February are identical as only a total for both months is provided in the source. A fitted exponential trend is shown.

Source: National Bureau of Statistics of China (data.stats.gov.cn/english/).

At the time of writing, nine months of data were available for 2024. Comparing these first nine months with the corresponding period in 2023, we can evaluate whether similar trends have persisted into 2024. According to Carbon Monitor, emissions fell 0.6 per cent in 2024 compared with 2023. Emissions from power generation rose by 1.6 per cent, with smaller increases in transport (0.7 per cent) and residential (0.8 per cent) emissions offset by a 4 per cent fall in industrial emissions. Does this mean that Chinese emissions are peaking?

Electricity output increased by 6.3 per cent between the first nine months of 2023 and 2024. The increase was supplied by roughly equal increases in thermal power, hydropower and the new renewables. The increase in hydropower is weather related. With less rainfall there would have been a more significant increase in thermal power.

Coal production was up only 0.7 per cent on 2023. In the first few months of the year, coal production was lower than in the previous year but by October it was running at 6 per cent above the level of 2023. This is identical to the compound trend between 2017 and 2023.

In conclusion, the fall in emissions in the first nine months of 2024 is probably partly due to the increase in hydroelectric output and slow economic activity. It probably does not yet constitute a sustainable peak in emissions.

Discussion

There could be catchup growth as well as economic stimulus as the economy comes out of the pandemic period of slow growth, though, as of the date of writing, news reports suggest a slow recovery exacerbated by a real estate bust. Previous research has found mixed results regarding the behaviour of carbon emissions after recessions. Burke et al. (2015) found that globally over the five decades from 1961 to 2010 emissions tended to grow more slowly relative to GDP following recessions than after economic expansions became established. Similarly, Bersalli et al. (2023) found that 26 of the 28 countries that had so far peaked their carbon dioxide emissions did so just before or during a recession. However, emissions grew strongly in 2010 following the Global Financial Crisis (GFC) mainly because energy intensity rose (Jotzo et al. 2012).

China suffers from energy insecurity as it consumes much more oil and coal than it produces. China also has the most to gain from reducing carbon emissions. Tol (2019: 188–89) estimates that around half the global benefits from climate mitigation accrue to China and, as the largest emitter, a given percentage of reduction in emissions in China translates to more avoided carbon than anywhere else. For these reasons, we might expect China to be a leader on climate action among developing countries. In the past, China has set seemingly ambitious climate mitigation goals, though these were not very strict compared with business as usual (Stern and Jotzo 2010). Under the Paris Agreement, China has pledged to peak carbon emissions by 2030 and reach net-zero emissions by 2060. However, after the energy crisis of 2022 and increased tensions between China and the West, China may have less incentive to continue to invest in the development and manufacture of renewable energy technologies (Goldthau and Tagliapietra 2022). Furthermore, the increasing political tensions between the United States and China are reducing the potential for cooperation on climate policy. For example, at the UN Climate Conference COP27 in Sharm El-Sheikh, Egypt, in November 2022, China did not join a pledge to curb methane emissions and refused to provide financial support as part of the Loss and Damage Fund. This stance and the data we have presented in this chapter cast some doubt on China following through on its previous pledges and certainly on peaking emissions in 2025.

Conclusion

To examine whether China is on a path to peak carbon emissions by 2030 and carbon neutrality by 2060 under the Paris Agreement, we compared China's carbon emissions and energy production trajectories between 2019 and 2023, defined as pre-pandemic and post-pandemic periods, respectively. I found that China's total emissions rose by 8 per cent, mainly due to an increase of 18 per cent in power sector emissions. The production of coal grew at a compound annual rate of 6 per cent. Moreover, we note that the share of coal in China's total energy production increased for the first time since 2007, from 56 per cent in 2021 to 56.2 per cent in 2022. Though emissions fell in the first nine months of 2024 relative to the same period in 2023, this was due to the slow pace of economic activity and wet weather that resulted in a big increase in hydropower generation.

Our new findings contrast with the research of Le Quéré et al. (2021: 197), who argued that 'the pervasive disruptions from the COVID-19 pandemic have radically altered the trajectory of global CO_2 emissions'. Emissions fell sharply globally because of the curtailment of passenger transport during the pandemic (Jiang and Stern 2023), but Le Quéré et al. (2021) believed there was a window of opportunity to continue the slowing of emissions growth that they had seen since 2015. Similarly, Chen et al. (2022) predicted that Chinese emissions would peak between 2021 and 2026 at a level between 11.7 and 13.1 gigatonnes, with more than 80 per cent probability, and Lui (2022) wrote that China is 'set to significantly overachieve [on] the targets it promised internationally for 2030, with emissions peaking by 2025'. Nonetheless, our findings suggest that China may find it challenging to achieve these targets within that time frame.

Acknowledgements

I thank Khalid Ahmed for his work on the original paper (Ahmed and Stern 2023), which contributed to this updated work.

References

Ahmed, Khalid. 2023. 'Perspective on China's Commitment to Carbon Neutrality Under the Innovation–Energy–Emissions Nexus.' *Journal of Cleaner Production* 390: 136202. doi.org/10.1016/j.jclepro.2023.136202.

Ahmed, Khalid, and David I. Stern. 2023. 'China's Carbon Emissions Trend After the Pandemic.' *Environmental Challenges* 13: 100787. doi.org/10.1016/j.envc.2023.100787.

Bersalli, Germán, Tim Tröndle, and Johan Lilliestam. 2023. 'Most Industrialised Countries Have Peaked Carbon Dioxide Emissions During Economic Crises Through Strengthened Structural Change.' *Communications Earth & Environment* 4, no. 1: 44. doi.org/10.1038/s43247-023-00687-8.

Burke, Paul J., Md Shahiduzzaman, and David I. Stern. 2015. 'Carbon Dioxide Emissions in the Short Run: The Rate and Sources of Economic Growth Matter.' *Global Environmental Change* 33: 109–21. doi.org/10.1016/j.gloenvcha.2015.04.012.

Chen, Jiandong, Chong Xu, Ming Gao, and Ding Li. 2022. 'Carbon Peak and its Mitigation Implications for China in the Post-Pandemic Era.' *Scientific Reports* 12, no. 1: 3473. doi.org/10.1038/s41598-022-07283-4.

Dai, Jing, Martin Kesternich, Andreas Löschel, and Andreas Ziegler. 2015. 'Extreme Weather Experiences and Climate Change Beliefs in China: An Econometric Analysis.' *Ecological Economics* 116: 310–21. doi.org/10.1016/j.ecolecon.2015.05.001.

Goldthau, Andreas, and Simone Tagliapietra. 2022. 'Energy Crisis: Five Questions That Must Be Answered in 2023.' *Nature* 612, no. 7941: 627–30. doi.org/10.1038/d41586-022-04467-w.

Jiang, Xueting, and David I. Stern. 2023. 'Asymmetric Business Cycle Changes in US Carbon Emissions and Oil Market Shocks.' *Climatic Change* 176: 147. doi.org/10.1007/s10584-023-03620-2.

Jotzo, Frank, Paul J. Burke, Peter J. Wood, Andrew Macintosh, and David I. Stern. 2012. 'Decomposing the 2010 Global Carbon Dioxide Emissions Rebound.' *Nature Climate Change* 2, no. 4: 213–14. doi.org/10.1038/nclimate1450.

Le Quéré, Corinne, Glen P. Peters, Pierre Friedlingstein, Robbie M. Andrew, Josep G. Canadell, Steven J. Davis, Robert B. Jackson, and Matthew W. Jones. 2021. 'Fossil CO_2 Emissions in the Post-COVID-19 Era.' *Nature Climate Change* 11, no. 3: 197–99. doi.org/10.1038/s41558-021-01001-0.

Li, Hu, Bo Zheng, Philippe Ciais, K. Folkert Boersma, T. Christoph V. Riess, Randall V. Martin, Gregoire Broquet, Ronald van der A, Haiyan Li, Chaopeng Hong, Yu Lei, Yawen Kong, Qiang Zhang, and Kebin He. 2023. 'Satellite Reveals a Steep Decline in China's CO_2 Emissions in Early 2022.' *Science Advances* 9, no. 29: eadg7429. doi.org/10.1126/sciadv.adg 7429.

Liu, Zhu, Philippe Ciais, Zhu Deng, Steven J. Davis, Bo Zheng, Yilong Wang, Duo Cui, Biqing Zhu, Xinyu Dou, Piyu Ke, Taochun Sun, Rui Guo, Haiwang Zhong, Olivier Boucher, François-Marie Bréon, Chenxi Lu, Runtao Guo, Jinjun Xue, Eulalie Boucher, Katsumasa Tanakam, and Frédéric Chevallier. 2020. 'Carbon Monitor, A Near-Real-Time Daily Dataset of Global CO_2 Emission from Fossil Fuel and Cement Production.' *Scientific Data* 7, no. 1: 392. doi.org/10.1038/s41597-020-00708-7.

Lu, Liang-Chun, Shih-Yung Chiu, Yung-ho Chiu, and Tzu-Han Chang. 2022. 'Sustainability Efficiency of Climate Change and Global Disasters Based on Greenhouse Gas Emissions from the Parallel Production Sectors—A Modified Dynamic Parallel Three-Stage Network DEA Model.' *Journal of Environmental Management* 317: 115401. doi.org/10.1016/j.jenvman. 2022.115401.

Lui, Swithin. 2022. 'Guest Post: Why China Is Set to Significantly Overachieve Its 2030 Climate Goals.' *Carbon Brief*, 19 May. www.carbonbrief.org/guest-post-why-china-is-set-to-significantly-overachieve-its-2030-climate-goals/.

Newey, Whitney K., and Kenneth D. West. 1987. 'A Simple, Positive Semi-Definite, Heteroskedasticity and Autocorrelation Consistent Covariance Matrix.' *Econometrica* 55, no. 3: 703–8. doi.org/10.2307/1913610.

Stern, David I., and Frank Jotzo. 2010. 'How Ambitious Are China and India's Emissions Intensity Targets?' *Energy Policy* 38, no. 11: 6776–83. doi.org/10.1016/j.enpol.2010.06.049.

Tol, Richard S.J. 2019. *Climate Economics: Economic Analysis of Climate, Climate Change and Climate Policy*. 2nd edn. Cheltenham: Edward Elgar.

9

China's local government debt: Causes, consequences, characteristics, governance and suggestions

Yang Yao and Ling Yu

China's local government debt has long been a critical engine of economic growth; however, its rapid expansion since 2009 has generated significant systemic risks to fiscal stability and macroeconomic security. Despite sustained policy efforts to curb debt accumulation, subnational liabilities have continued to grow. In this chapter, we employ a multidimensional framework to investigate: 1) the institutional drivers of debt escalation, including imbalances in fiscal decentralisation, subnational development imperatives, the characteristics of a unitary political system, frequent official staff turnover and regulatory gaps in off-budget financing under the Budget Law; 2) the dual consequences of local government debt as both a catalyst for growth and an amplifier of fiscal vulnerability; 3) the dynamics of debt accumulation from 2015 to 2023, as quantified through panel data analysis; and 4) the central government's debt governance mechanisms, particularly in the context of its debt-swap programs. Based on our findings, we propose a comprehensive three-step debt resolution strategy designed to mitigate both current and future debts, thereby enhancing fiscal sustainability and promoting balanced economic development.

Global debt has escalated sharply in recent years. According to a UN report, global public debt reached a record high of US$97 trillion in 2023, with public debt in developing countries increasing at twice the rate of that in developed nations (UNCTAD 2024). According to the International Monetary Fund (IMF 2024), global public debt was expected to exceed US$100 trillion (93 per cent of global GDP) by 2024. As the second-largest economy in the world, China's central government debt is relatively light compared with other large countries, but the debt borne by its local governments has reached an alarming level. According to the Ministry of Finance (MoF), China's official on-budget local government debt was RMB40.74 trillion in 2023 (State Council 2023). However, this figure fails to account for local governments' substantial off-budget borrowings that are disguised as commercial debts borrowed by government-owned commercial entities. Given the profound implications of local government debt for economic development and financial stability, a comprehensive understanding of its causes, consequences and governance is of great value to both scholars and policymakers.

The off-budget debt is primarily borrowed through local government financing vehicles (LGFVs). In response to the 2008 GFC, the Chinese Government launched a RMB4-trillion (US$586-billion) infrastructure-centric investment stimulus plan to stabilise economic growth (Fan et al. 2022). To address the substantial funding gap between local government fiscal revenue and the expenditure required for the stimulus plan, in 2009, the central government permitted local governments to establish LGFVs to raise off-budget funds for investment projects (Huang et al. 2020; PBOC and CBRC 2009; MoF 2009). In legal terms, LGFVs are commercial entities, just like other kinds of state-owned enterprises (SOEs). However, unlike other kinds of SOEs that operate for profit, their major task is to raise funds from the debt market to finance local government projects. Subsequently, LGFV-induced off-budget financing became a critical funding source, leading to rapid off-budget debt accumulation. By 2023, off-budget debt through LGFVs had ballooned to RMB61.56 trillion, accounting for 87 per cent of local government fiscal revenues, or 70 per cent of the total government debt balance (RMB70.77 trillion) and 151 per cent of the official on-budget local government debt balance (RMB40.74 trillion).[1]

The accumulation of local government debt in China is influenced by several institutional factors. The 1994 tax-sharing reform restructured fiscal relations by shifting revenue authority upwards to the central government while decentralising administrative responsibilities to local governments. This restructuring led to increased fiscal pressures on local governments to finance their expenditure needs, thereby incentivising off-budget borrowing. Additionally, China's rapid urbanisation has intensified demands for local economic development, further straining fiscal resources and necessitating debt-financed investment. China's unitary state structure implies that the central government ultimately bears responsibility for local debt, creating moral hazard as local governments anticipate potential bailouts and thus dare to engage in excessive borrowing. The frequent rotation of local officials, combined with the pressures of the 'promotion tournament', exacerbates the issue, as officials' performance evaluations are heavily tied to short-term GDP growth. This system incentivises large-scale infrastructure and industrial investment through excessive borrowing. Moreover, weak budgetary constraints and an underdeveloped municipal bond market further exacerbate the reliance on off-budget financing, amplifying risks associated with local government debt accumulation.

---

1   The calculation is based on data from the Wind database (www.wind.com.cn/).

While local government debt has played a positive role in supporting infrastructure investment (Aschauer 1989; Ludvigson 1996), enhancing market liquidity (Yakita 2008) and promoting economic growth (Adelino et al. 2017), its mismanagement has led to significant risks. Excessive debt accumulation has heightened repayment pressures, weakened fiscal discipline and resulted in inefficient resource allocation (Huang et al. 2020). Low-return investments have put financial distress on LGFVs, many of which are on the verge of bankruptcy, necessitating central government intervention. Furthermore, mounting debt burdens have exacerbated financial market volatility (Huang and Du 2018), triggered inflation (Cochrane 2011) and heightened systemic risks (Chen et al. 2020), particularly in the real estate and financial sectors, where local government debt is often intertwined with land markets and banking institutions. This poses a dilemma for policymakers in balancing perceived trade-offs.

To address these challenges, the central government has undertaken a series of governance reforms, including the introduction of a quota-based local government bond issuance system, strengthened monitoring and enhanced transparency requirements. In 2015, the central government launched a three-year debt-swap program that converted RMB14.34 trillion of LGFV debt into local government bonds to lessen local governments' short-term payment pressure. However, this program did not stop local governments borrowing through their LGFVs. In 2019, pilot resolution programs in six highly indebted provinces facilitated RMB142.92 billion for debt swap again.[2] The Covid-19 pandemic further strained fiscal conditions, prompting a temporary relaxation of LGFV bond issuance in 2020. Recognising the mounting risks, the central government expanded debt-swap efforts to 26 provinces, reaffirming its commitment to long-term fiscal sustainability.

What has happened in China in recent years is paradoxical: while the central government has geared up its efforts to resolve the debt issue, local government debt has surged and exhibited sustained growth. To this end, we present a thorough examination of the causes, functions and problems of local debt, alongside a review of the central government's strategies for managing the risks. We seek to provide insights into how to reduce financial vulnerabilities, improve debt management practices and ensure that local government debt contributes to sustainable economic development without posing systemic risks to the broader economy.

The chapter is structured as follows: section two presents the categorisation of local debts; section three examines the factors driving the accumulation of local government debt; section four explores the economic consequences of local government borrowing; section five outlines the characteristics and scale of local government debt; section six evaluates the central government's approaches to managing local government debt; and section seven proposes systematic solutions for mitigating the risks associated with local government debt.

---

2   ibid.

Categories of local government debt

Before setting out our analysis, it is useful to present the categories of local government debt in China, which have evolved over time. Currently, local government debt is classified into on-budget and off-budget debt based on whether the obligations are recorded in the government's balance sheet and whether they constitute legally mandated liabilities.

Categories of on-budget local government debt

On-budget debt includes obligations explicitly recorded in government accounts and legally required to be repaid using fiscal revenues within statutory debt ceilings. This category primarily comprises general and special debts raised through the issuance of local government bonds (LGBs). General debt is typically financed through the issuance of general LGBs and is allocated to non–revenue-generating public welfare projects, such as education, health care and social security. These debts are primarily repaid using general budget revenues. In contrast, special debt is financed through the issuance of special LGBs, which are directed towards revenue-generating projects, such as infrastructure development in transportation, telecommunications and housing. Special debts are predominantly repaid using corresponding government funds or designated revenues. On-budget debt is characterised by greater transparency and more stringent monitoring and is commonly referred to as explicit debt.

Categories of off-budget local government debt

Off-budget local government debt refers to obligations that are not formally recorded in government accounts and fall outside the statutory debt ceiling or central government–authorised debt frameworks. Some of these liabilities are guaranteed by fiscal funds or may involve illegal guarantees (Xu et al. 2020). This category primarily includes the following forms.

Direct implicit local government debt

This includes three categories: 1) implicit debt arising from social security obligations, such as pension deficits and healthcare insurance shortfalls. In cases where social security funds fall short, the responsibility for bridging the payment gap ultimately rests with local governments. 2) Debt related to future capital and recurring expenditures arising from public investment projects, notably under government-paid public–private partnership (PPP) models and viability gap-funded PPPs. The reduction in financing costs for such PPP projects is heavily dependent on credit enhancements provided by government-issued approval documents. Furthermore, debt repayment for these PPP projects is contingent on government payments or subsidies. When the total scale of these projects becomes unmanageable or when payment delays, defaults or project operation failures occur, the local government assumes the responsibility for repayment. 3) Implicit debt resulting from arrears in various on-budget expenditures, such as unpaid project costs, outstanding procurement payments and refundable deposits that have been collected from enterprises but remain unpaid. These are obligations that the government or its departments are required to fulfil once the payment conditions are met.

Contingent implicit local government debt

There are two categories here: 1) illegal or irregular implicit debt, such as the issuance of unlawful or irregular commitment letters, guarantees or similar documents that are often provided for LGFV bonds, the execution of construction projects disguised as government service procurement, PPP projects structured as equity investments with government repurchase agreements or guaranteed fixed returns, effectively converting equity into disguised debt, guarantees provided by industrial guidance funds to other partners and local governments raising debt in the name of SOEs or public institutions; and 2) debt arising from local governments assuming rescue responsibilities, which manifests in expenditures aimed at addressing liabilities such as debts of public institutions, debts and losses of local SOEs, non-performing assets of local financial institutions, obligations of LGFVs, fiscal shortfalls in subordinate governments and defaults on non-guaranteed debts. Since these debts are typically excluded from official government statistics, they are characterised by a high degree of opacity and are commonly referred to as implicit debt.

General LGFV debt

This kind of debt is issued by the LGFVs that are commercial identities legally separated from the government. However, it is widely acknowledged that LGFV debts are a convenient tool for local governments to finance their infrastructural projects. The market also believes that those debts are government debts and will never be defaulted even if they do not bear government guarantees—and there has never been a single default. LGFV debt can be bank loans, municipal bonds, trusted loans and other forms of shadow bank finance. There are no reliable data on the size of LGFV debts, but, by most estimations, they make up the largest part of local government debts.

Figure 9.1 provides a visual presentation of the categorisation. The central government only recognises on-budget debts and direct and contingent implicit debts among the off-budget debts. In theory, this leaves most of the LGFV debts to the market, which contradicts the widespread belief held by the market that those debts are also government debts.

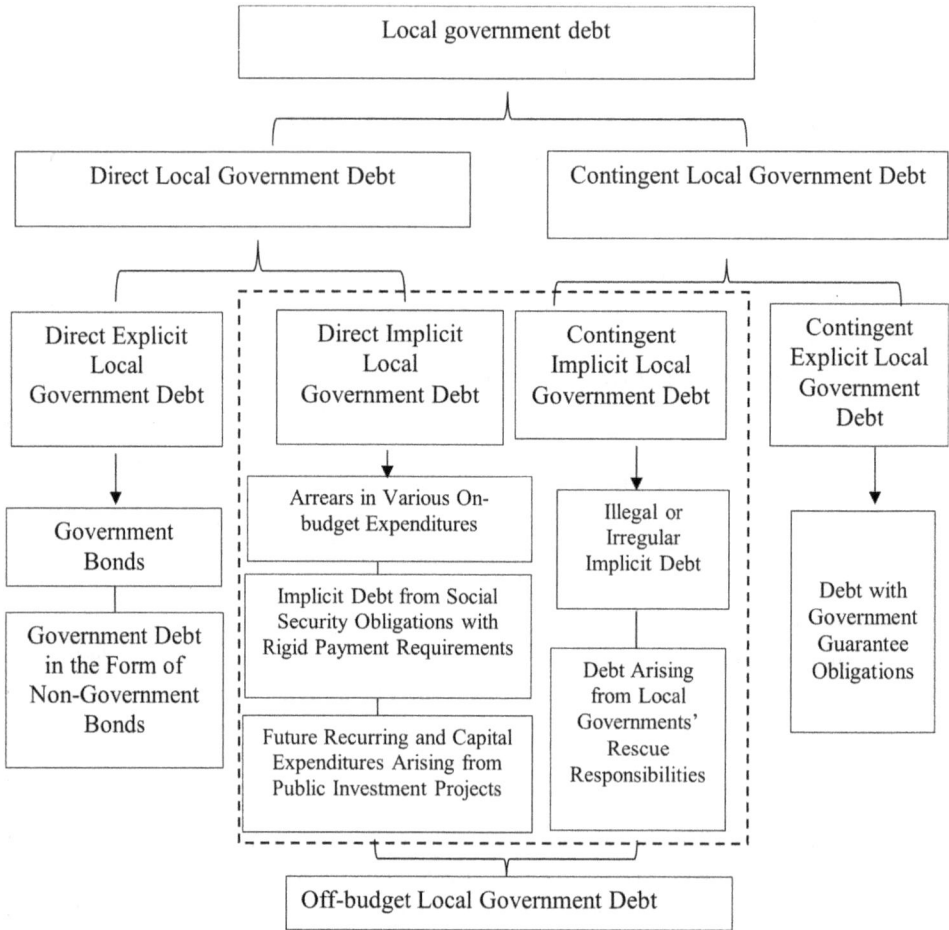

Figure 9.1: Categories of local government debt in China
Source: Authors' original illustration.

Causes of rising local government debt

The 1994 tax-sharing reform

From the early 1980s to early 1990s, China had a contract system for central–local fiscal relationships, which greatly tilted towards local governments. The rapid expansion of rural collective industries in the 1980s significantly propelled China's early economic growth, generating a substantial surplus that became a primary revenue source for local governments (Tian 2000; Huang 2012; Banerjee et al. 2020). However, this led to a fiscal imbalance, with local government revenues significantly surpassing those of the central government. By 1993, the central government's share of total on-budget revenues was merely 22.02 per cent, while local governments accounted for 77.98 per cent (see Figure 9.2).

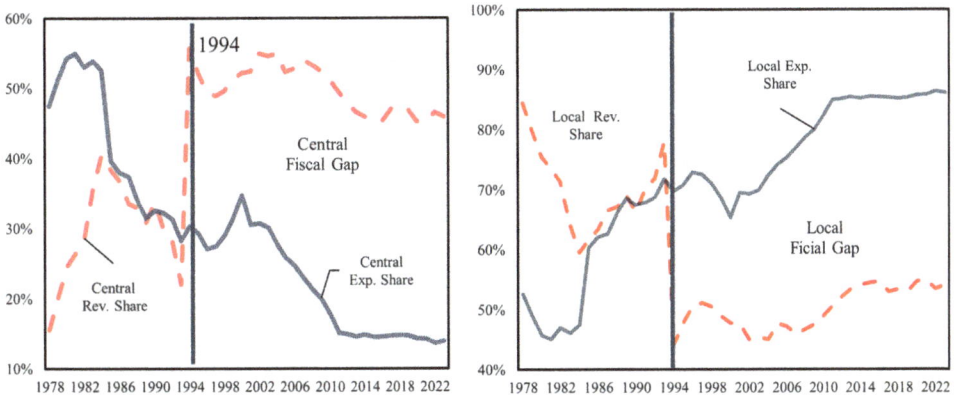

Figure 9.2: An effect of China's 1994 tax-sharing reform

Notes: Central government (left) and local government (right).

Source: National Bureau of Statistics of China (data.stats.gov.cn/english/).

To address this imbalance, the central government implemented the tax-sharing reform in 1994, categorising tax revenue into central, local and shared taxes, thereby fundamentally altering the revenue-sharing arrangement between the central and local governments (Liu et al. 2017; Tsui 2005). Despite its intentions, the reform primarily resulted in an upward shift of fiscal resources to the central government (Chang et al. 2013). Stable and high-growth tax sources were designated as exclusive central government revenue or shared revenue. Consequently, local governments became increasingly reliant on smaller, less stable tax bases characterised by fragmented sources, inefficient collection mechanisms and high administrative costs. Notably, the value-added tax, the most significant shared tax, allocated 75 per cent to the central government and only 25 per cent to local governments (Liu et al. 2022). This redistribution effectively recentralised fiscal control, substantially reducing local government revenues.

Compounding the issue, the reform rectified the decentralised administrative system that allocated the bulk of financial responsibilities to local governments, including social security, basic education, health care and public safety (See Table A9.1). Wong (2025) observes that in China's fiscal structure, prefectures and counties, constituting the third and fourth tiers of the government, respectively, are primarily responsible for nearly all social security expenditures, encompassing old-age pensions, unemployment insurance and other welfare programs. Counties and townships, representing the fourth and fifth tiers, respectively, account for approximately 70 per cent of total education spending and between 55 and 60 per cent of healthcare expenditures. As a result, local governments faced heightened fiscal pressures, with expanding expenditure obligations and contracting revenue bases. The decline in revenues coupled with rising expenditures produced a persistent and widening fiscal gap for local governments (see Figure 9.2). In the post-reform period, although the central government has secured more than 50 per cent of total on-budget fiscal revenues, local governments have been tasked with nearly 80 per cent of on-budget fiscal expenditures. Although central government transfers fill most of the gap, fiscal independence forces local governments to increasingly resort to borrowing, thereby contributing to the expansion of local government debt.

Local development imperatives

China's urbanisation has experienced a sustained upwards trend. In 1994, urban residents as a share of the national population exceeded 30 per cent and, by the end of 2023, the share reached 66 per cent, reflecting an average annual increase of 0.93 per cent over the previous five years.[3] This rapid urbanisation has amplified local development demands; however, suboptimal infrastructure standards, insufficient coordination of infrastructure and service facilities between regions (for example, poor functional integration between transportation, utilities and public service systems) and deficiencies in urban governance have constrained local economic growth. Consequently, there is an urgent need for local governments to engage in municipal construction to stimulate GDP growth as well as to provide necessary public goods for the newly added urban population. Most government-invested projects are the responsibility of local governments rather than the central government, intensifying the urgency for local authorities to secure substantial financing resources within a short time.

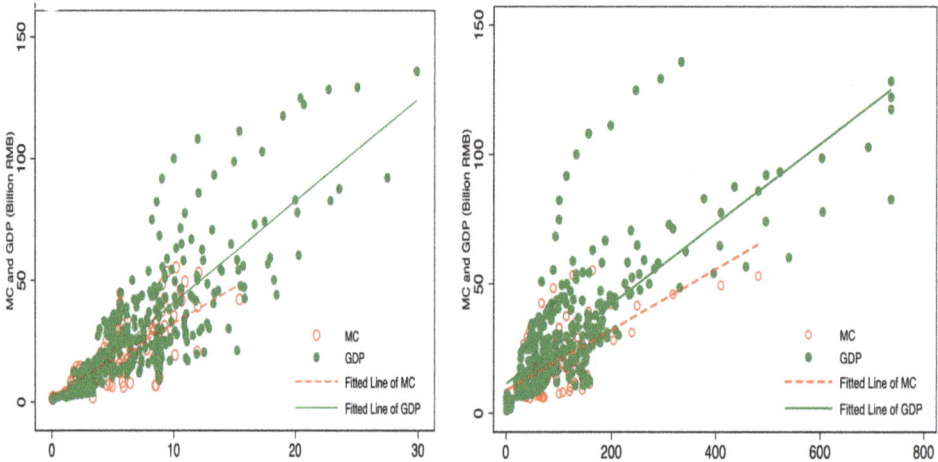

Figure 9.3: The relationship between municipal construction, GDP and local government debt in China, 2015–23

Notes: On-budget local government debt (left) and LGFV-induced off-budget local government debt (right). MC = municipal construction. LGD = local government debt.

Sources: National Statistical Yearbooks (www.stats.gov.cn/english/Statisticaldata/yearbook/); Local Government Bond Information Disclosure Platform (www.celma.org.cn); and Enterprise Early Warning System.

---

3   Data are sourced from the National Bureau of Statistics of China (data.stats.gov.cn/english/).

Local governments' fiscal resources comprise four primary components: formal budgetary revenues, transfers from the central government, off-budget revenues and proceeds from land-related transactions (Wu and Feng 2014). Revenues from the first two components are strictly controlled under the budgetary arrangements and are predominantly allocated to daily government operations, education, health care and social security, leaving limited funds available for public projects. Notably, municipal construction—the principal component of infrastructure investments—is generally not included in these budgetary allocations (Xu and Zhang 2013). As a result, only the last two revenue sources can fund municipal construction, and any financing shortfall must be met through alternative channels—primarily debt borrowings. Empirical analysis demonstrates that both municipal construction and GDP had a positive linear relationship with local government debt from 2015 to 2023 (see Figure 9.3).

A unitary political system

Since the fall of Qing Dynasty, China has adopted a unitary political system (as shown in Figure 9.4). Under this system, the country is divided into administrative regions (provinces) that operate under a unified sovereign authority (Xu 2011). This structural arrangement has two key implications. First, the authority of local governments is entirely delegated by the central government as a direct consequence of the hierarchical political structure. As a result, local governments primarily function as administrative extensions of the central government rather than as autonomous political entities. Consequently, when local governments incur debt, the ultimate responsibility for repayment rests with the central government. This arrangement creates strong incentives for local governments to engage in excessive borrowing, as they can effectively shift the financial burden upwards, relying on the central government to absorb the risks associated with their fiscal imbalances. Second, the unitary political system precludes local governments from utilising formal bankruptcy mechanisms, such as Chapter 9 of the US Bankruptcy Code, to discharge their debt obligations. In this legal framework, any bankruptcy or default by a local government would undermine the creditworthiness of the central government and potentially trigger a systemic financial crisis (Gao et al. 2021).[4] As a result, the central government is compelled to assume responsibility for local government debt, which further incentivises local governments to borrow excessively to finance their development projects.

---

4   Under this legal framework, Detroit's bankruptcy proceedings facilitated the discharge of its outstanding debt, thereby laying the groundwork for its subsequent economic recovery.

Politburo of CPC
(National-level Leaders,
e.g., Party Secretary General and Premier)

Provincial Secretary
(Ministerial Level)

Central Committee
(Ministerial Level)

Vice Provincial secretary
(Deputy Ministerial Level)

Central Committee (Alternate)
(Deputy Ministerial Level)

City secretary/mayor
(Department Level)

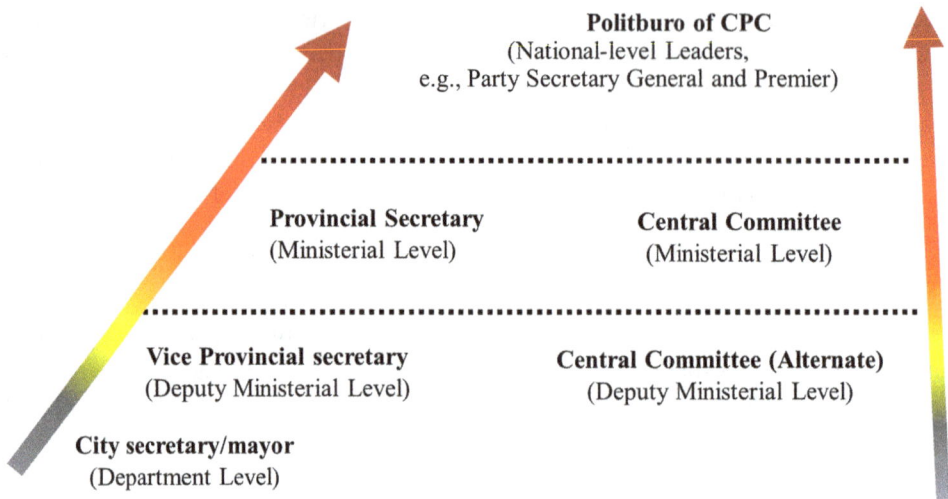

Figure 9.4: The hierarchical structure of the unitary political system in China

Notes: At the top of the hierarchy are the members of the Politburo of the Communist Party of China (CPC), including the party General Secretary and the Premier, who hold national rank. Directly beneath them is the Central Committee of the CPC, which comprises approximately 200 full members at the ministerial level and about 170 alternative members. Provincial party secretaries and governors, typically ranked at the ministerial level, also serve as members of the Central Committee. Notably, some provincial leaders, such as the party secretaries of Guangdong and Xinjiang, hold positions within the Politburo. Below this level are city secretaries and mayors, who are generally classified at the department level within the hierarchy.

Source: Authors' description.

Frequent rotation of officials

Chinese officials are engaged in a promotion tournament. Frequent rotation is a distinctive feature of this system (Li and Zhou 2005; Chen and Kung 2019).[5] According to the *Interim Provisions on Tenure of Leading Party and Government Cadres* (General Office of the CPC Central Committee 2006), once an official has held a position for two consecutive terms, they will be moved to a new position if they have not reached retirement age. The performance of local officials is primarily evaluated based on short-term economic growth, which is crucial for advancing their political careers (Xu 2011). In pursuit of economic growth, local officials often engage in building physical infrastructure (Bai and Qian 2010; Ru 2018), establishing special economic zones to attract investment (Wang 2013; Lu et al. 2019) and appointing capable bureaucrats (Xu 2011; Yao and Zhang 2015). Among them, public investment requires significant financial resources and raising taxes may adversely affect officials'

---

5   While the standard tenure for government officials is generally set at five years, early appointments to new positions, regardless of whether they constitute a promotion, may occur based on specific exigencies or directives from higher levels of government. Promotions typically proceed incrementally, with officials advancing one rank at a time, such as from the prefectural to the subprovincial level. Each political rank is subject to an age threshold—referred to as the 'age dilemma'—above which officials become ineligible for promotion. This age limit varies across ranks, typically ranging from 40 to 67 years, as specified by Kou and Tsai (2014).

reputations and prospects for promotion. As a result, borrowing becomes a preferred method for local officials to mobilise the necessary funds to achieve their economic objectives (Fan et al. 2022).

Frequent political rotations also complicate audits for short-tenure local officials, presenting challenges in assessing the legitimacy of borrowing decisions for public investments in the short term. This situation exacerbates the moral hazard associated with borrowing, as local officials may exploit this ambiguity to meet their economic objectives. Consequently, fiscal resources obtained through borrowing are frequently allocated to short-term and low-return projects, often serving to save face for the incumbent officials.

Insufficient regulation for off-budget debt under the Budget Law

The persistent lack of transparency and accountability in fiscal budgeting has significantly exacerbated the challenges inherent in soft budget constraints. In response, the Budget Law was introduced in 1995, banning local governments from running deficits in their fiscal budgets. However, the law did not ban local governments' off-budget borrowing. Consequently, local off-budget expenditures have grown considerably yet remained largely unreported. The RMB4-trillion stimulus in 2009 accelerated this process.

To reflect the reality that off-budget borrowing had become an important source of finance for local governments, the Budget Law was revised in 2014 to allow local governments to borrow within their budget. The law requires that the drafting and implementation of both central and local government budgets be reviewed and approved by the National People's Congress (NPC). However, the law still does not provide a framework for the central government to regulate local governments' off-budget borrowing, which has allowed local governments to circumvent formal debt constraints. In particular, local governments have increasingly relied on massive borrowing through various financing vehicles, such as LGFVs. A large part of these borrowings is secured by pledges of land or other assets, which already exposes local governments to significant debt-related risks (for example, Ang et al. 2018; Chen et al. 2018, 2020). Worse, to circumvent the borrowing constraints, local governments have progressively adopted more sophisticated and intricate financial instruments to run funds, including borrowing from trust funds, issuing LGFV bonds and borrowing from enterprises. Many of these are not reported, complicating the accurate estimation of their size (Chang et al. 2013).

The dual consequences of local government debt

The developmental role of local government debt

Enhancing infrastructure investment

Local government debt (LGD) functions as a pivotal fiscal mechanism in stimulating aggregate demand, with a pronounced emphasis on infrastructure investment. On-budget debt mechanisms, especially the special LGBs, have become the dominant financing channel for public infrastructure projects. Between 2015 and 2023, the infrastructure-targeted special LGBs accounted for 40.39 per cent (RMB9.42 trillion) of total special bond issuance. Within this allocation, infrastructure construction for municipal facilities and industrial parks comprised 20.78 per cent, transportation infrastructure 18.39 per cent and cold chain logistics and new infrastructure projects 0.71 per cent and 0.51 per cent, respectively. Additionally, a substantial share was allocated to infrastructure-adjacent areas, including shantytown renovation (12.03 per cent), agricultural, forestry and water projects (3.49 per cent), ecological and environmental protection (2.58 per cent), rural revitalisation (0.84 per cent), urban old residential community renovation (0.64 per cent) and energy projects (0.33 per cent) (see Figure 9.5-A).

Complementarily, off-budget financing via LGFVs addresses diverse infrastructure needs (Chen et al. 2020). From 2015 to 2023, the bond issuance of LGFVs was distributed across urban infrastructure construction (36.09 per cent), industrial investment (27.77 per cent), transportation construction and operation (12.57 per cent), land development and consolidation (8.72 per cent), public utilities (5.76 per cent) and other areas such as shantytown renovation and affordable housing development (3.79 per cent), comprehensive platforms (3.39 per cent),[6] industrial park development and operation (1.71 per cent) and cultural tourism (0.21 per cent) (see Figure 9.5-B). Correspondingly, the distribution of LGFVs' total debt reflected a comparable pattern, with infrastructure construction (40.05 per cent), industrial investment (26.77 per cent), transportation construction and operation (7.88 per cent), land development and consolidation (9.24 per cent), public utilities (6.44 per cent), shantytown renovation and affordable housing development (4.44 per cent), comprehensive platforms (3.26 per cent), industrial park development and operation (1.66 per cent) and cultural tourism (0.26 per cent) (see Figure 9.5-C). The investment carried out by the LGFVs has significantly enhanced the functionality of the urban system and addressed critical infrastructure gaps while providing a robust foundation for sustainable economic development and improved quality of life for ordinary people.

---

6   Comprehensive platforms are a type of LGFV the operations of which span multiple domains of urban development and serve dual purposes: financing and investing in public-interest projects and engaging in market-oriented operations.

Figure 5-A: Allocation of Funds Raised Through Special Bond Issuance

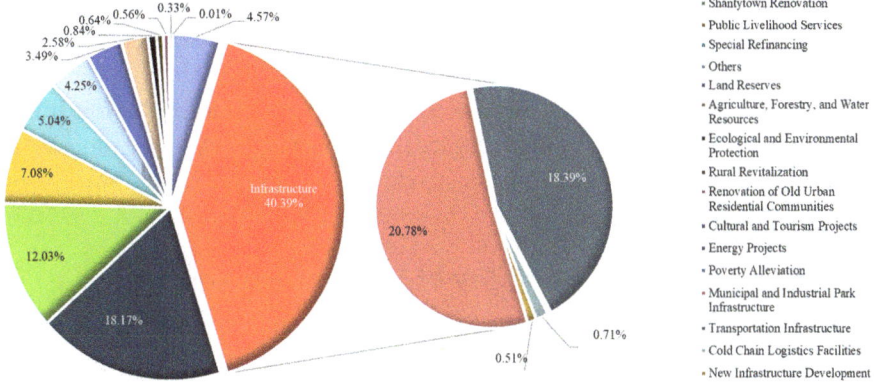

- Borrowing New Debt to Repay Old Debt
- Shantytown Renovation
- Public Livelihood Services
- Special Refinancing
- Others
- Land Reserves
- Agriculture, Forestry, and Water Resources
- Ecological and Environmental Protection
- Rural Revitalization
- Renovation of Old Urban Residential Communities
- Cultural and Tourism Projects
- Energy Projects
- Poverty Alleviation
- Municipal and Industrial Park Infrastructure
- Transportation Infrastructure
- Cold Chain Logistics Facilities
- New Infrastructure Development

Figure 5-B: Share of LGFV Bond Issuance by Business Type   Figure 5-C: Proportion of LGFVs by Business Type

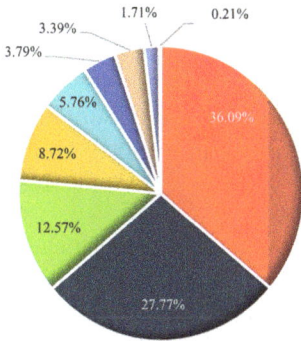

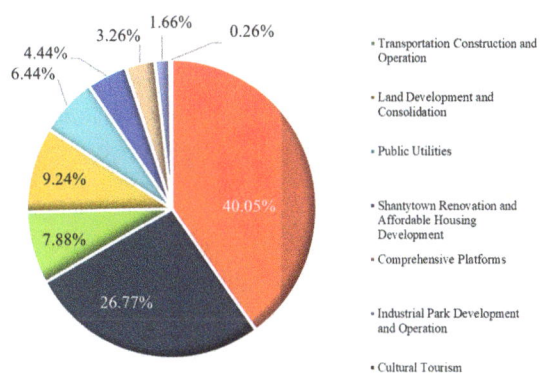

- Infrastructure Construction
- Industrial Investment Platforms
- Transportation Construction and Operation
- Land Development and Consolidation
- Public Utilities
- Shantytown Renovation and Affordable Housing Development
- Comprehensive Platforms
- Industrial Park Development and Operation
- Cultural Tourism

Figure 9.5: Distribution of special bond funds and LGFV bond issuance by business type

Notes: Allocation of funds raised through special bond issuance (A), share of LGFV bond issuance by business type (B) and proportion of LGFVs by business type (C).

Source: Wind database (www.wind.com.cn/).

Financial repurposing: From savings glut to productive capital

The rapid accumulation of savings in China has created a substantial pool of idle capital (Chen and Wen 2017; Piketty et al. 2019; Zhang et al. 2021). By the end of 2015, household deposit balances stood at RMB54.6 trillion, rising sharply to RMB137 trillion by 2023 (PBOC 2024). However, this capital surplus has led to a decline in its utilisation efficiency, as substantial savings remain underemployed in productive economic activities.

Local governments have utilised financing platforms and related mechanisms to mobilise low-yield idle funds, redirecting them towards high-yield capital expenditures (Zhang and Tsai 2024). Strategically, these funds have been allocated to sectors critical to economic development, including transportation, agriculture and water conservancy, energy and ecological and environmental protection (Zhu 2021). Investments in these areas not only strengthen infrastructure but also promote industrial chain coordination, fostering regional economic growth. Furthermore, many infrastructure projects generate sustainable revenue streams, such as toll fees from highways and revenues from utility projects. These revenue flows mitigate local fiscal shortfalls and contribute to long-term GDP growth, thereby reinforcing the foundation for high-quality economic development. Through this dual mechanism of resource mobilisation and revenue generation, idle funds are effectively utilised to advance regional and national economic objectives.

Enhancing the fiscal flexibility of local governments

China has made significant progress in developing a market-oriented economy (Chen et al. 2020). Market-based financing instruments, characterised by high operational flexibility and risk-sharing mechanisms, have systematically enhanced local governments' fiscal flexibility through three channels: 1) diversification of funding sources beyond traditional fiscal transfers; 2) optimisation of intertemporal resource allocation via capital market integration; and 3) establishment of shock-absorbing mechanisms for fiscal resilience. Through these mechanisms, local governments have strengthened fiscal stability, ensured expenditure continuity and improved resource allocation efficiency. This institutional innovation has enabled provincial and municipal authorities to maintain fiscal stability while ensuring expenditure continuity and improving allocative efficiency.

Between 2015 and 2023, a confluence of external and internal pressures, including escalating United States–China trade frictions, the transition to new growth drivers, tax and fee reductions and the economic disruptions caused by Covid-19, intensified fiscal imbalances. During this period, the national public budget self-sufficiency rate declined from 86.60 per cent in 2015 to 76.76 per cent in 2023.[7] In response, local governments issued RMB26.91 trillion in new bonds, with the proportion of on-budget bond issuance relative to fiscal resources rising from 3.88 per cent in 2015 to 21.15 per cent in 2023.[8] Additionally, LGFVs issued RMB33.44 trillion in bonds, with the proportion of LGFV bonds relative to fiscal resources increasing from 9.75 per cent in 2015 to 30.36 per cent in 2023.[9] These bond issuances not only mitigated fiscal shortfalls but also strengthened local governments' financial autonomy, enhancing their capacity to address regional economic challenges and implement development priorities more effectively.

The crowding-in effects: Stimulating private investment

While the conventional literature finds that government debt crowds out private investment, there are several ways for government debt to have a crowding-in effect for private investment.

---

7   The fiscal self-sufficiency rate is the ratio of general budget revenue to general budget expenditure. The calculation is based on data from the official website of the Chinese Government (english.www.gov.cn/).

8   The data are sourced from the local government bond information disclosure platform. See CELMA (n.d.).

9   The calculation is based on data from the Wind database (www.wind.com.cn/).

The first is the trade credit channel. A portion of local government debt arises from repayment obligations to private suppliers through procurement activities (Hu et al. 2022). Delays in payments or the accumulation of accounts receivable often impose liquidity constraints on private enterprises, adversely affecting their financial stability. To mitigate these challenges, private enterprises have increasingly used high-credit government receivables as collateral to secure external financing, thereby replenishing working capital. In this context, local government debt may serve as a catalyst for private enterprise financing by enhancing access to financial resources that might otherwise be restricted. Contrary to concerns about crowding out private financing, government debt may stimulate additional funding opportunities, enabling private firms to address liquidity challenges and support their operational growth. Thus, local government debt not only stabilises private enterprise financing but also contributes to broader economic resilience and growth.

The second is the reputation channel. Private firms engaged in government procurement or investment projects must go through a screening process. As a result, they are often better than the average private firm in the market. In this case, local governments play the role of quality screening for the entire market. Private firms passing the screening process thus build a reputation that can facilitate their borrowing in the market.

The third is the channel of financial acceleration. Much local government borrowing has been spent on building infrastructure, which has accumulated a large number of assets, facilitating further local government borrowing. This is a kind of financial acceleration like the one created by investment in the real estate sector. Moreover, private firms can also benefit. Better infrastructure increases the value of assets held by private firms. As a result, private firms can borrow more from banks and other financial institutions because they have large amounts of collateral to offer.

The challenges of local government debt

Escalating debt servicing risks

The exponential growth of local government debt issuance and its cumulative stock has imposed severe repayment obligations, thereby amplifying local government debt risks. The off-budget debts are especially concerning. Traditionally, repayment of those debts has relied on revenues from government-managed funds, primarily coming from land sales, in addition to the revenue generated by the projects financed by those debts. However, in recent years, economic slowdowns, adjustments to real estate policies and other structural factors have constrained the growth of revenues from the sale of state-owned land use rights. From 2015 to 2023, the average annual growth rate of revenues from government-managed funds stood at merely 5.07 per cent,[10] significantly lagging the average annual growth rate of local debt balances. This widening gap has grown into a huge risk.

---

10 The data are sourced from CELMA (n.d.).

To make up the gap, local governments have increasingly relied on borrowing new debts to roll over the old debts, exacerbating the risks. As we will show later, more than 90 per cent of the LGFV bonds have been refinancing borrowings in recent years. However, this is not enough to cover all the maturing LGFV bonds. Local governments, under the approval of the central government, must issue on-budget refinancing special LGBs to swap maturing LGFV bonds. Since 2018, local governments have increasingly resorted to refinancing special LGBs to alleviate their repayment pressures. The issuance of such bonds surged from RMB135.8 billion in 2018 to RMB2.03 trillion in 2023, with their share of total special LGBs rising from 6.98 per cent to 33.96 per cent.[11] While the swaps provided short-term relief, their rapid growth has created long-term challenges. Under the constraints of the official debt ceilings, the expansion of refinancing special LGBs has reduced the available quota for new bond issuances, thereby limiting the financing capacity for new projects. In addition, by moving local governments' repayment pressures from their off-budget balance sheets to their on-budget balance sheets, debt swaps have increased local governments' fiscal stress over time.

Return and term mismatches

Local government financing is predominantly directed towards public infrastructure projects. The heavy reliance on market finance leads to dual mismatches. First, there is a mismatch between the public nature of the projects and the cost of finance. Infrastructural projects bring low returns and many generate no return. But the interest rates incurred by LGFVs, whether through bond issuance (for example, 4.07 per cent in 2023) or other mechanisms,[12] often exceed the financial returns generated by infrastructure projects. As of 31 December 2023, the average coupon rate of LGFV bond issuance was 4.07 per cent. The resulting cost–return imbalances have led to persistent financial losses for LGFVs, undermining their long-term operational sustainability. Second, a mismatch also exists between the timespan for infrastructural projects to recoup their costs and the term of most off-budget debts. As Kroeber (2016) describes, 'localities often used three to five-year bank loans to finance infrastructure projects whose economic benefits (and revenue streams, if any) would only materialize over two or three decades'. The maturity of LGFV bonds is like that of bank loans. This disparity implies that debt obligations often mature before the infrastructure projects can generate sufficient returns, thereby exacerbating fiscal pressures and increasing financial risks.

Underutilisation of debt

Despite substantial funds raised through debt issuance, many local governments fail to effectively channel these resources into productive investments, leading to resource misallocation and the underutilisation of local economic development potential.

First, the prevalent issue of 'over-packaging' of investment projects significantly undermines the efficiency of fund allocation. To secure approval and additional financial support, some local governments artificially enhance project feasibility by

---

11   The data are sourced from the Wind database (www.wind.com.cn/).
12   For details, see the subsection titled 'Overview of the LGFVs' bond issuance'.

inflating projected revenues or underestimating costs. In 2022, 24 regions issued a total of RMB198.21 billion in special LGBs through such practices. Of this amount, RMB5.03 billion was illegally allocated to prohibited areas, including landscaping and commercial projects. Moreover, 47 regions misappropriated RMB15.80 billion, while five regions falsified expenditure progress for 33 projects, leaving RMB6.03 billion in unused funds by the end of the year. These actions not only led to the inefficient use of financial resources but also increased repayment risks, particularly for projects with low or negligible returns.

Second, the persistence of idle funds exacerbates inefficiencies. Poor project preparation, inadequate planning and delays in implementation often leave allocated funds unutilised for extended periods. For example, the National Audit Office reported that the value of special bond funds left idle for more than one year was RMB13.2 billion in 2020, rising to RMB21.7 billion in 2021. Such prolonged underutilisation reflects structural weaknesses in project execution and resource management.

Third, the misuse of funds remains widespread, undermining the intended objectives of local government debt financing. While these funds are designated for public projects within specific sectors, some regions have diverted them to unauthorised purposes, such as projects on the negative list or recurrent expenditures such as employee salaries. From 2020 to 2022, illegal fund usage amounted to RMB41.32 billion, RMB13.66 billion and RMB20.80 billion, respectively. Such mismanagement not only distorts resource allocation but also exacerbates fiscal risks, ultimately compromising the sustainability of local government finances (NAO 2023).

How big is Chinese local government debt?

On-budget local government debt

Over the years, the balance of on-budget debt has grown substantially while it generally remaining within the statutory limits. Figure 9.6 illustrates the trend in on-budget local government debt from 2015 to 2023. During this period, the total debt balance increased from RMB14.21 trillion in 2015 to RMB40.74 trillion in 2023, staying below the statutory debt ceiling of RMB42.17 trillion. However, by 31 October 2024, the total on-budget debt had risen to RMB45.31 trillion. In terms of debt composition, special debt has seen accelerated growth since 2019, surpassing general debt in terms of total balance. As of 31 October 2024, the balance of special debt reached RMB28.71 trillion, compared with RMB16.60 trillion for general debt. Notably, the balance of special debt remains within the prescribed ceiling, with the ceiling projected to increase from RMB29.52 trillion to RMB35.52 trillion by the end of 2024.[13]

---

13 The Standing Committee of the NPC (2024) approved an increase of RMB6 trillion in the local government debt ceiling to replace existing implicit debt.

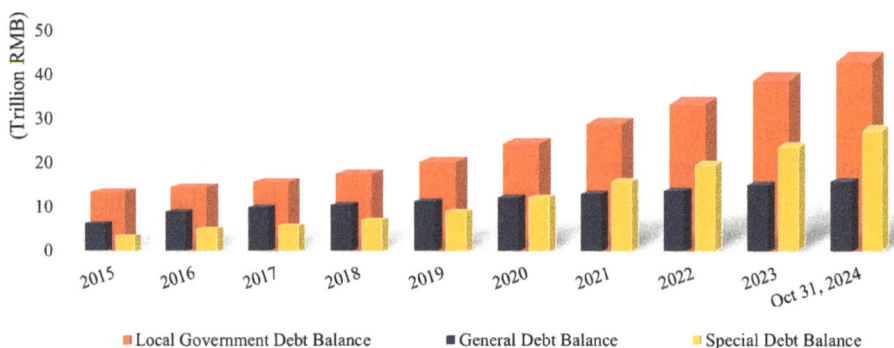

Figure 9.6: Local government debt balance in China, 2015–24

Source: Local Government Bond Information Disclosure Platform (www.celma.org.cn).

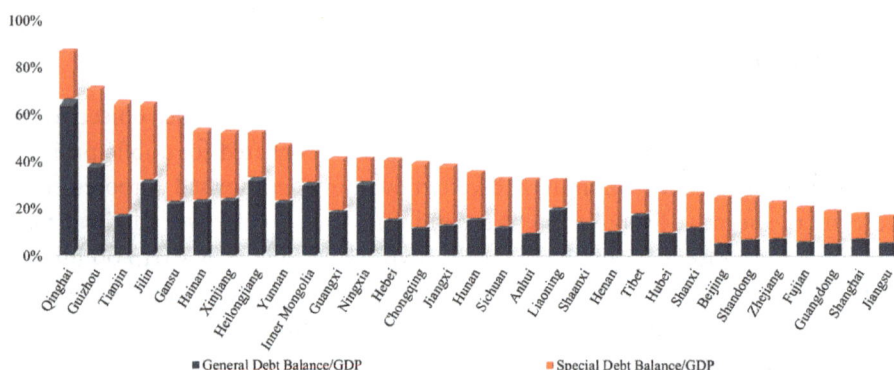

Figure 9.7: Local government bond balances in Chinese provinces in 2023

Source: Local Government Bond Information Disclosure Platform (www.celma.org.cn).

Significant regional disparities exist. Figure 9.7 presents the regional trends in on-budget local government debt ratios in 2023. The provinces with the highest debt ratios were Qinghai, Guizhou and Tianjin, with outstanding balances of RMB3.33 trillion, RMB15.12 trillion and RMB11.12 trillion, respectively, accounting for 87.87 per cent, 72.32 per cent and 66.42 per cent of their regional GDP. Within these provinces, general debt stood at RMB2.54 trillion in Qinghai, accounting for 66.78 per cent of GDP; RMB8.23 trillion in Guizhou, accounting for 39.36 per cent of GDP; and RMB2.93 trillion in Tianjin, accounting for 17.51 per cent of GDP. Meanwhile, special debt reached RMB800 billion in Qinghai, representing 21.08 per cent of GDP; RMB6.89 trillion in Guizhou, or 32.96 per cent of GDP; and RMB8.18 trillion in Tianjin, or 48.91 per cent of GDP. In contrast, the provinces with the lowest local government debt ratios were Guangdong, with outstanding balances of RMB27.19 trillion, accounting for 20.04 per cent of GDP; Shanghai, RMB8.83 trillion, or 18.70 per cent of GDP; and Jiangsu, RMB22.73 trillion, or 17.73 per cent of GDP. General debt in these regions amounted to RMB7.50 trillion in Guangdong, representing 5.53 per cent of GDP; RMB3.68 trillion in Shanghai, or 7.78 per cent of GDP; and RMB7.66 trillion in Jiangsu, or 5.98 per cent of GDP. Meanwhile, special debt totalled RMB19.69 trillion in Guangdong, or 14.51 per cent of GDP; RMB5.16 trillion in Shanghai, or 10.92 per cent of GDP; and RMB15.06 trillion in Jiangsu, or 11.75 per cent of GDP.

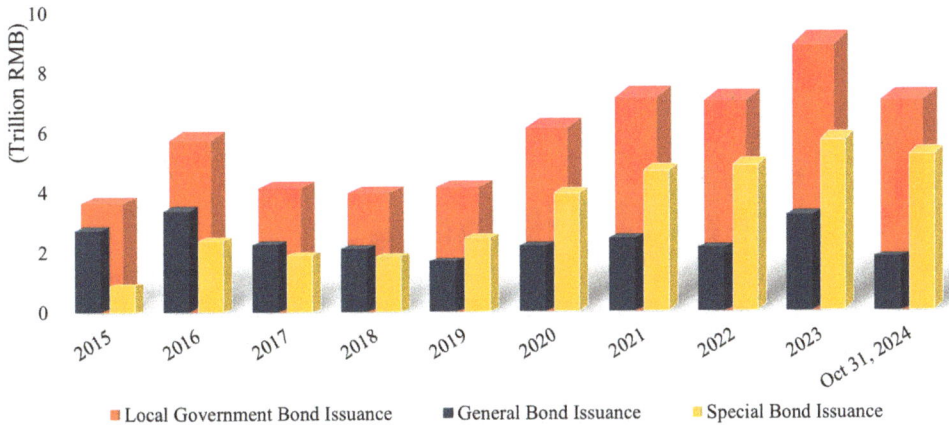

Figure 9.8: Local government bond issuance in China, 2015–24

Source: Local Government Bond Information Disclosure Platform (www.celma.org.cn).

LGB issuance in China has shown a generally increasing trend, albeit with fluctuations. Figure 9.8 illustrates LGB issuance since 2015, revealing an upward trajectory. In 2015, the total LGB issuance amounted to RMB3.84 trillion, including RMB2.86 trillion in general LGBs and RMB970 billion in special LGBs. By 2023, total LGB issuance had risen to RMB9.34 trillion, comprising RMB3.35 trillion in general LGBs and RMB5.99 trillion in special LGBs. From January to October 2024, total LGB issuance amounted to RMB7.38 trillion, including RMB1.90 trillion in general LGBs and RMB5.49 trillion in special LGBs.

Regional disparities in LGB issuance are evident across China. Figure 9.9 illustrates the regional variations in the on-budget LGB issuance ratio for 2023. The provinces with the highest LGB issuance ratios were Tianjin, Jilin and Qinghai, with total LGB issuance of RMB361.2 billion in Tianjin, accounting for 21.58 per cent of regional GDP; RMB2.38 trillion in Jilin, or 17.58 per cent of regional GDP; and RMB57.0 billion in Qinghai, or 15.00 per cent of regional GDP. In these provinces, general bond issuance reached RMB116.5 billion in Tianjin, or 6.96 per cent of GDP; RMB116.2 billion in Jilin, or 8.59 per cent of GDP; and RMB46.0 billion in Qinghai, or 12.11 per cent of GDP. Meanwhile, special bond issuance amounted to RMB244.7 billion in Tianjin, or 14.62 per cent of GDP; RMB8.99 billion in Jilin (2.90 per cent); and RMB11.0 billion in Qinghai (2.89 per cent). In contrast the provinces with the lowest LGB issuance ratios were Beijing, RMB171.3 billion, or 3.91 per cent of GDP; Jiangsu, RMB458.0 billion (3.57 per cent); and Shanghai, RMB112.0 billion (2.37 per cent). General bond issuance accounted for RMB37.4 billion in Beijing, representing 0.85 per cent of GDP; RMB134.7 billion in Jiangsu (1.05 per cent); and RMB52.1 billion in Shanghai (1.10 per cent). Meanwhile, special bond issuance totalled RMB133.9 billion in Beijing, or 3.06 per cent of GDP; RMB323.3 billion in Jiangsu (2.52 per cent); and RMB59.8 billion in Shanghai (1.27 per cent).

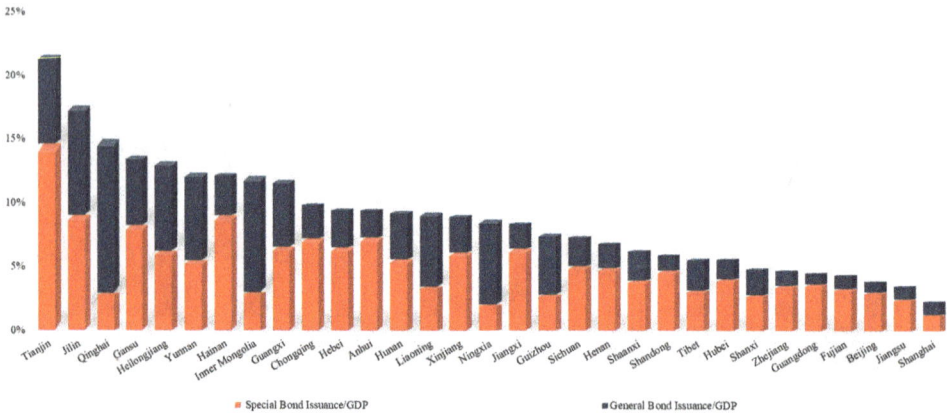

Figure 9.9: Regional distribution of local government bond issuance in China in 2023
Source: Local Government Bond Information Disclosure Platform (www.celma.org.cn).

In addition to general and special debt, local government fiscal deficits also constitute a component of on-budget local government debt, as these deficits are formally recorded in government accounts and reflect structural funding gaps where fiscal revenues are insufficient to meet expenditure demands, thus are legally required to be repaid through government debt instruments or other financing mechanisms. The primary drivers of local fiscal deficits can be summarised as follows: first, the reduction in tax revenue. As a fundamental pillar of fiscal income, tax revenue has undergone a significant contraction due to the sustained implementation of tax cuts and fee reduction policies in recent years. The ratio of tax revenue to GDP in China declined from 18.45 per cent in 2015 to 14.36 per cent in 2023.[14] This downward trend has substantially amplified local fiscal deficits, undermining the ability of local governments to meet their financial obligations. Second is the decline in revenue from land-use rights transfers. Traditionally, revenue from this source has served as an essential supplementary income stream for local governments, alleviating fiscal pressures and supporting expenditure needs. However, this revenue has experienced a sharp decline in recent years. In 2023, revenue from state-owned land-use rights amounted to RMB5.8 trillion, representing a 31.8 per cent decrease from its peak in 2021.[15] This contraction has further exacerbated fiscal shortfalls, reflecting both a cooling real estate market and the diminishing efficacy of land-based financing strategies. Third, the significant increase in government expenditure due to the Covid-19 pandemic. According to estimates by Chong'en Bai, the pandemic resulted in a cumulative local government fiscal deficit exceeding RMB4 trillion over three years. Specifically, local deficits were RMB2.08 trillion in 2020, RMB988.8 billion in 2021 and RMB1.19 trillion in 2022. The financial pressures stemming from pandemic-related expenditures have directly contributed to the substantial growth of local fiscal deficits, ultimately increasing on-budget local government debt.[16]

---

14   The data are sourced from the official website of the Chinese Government (english.www.gov.cn/).

15   ibid.

16   The data were presented by Chong'en Bai, a member of the Academic Committee of the China Economist 50 Forum, during the 403rd session of the Chang'an Forum on 21 September 2023.

In recent years, although the scale of on-budget local government debt has increased, it remains within a manageable range of risk. This is primarily due to the high transparency of on-budget debt, as the Chinese Government has established clear regulations governing the scale of on-budget debt financing, which stipulate that local governments may only borrow within prescribed limits, as well as the arrangements for debt repayment.

Off-budget local government debt

As indicated in section two, off-budget debt includes implicit local government debt and general LGFV debt. They pose a significant financial risk within China's local government debt framework. They are not included in official government debt statistics and there is a lack transparency (Bai et al. 2016; Chen et al. 2020), creating substantial challenges for government departments to effectively supervise and manage them. Their concealed nature makes it difficult to accurately assess their scale.

Several studies have sought to estimate the scale of off-budget implicit local government debt using various methodologies, yielding significantly varied results. A common approach involves using the debt of LGFVs as a proxy for implicit debt. For instance, B. Zhang et al. (2018) and X.J. Zhang et al. (2018) estimated the off-budget implicit local government debt to be approximately RMB39.25 trillion by the end of 2017. Similarly, Liang and Liu (2019) reported that, by the end of 2018, off-budget implicit local government debt had surpassed RMB30 trillion. Another methodology for estimating this debt involves analysing various debt sources (for example, LGFV debt, debt from local SOEs, government-paid PPP projects and collateralised supplementary loans) and debt allocation. Mao et al. (2018) applied this approach and estimated that, as of 2016, the scale of off-budget implicit local government debt ranged between RMB21 trillion and RMB30.5 trillion, based on these debt sources. Li and Zhang (2024) refined this estimate further, projecting that implicit local government debt reached RMB59.83 trillion by 2020, considering debt allocation patterns. In addition to these domestic estimates, the IMF (2022) provided its own evaluation, estimating that the scale of implicit local government debt stood at RMB48.6 trillion by the end of 2020. Furthermore, data from the Bank for International Settlements reveal that the nominal value of government sector credit in China reached RMB68.05 trillion by the end of 2020. Taking into account the Ministry of Finance's report that China's government debt balance stood at RMB46.55 trillion by the same period, the scale of implicit local government debt for 2020 can be further estimated at RMB21.50 trillion (BIS n.d.). Additionally, after comprehensive project identification and multi-level reviews, the Ministry of Finance reported that the national balance of implicit local government debt stood at RMB14.3 trillion by the end of 2023.[17]

---

17 These implicit debt data include obligations explicitly guaranteed and acknowledged by the government, while numerous implicit debts lacking government guarantees remain unrecognised. The data are sourced from NPC Finance and Economic Committee (2024).

The implicit local government debt is not the total of local governments' potential off-budget debt; all the debt incurred by LGFVs, regardless of whether it is guaranteed by local governments, could have to be paid back by local governments. This debt mainly comprises loans obtained from commercial banks, corporate bonds issued by LGFVs in the capital markets and funds obtained by non-standard financing tools (for example, trust loans financing, direct financing from non-bank financial institutions or investors, wealth management direct financing, factoring and financing leases). Here, we focus on the interest-bearing liabilities of LGFVs because relatively complete data exist for them. Note that these liabilities overlap with the implicit government debt.

Overview of the interest-bearing debt of LGFVs

The interest-bearing debt of LGFVs has experienced significant and sustained growth. As depicted in Figure 9.10, the total interest-bearing debt of LGFVs rose from RMB20.09 trillion in 2015 to RMB61.56 trillion in 2023.[18] In terms of financing structure, bank loans have consistently remained the primary source of capital for LGFVs. From 2015 to 2023, bank loans accounted for more than 50 per cent of the total financing. Specifically, in 2023, bank loans to LGFVs amounted to RMB41.32 trillion, representing 63.92 per cent of the total interest-bearing debt. Bond financing has emerged as the second-largest source of capital, contributing more than 27 per cent of the total interest-bearing debt since 2015. In 2023, bond financing reached RMB15.55 trillion, accounting for 27 per cent of the total debt. Non-standard financing had consistently accounted for more than 5 per cent of the total interest-bearing debt between 2015 and 2023. In 2023, non-standard financing totalled RMB4.69 trillion, constituting 5.72 per cent of the overall debt.

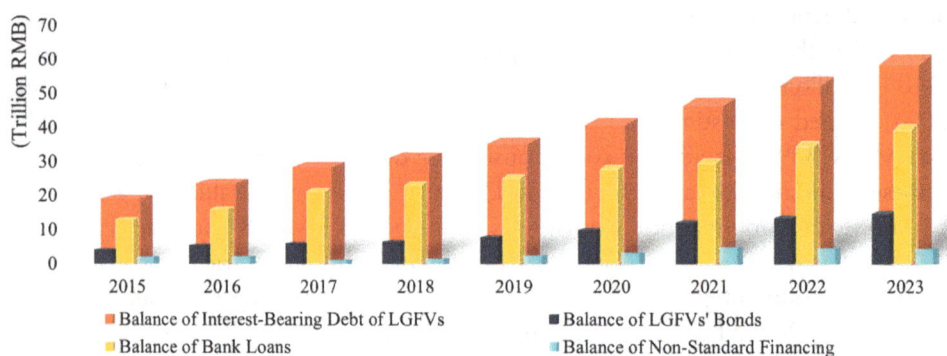

Figure 9.10: Interest-bearing debt balance of LGFVs, 2015–23
Source: Enterprise Early Warning System (www.qyyjt.cn).

---

18   The interest-bearing debt of LGFVs here is based on the debt data disclosed by enterprises through the Enterprise Early Warning Platform. However, a significant portion of LGFVs' interest-bearing debt remains unreported, primarily due to the opacity of many debt obligations. This includes debts incurred through PPP projects, entrusted loans, financing leases and other similar financial arrangements.

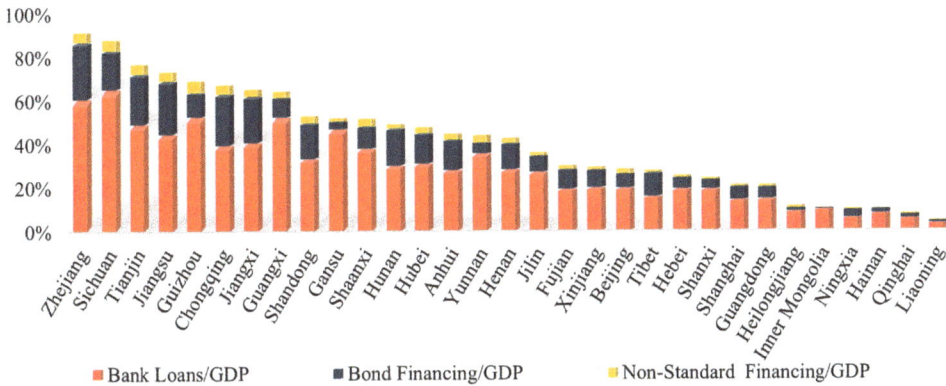

Figure 9.11: Interest-bearing debt balances of regional LGFVs in China in 2023

Source: Wind database (www.wind.com.cn/).

The interest-bearing debt of LGFVs also shows considerable regional variation. As depicted in Figure 9.11, in 2023, the provinces of Zhejiang, Guangdong and Tianjin recorded the highest interest-bearing debt ratios, with total debt of RMB7.57 trillion, RMB5.31 trillion and RMB1.29 trillion, respectively. These figures represented 91.75 per cent, 88.28 per cent and 76.96 per cent of their respective regional GDPs. Financing through bank loans accounted for RMB5.02 trillion in Zhejiang, or 60.88 per cent of regional GDP; RMB3.93 trillion in Guangdong (65.45 per cent); and RMB820 billion in Tianjin (49.31 per cent). Bond financing amounted to RMB2.14 trillion in Zhejiang (25.97 per cent of regional GDP); RMB1.06 trillion in Guangdong (17.57 per cent); and RMB390 billion in Tianjin (23.12 per cent). Non-standard financing totalled RMB405.19 billion in Zhejiang (4.91 per cent of regional GDP); RMB315.65 billion in Guangdong (5.25 per cent); and RMB75.84 billion in Tianjin (4.53 per cent). In contrast, the provinces exhibiting the lowest interest-bearing debt ratios were Hainan, with total debt of RMB75.70 billion, or 10.02 per cent of regional GDP; Qinghai, with RMB29.89 billion (7.87 per cent); and Liaoning, with RMB141.66 billion (4.69 per cent). Bank loans amounted to RMB59.49 billion in Hainan, RMB20.71 billion in Qinghai and RMB99.53 billion in Liaoning, representing 7.88 per cent, 5.45 per cent and 3.29 per cent of regional GDP, respectively. Bond financing in these provinces totalled RMB15.46 billion in Hainan (2.05 per cent); RMB6.41 billion in Qinghai (1.69 per cent); and RMB23.28 billion in Liaoning (0.77 per cent). Non-standard financing in these regions amounted to RMB740 million in Hainan (0.10 per cent); RMB2.76 billion in Qinghai (0.73 per cent); and RMB18.84 billion in Liaoning (0.62 per cent).

Overview of LGFV bond issuance

LGFV bonds are corporate bonds issued by LGFVs that have implicit government guarantees (Liu et al. 2017). Bond issuances of LGFVs in China have exhibited a consistent upward trend, albeit with minor fluctuations over time. As depicted in Figure 9.12, the issuance trajectory of LGFV bonds from 2015 to 2023 demonstrates sustained growth, except for declines in 2017 and 2022. The total bond issuance increased from RMB1.48 trillion in 2015 to a peak of RMB6.68 trillion in 2023.

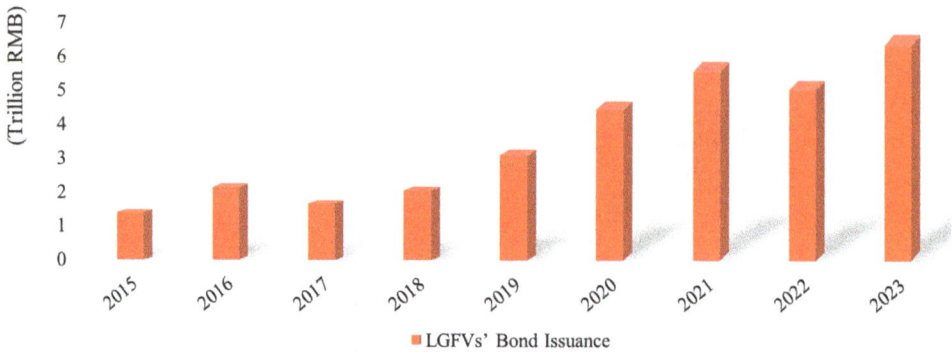

Figure 9.12: LGFV bond issuance, 2015–23

Source: Enterprise Early Warning System.

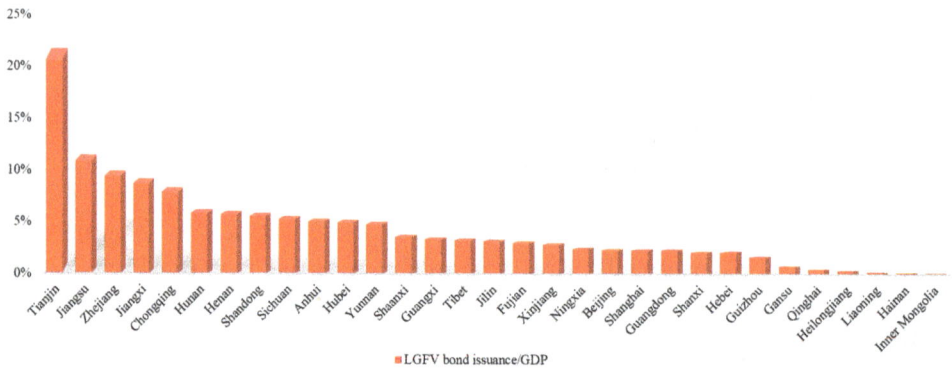

Figure 9.13: Regional LGFV bond issuance in 2023

Source: Enterprise Early Warning System.

From a regional perspective, provinces with stronger economies and more robust fiscal capacities tend to exhibit higher LGFV bond issuance volumes, whereas less-developed provinces generally show lower issuance volumes. As shown in Figure 9.13, in 2023, Tianjin, Jiangsu and Zhejiang ranked as the top three provinces in terms of LGFV bond issuance relative to regional GDP, with issuance amounts of RMB360 billion, RMB1.47 trillion and RMB820 billion, respectively. These amounts accounted for 5.43 per cent, 21.94 per cent and 12.24 per cent of the national total LGFV bond issuance (RMB6.68 trillion) and represented 21.70 per cent, 11.43 per cent and 9.91 per cent of their respective regional GDPs. In contrast, Liaoning recorded issuances of RMB69.60 billion, accounting for only 0.10 per cent of the national total LGFV bond issuance; Hainan recorded RMB12.00 billion (0.02 per cent); and Inner Mongolia recorded RMB35.00 billion (0.05 per cent). These amounts represented 21.10 per cent, 11.43 per cent and 9.91 per cent of their respective regional GDPs.[19]

---

19 In 2023, the GDP of Tianjin was RMB17.21 trillion; Jiangsu, RMB128.22 trillion; Zhejiang, RMB82.55 trillion; Liaoning, RMB30.21 trillion; Hainan, RMB7.55 trillion; and Inner Mongolia, RMB24.63 trillion.

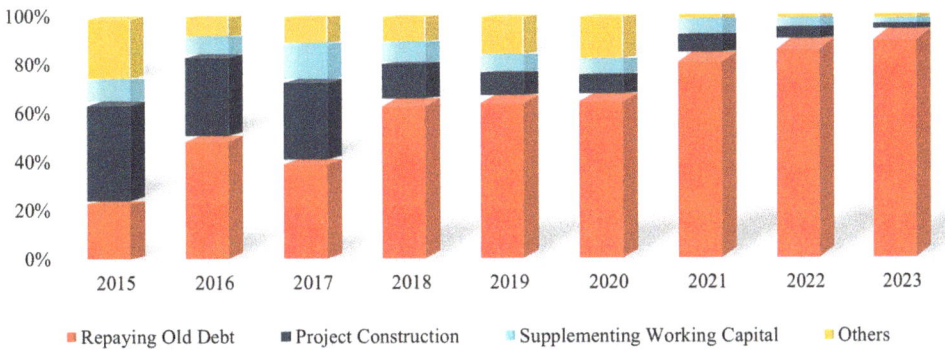

Figure 9.14: Fund allocation of LGFV bond issuance, 2015–23
Source: Enterprise Early Warning System.

The allocation of funds raised through LGFV bond issuance has undergone substantial changes in recent years, with a notable shift towards increasing the proportion directed to debt repayment. LGFVs reveal the fund allocation of bond proceeds in their prospectuses, and these funds can be classified into four primary categories: 1) debt repayment, which includes the settlement of maturing LGFV bonds, as well as the principal and interest on bank loans, non-standard borrowings and other financial obligations of LGFVs; 2) project construction such as rebuilding shanty areas or constructing infrastructure; 3) supplementing operating capital; and 4) others such as financing for other entities through entrusted loans. Figure 9.14 provides a detailed analysis of the allocation of funds from LGFV bond issuances between 2015 and 2023. The data clearly indicate a shift towards debt refinancing, with the proportion of funds allocated to this purpose rising from 23.91 per cent in 2015 to 94.00 per cent in 2023. In contrast, the share of funds allocated to project construction has steadily declined, from 40.87 per cent in 2015 to only 2.31 per cent in 2023. Similarly, the proportion of funds used to supplement operating capital has decreased from 9.13 per cent in 2015 to 1.93 per cent in 2023. Moreover, the allocation of funds to other purposes has decreased significantly, from 26.09 per cent in 2015 to 1.76 per cent in 2023. These trends highlight a growing reliance on bond issuance to service existing liabilities, accompanied by a substantial reduction in funding for new projects and operational needs.

Regionally, certain provinces allocate a larger proportion of the funds raised through LGFV bond issuance to the repayment of existing debt. As shown in Figure 9.15, all provinces allocate at least 70.03 per cent of the raised funds for debt repayment. Among them, provinces such as Ningxia, Jilin, Qinghai, Hainan, Tianjin, Guizhou, Liaoning, Yunnan and Tibet allocate close to 100 per cent of the raised funds to debt repayment. Additionally, Guangxi allocates 99.27 per cent to debt repayment; Hebei, 97.91 per cent; Jiangsu, 97.71 per cent; Hunan, 97.25 per cent; and Henan, 95.74 per cent. This pattern highlights the significant debt burdens in these regions and emphasises their increasing reliance on bond issuance as a key mechanism for managing and servicing growing debt obligations.

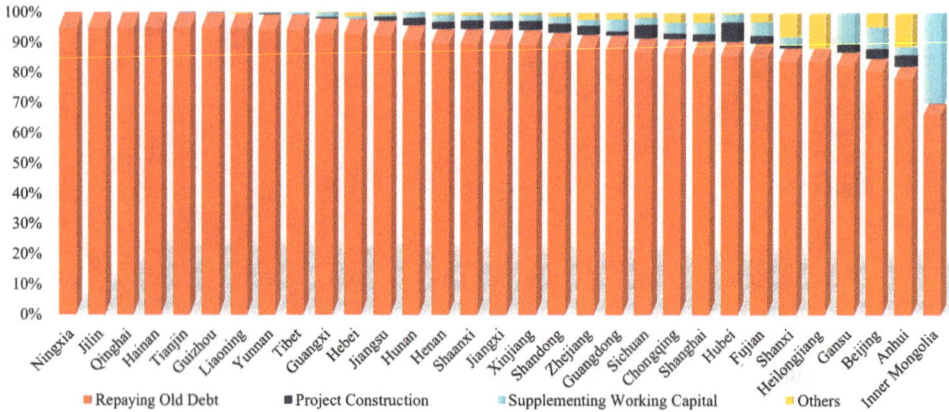

Figure 9.15: Regional fund allocation of LGFV bond issuances in 2023

Source: Enterprise Early Warning System.

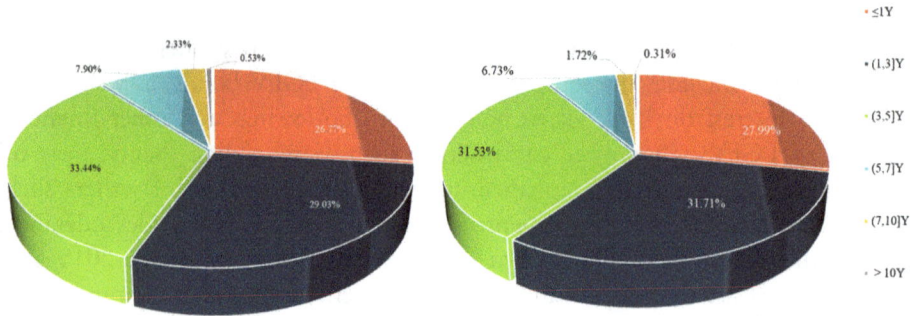

Figure 9.16: Distribution of LGFV bond maturity in 2023

Notes: Funds distribution by LGFVs bond issuance maturity (left) and distribution of LGFVs by issuance maturity (right).

Source: Enterprise Early Warning System.

In terms of maturity, most bonds are concentrated in the range of three to five years. As shown in Figure 9.16, in 2023, bonds with a maturity of three to five years accounted for the largest proportion of funds raised, representing 33.44 per cent of the total issuance amount. The corresponding issuance quantity was 31.53 per cent of the total number of issuances. Bonds with a maturity of one to three years ranked second, raising 29.03 per cent of the total amount, with their issuance quantity making up 31.71 per cent of the total issuances. Bonds with a maturity of less than one year followed, representing 26.77 per cent of the total issuance value and 27.99 per cent of the total number of issuances. Bonds with maturities between five and seven years accounted for 7.99 per cent of the total issuance value, with their corresponding issuance quantity representing 6.73 per cent of the total. Bonds with maturities exceeding seven years represented the smallest share, accounting for only 2.86 per cent of the total issuance value and 2.03 per cent of the total issuance quantity.

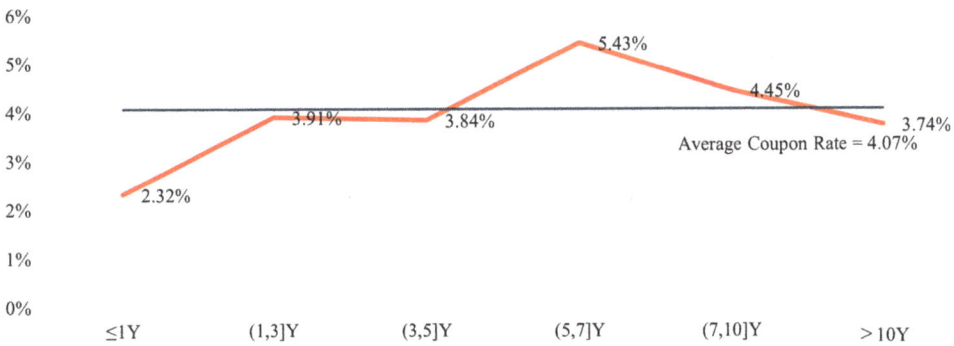

Figure 9.17: Distribution of coupon rates for LGFV bond issuance in 2023
Source: Enterprise Early Warning System.

The coupon rate for LGFV bond issuance has consistently remained above 2.3 per cent. As shown in Figure 9.17, the average coupon rate for LGFV bond issuance in 2023 was 4.07 per cent. A breakdown by maturity indicates that the weighted average coupon rate is 2.32 per cent for bonds with maturities of less than one year, 3.91 per cent for those with maturities of one to three years, 3.84 per cent for three to five years, 5.43 per cent for five to seven years, 4.45 per cent for seven to 10 years and 3.74 per cent for those exceeding 10 years.

The central government's debt-swap programs

The indebtedness of local governments has profound implications for the country's financial security and economic stability. To assess the scale of local government debt, the central government conducted two comprehensive audits between 2011 and 2013. The first audit, conducted in 2011, revealed that the total on-budget explicit local government debt amounted to RMB10.72 trillion. This figure included RMB6.71 trillion with direct repayment obligations, RMB2.34 trillion related to guarantee obligations and RMB1.67 trillion potentially subject to bailout measures. In parallel, the off-budget implicit local government debt induced by LGFVs was quantified at RMB4.97 trillion. Of this, RMB3.14 trillion was classified as debt with repayment obligations, while RMB1.83 trillion was considered contingent liabilities, collectively representing 46.38 per cent of the on-budget explicit local government debt (NAO 2011).

The second audit, conducted from November 2012 to February 2013, revealed a substantial increase in both on-budget and off-budget local government debt. On-budget explicit debt had surged to RMB17.89 trillion, with RMB10.89 trillion classified as debt with repayment obligations, RMB2.67 trillion as debt with guarantee obligations and RMB4.34 trillion as debt that could potentially require government intervention. Similarly, the off-budget implicit debt induced by LGFVs had risen to RMB6.97 trillion, with RMB4.08 trillion categorised as debt with repayment obligations and RMB2.89 trillion as contingent liabilities. This off-budget debt accounted for 38.96 per cent of the on-budget explicit debt (NAO 2013).

These audit results highlight the persistent and accelerating rise in local government debt, particularly the growing volume of off-budget implicit debt generated through LGFVs. In response to these mounting challenges, the central government conducted several rounds of debt swaps.

Debt screening and the first debt-swap program from 2015 to 2018

In October 2014, the central government introduced the *Measures for the Clearance and Screening of Local Government Outstanding Debt into Budget Management*, which mandated a comprehensive assessment and classification of local government debt (MoF 2014). This initiative required local governments to integrate their outstanding debt into formal budgetary management frameworks. By the end of 2014, the total outstanding debt of local governments nationwide amounted to RMB15.40 trillion. Of this, RMB1.06 trillion had been authorised for LGB issuance, while RMB14.34 trillion was raised through nongovernment financing channels, such as bank loans, trust financing and LGFV bond issuance (General Office of the State Council 2018).

To address the escalating debt burden, the revised Budget Law, effective 1 January 2015, authorised local governments to issue LGBs within the central government's prescribed debt limits. This revision allowed local governments to swap off-budget implicit debt with more transparent on-budget debt instruments, primarily LGBs. As shown in Figure 9.18, for each year between 2015 and 2018, the amount of LGBs for the swap purposes was RMB3.24 trillion, RMB4.88 trillion, RMB2.77 trillion and RMB1.32 trillion, respectively, accumulating to a total of RMB12.2 trillion. This debt-swap program converted virtually all off-budget implicit local government debt into LGBs with higher credit ratings, thereby reducing the interest burden on local governments. Moreover, this restructuring extended the maturity profile of the outstanding debt, further lessening local governments' repayment burden. By the end of 2018, the outstanding balance of off-budget implicit local government debt had been reduced to RMB315.1 billion, successfully achieving the key objectives of the debt-swap initiative (Xinhua 2016).

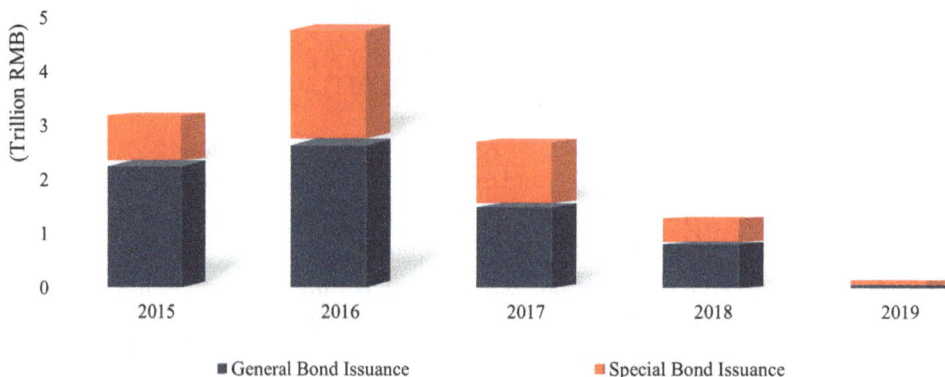

Figure 9.18: Local government bond issuance for debt swaps, 2015–19
Source: Wind database (www.wind.com.cn/).

The pilot program for small-scale implicit debt resolution

Despite the effort of the first round of debt swaps, local government debt continued to rise significantly during this period. By the close of 2016, the outstanding debt of local governments had reached RMB15.32 trillion, marking an increase of more than RMB4 trillion compared with 2011 (NAO 2017). Two primary factors contributed to this upward trend. First, although a comprehensive debt screening was conducted in late 2014, certain local government debts were excluded from official statistics. These uncounted debts remained the responsibility of local governments, thus creating potential future repayment or bailout obligations. Second, while the revised Budget Law established clear regulations governing government borrowing, local governments continued to engage in unregulated borrowing through newly established public institutions and LGFVs. This led to further accumulation of off-budget implicit debt, exacerbating the fiscal risk.

In response to the escalating levels of local government debt, a pilot program was launched in November 2019 to resolve implicit debt again by way of debt swaps. The program specifically targeted fiscally vulnerable counties in six provinces: Guizhou, Yunnan, Hunan, Gansu, Inner Mongolia and Liaoning. Operating within the central debt issuance limits, the program resulted in the issuance of RMB157.92 billion in LGBs across seven provinces (see Figure 9.19). Of this total, Jiangsu Province—although not initially included in the program—accounted for RMB15 billion, representing 24.27 per cent of the total issuance, while the remaining six provinces collectively issued RMB142.92 billion. The provincial breakdown is as follows: Guizhou, RMB38.32 billion (23.15 per cent); Yunnan, RMB22.62 billion (14.38 per cent); Hunan, RMB36.56 billion (14.32 per cent); Gansu, RMB5.97 billion (10.60 per cent); Inner Mongolia, RMB22.71 billion (9.50 per cent); and Liaoning, RMB16.74 billion (3.78 per cent) (see Figure 9.19).

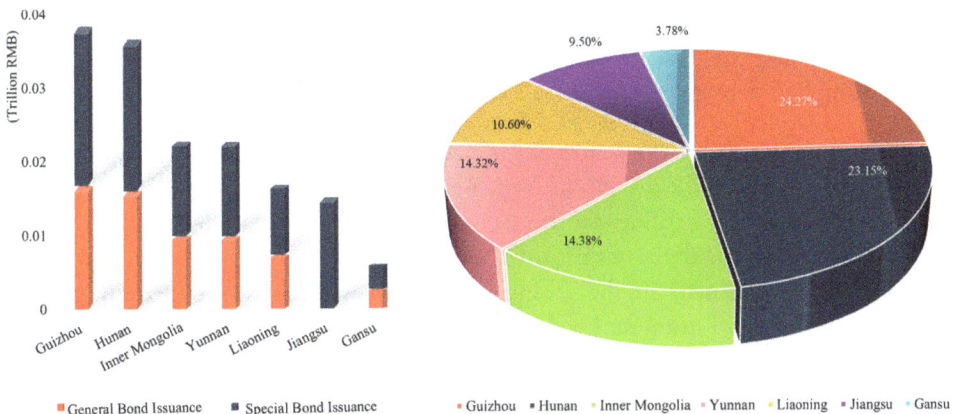

Figure 9.19: Local government bond issuance for the 2019 debt-swap program
Source: Wind database (www.wind.com.cn/).

The primary objective of this pilot program was to alleviate the implicit debt burden faced by financially vulnerable counties and cities, thereby mitigating fiscal risks and promoting regional economic stability. By converting implicit debt into more transparent and manageable LGBs, the program aimed to enhance fiscal transparency, improve debt sustainability and facilitate more efficient resource allocation within these regions.

Expansion of debt swaps and the 'no implicit debt' initiative

The Covid-19 pandemic significantly disrupted Chinese enterprises, leading to widespread production halts and, in some cases, permanent closures. This disruption resulted in a substantial decline in corporate income tax, value-added tax and other business-related revenues, thereby causing a sharp reduction in the overall tax revenue collected by the Chinese Government. Concurrently, the government increased public spending to ensure the wellbeing of citizens, including investments in nationwide nucleic acid testing, vaccination campaigns and the establishment and management of quarantine facilities. As a result, local governments faced mounting fiscal pressures, compelling them to resort to borrowing to address these financial difficulties, which contributed to a rise in local government debt (Oi et al. 2025; Wong 2025).

In response to these challenges, between December 2020 and September 2021, the government expanded its debt resolution pilot program, extending it from incorporated counties to 26 regions, and issuing special refinancing bonds (a type of LGB) to swap implicit debt.[20] During this period, RMB612.8 billion in special refinancing bonds were issued (see Figure 9.20). From October 2021 to June 2022, the government further advanced its efforts by implementing 'no implicit debt' pilot programs across entire regions, beginning with Beijing, Shanghai and Guangdong. In this phase, special refinancing bonds amounting to RMB325.23 billion were issued in Beijing, RMB65.48 billion in Shanghai and RMB113.48 billion in Guangdong, totalling RMB504.18 billion across the three areas (see Figure 9.20). By the end of the initiative, several regions, including the initial three, formally announced the successful completion of their 'no implicit debt' pilot programs.

This round saw a notable expansion in both the scope and the scale of debt resolution efforts. The introduction of regional 'no implicit debt' pilots in major cities marked a critical milestone, signalling the beginning of a new stage of debt management. This phase represents a decisive shift towards achieving a state of 'zero implicit debt'— a significant achievement in the Chinese Government's ongoing efforts to address implicit debt risks and promote fiscal transparency.

---

20 The distinctive feature of special refinancing bonds lies in the fact that the raised funds are not only allocated for the repayment of maturing debt, but also extended to address local governments' existing debt stock, including non-standard forms of off-budget implicit local government debt, such as debt associated with LGFVs and arrears owed to enterprises.

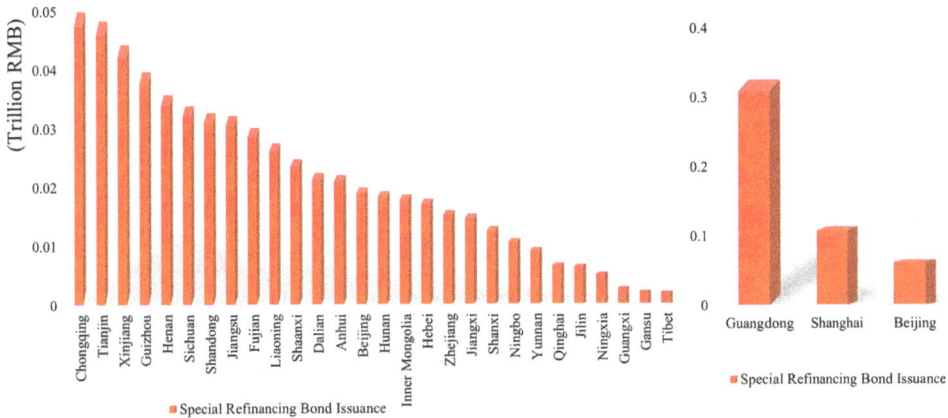

Figure 9.20: Local government bond issuance for debt swaps, December 2020 – June 2022

Source: Wind database (www.wind.com.cn/).

Strengthening of implicit debt resolution

Since 2022, local governments in China have faced severe financial challenges due to the combined effects of the Covid-19 pandemic and the downturn in the real estate sector. This has resulted in significant declines in tax revenues and land transfer income, severely reducing financial resources and solvency. Concurrently, local governments and LGFVs have accumulated substantial unpaid, interest-free debt, including supplier accounts, construction deposits, salaries and public sector allocations, exacerbating implicit debt risks and making its resolution a priority for both central and local authorities.

In response, the central government introduced a 'comprehensive debt reduction plan' in July 2023, marking the beginning of a new round of debt resolution efforts, focusing on three key areas. First, from October 2023 to September 2024, local governments issued special refinancing bonds and new special LGBs (as supplementary debt instruments) to address outstanding debt.[21] High-debt (that is, priority) provinces predominantly issued special refinancing bonds to resolve outstanding debt, while low-debt (non-priority) provinces primarily issued new special LGBs. The issuance of special refinancing bonds totalled RMB1.1 trillion in high-debt provinces and RMB400.5 billion in low-debt provinces, while the issuance of new special LGBs amounted to RMB294.4 billion in high-debt provinces and RMB819.9 billion in low-debt provinces (see Figure 9.21). Second, the central government increased the debt ceiling by more than RMB2.2 trillion for 2023 and committed to an additional RMB1.2 trillion for 2024. These

---

21 The new special government bonds represent a distinct category of local government bonds, characterised by the absence of detailed disclosures regarding the specific allocation of the funds raised for investment projects. Furthermore, these bonds do not provide key documents typically required for transparency, such as the project implementation plan, financial audit report and legal opinion. This lack of disclosure raises concerns about the level of transparency and accountability in the use of the funds.

measures aim to facilitate the issuance of LGBs to swap implicit debts and address outstanding arrears owed to enterprises (Shen 2024). Third, in November 2024, the government introduced the '6+4+2' collaborative debt resolution plan, targeting the replacement of RMB14.3 trillion in implicit debt and achieving 'zero implicit debt'. This includes: 1) a RMB6-trillion increase in the local debt ceiling from 2024 to 2026, with RMB2 trillion issued annually to swap existing implicit debt; 2) a RMB4-trillion allocation over five years to replace implicit debt with RMB800 billion in new special LGBs annually; and 3) RMB2 trillion in continued repayment of debt related to shantytown renovation, due after 2029, in accordance with the original contractual terms. It is projected that, by 2028, the total implicit debt of local governments will decline substantially, from RMB14.3 trillion to RMB2.3 trillion.

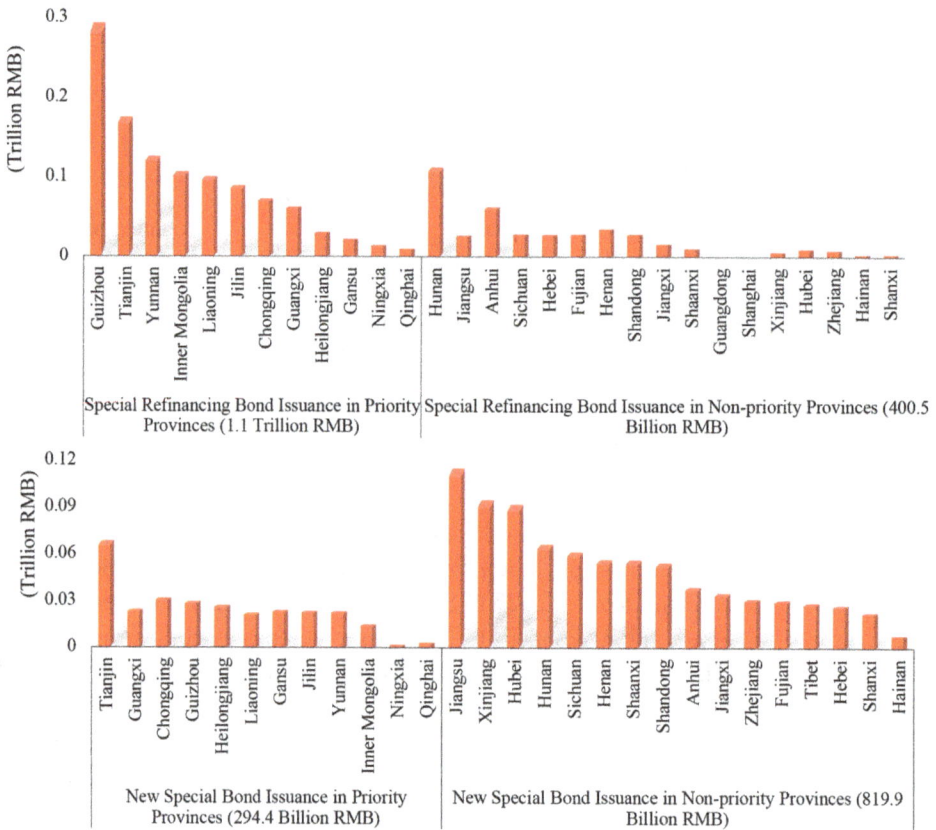

Figure 9.21: Local government bond issuance for debt swaps in priority provinces versus non-priority provinces, October 2023 to September 2024

Source: Wind database (www.wind.com.cn/).

Conclusion and policy implications

Conclusion

Local government debt has played an indispensable role in driving China's economic growth. However, the rapid accumulation of debt, particularly through off-budget financing mechanisms such as LGFVs, has led to significant systemic risks. The issue of local government debt in China has been the focus of scholarly attention. To this end, we have systematically examined the drivers and impacts of local debt expansion, provided a detailed overview of its current developmental characteristics and scale and reviewed the central government's management of local debt. Our analysis reveals that the current debt resolution paradigm inadequately addresses the intertemporal fiscal stress. To address the challenges posed by local government debt, we propose a comprehensive three-step debt resolution strategy.

Policy recommendations

Step 1: Enhancing fiscal relief for local governments

The fiscal conditions of local governments have significantly deteriorated due to the Covid-19 pandemic, posing a major threat to regional economic stability. Between 2019 and 2022, local governments in China recorded a cumulative fiscal deficit of RMB4 trillion. However, the scale of existing debt resolution plans is inadequate to effectively alleviate the growing fiscal pressure, leading to sharp reductions in public expenditures and declining quality of public service provision. This, in turn, may distort local government behaviour, such as increasing burdens on enterprises or adopting non-market mechanisms to fill fiscal gaps, exacerbating market uncertainty. To address the issue, the central government should issue special bonds to bail out local governments, easing their short-term fiscal difficulties, ensuring the fulfilment of public service responsibilities and stabilising market confidence. The central government has taken steps to bail out local governments. Premier Li Qiang announced a plan of RMB4.4 trillion in special bonds for local governments at the 2025 National People's Congress. Local governments can use the funds to pay their wage and business arrears.

Step 2: Resolving the stock of local government debt

To deal with the mounting stock of local government debt, the central government should move all LGFV debts to new or existing asset management companies (AMCs). For distressed but promising projects or platform company debts, AMCs could facilitate debt restructuring by negotiating adjustments to debt terms, such as extending repayment periods, lowering interest rates or converting debt to equity. For debts backed by liquid assets, public auctions, listings or private sales could be employed to monetise assets and repay the debts. For assets generating stable cash flows (for example, toll roads, bridges, wastewater treatment plants), AMCs could securitise future revenue streams into marketable financial products, raising funds in advance. AMCs could revitalise large fixed assets such as factories or equipment through leases or joint operations, securing rental income or profit-sharing agreements with capable enterprises.

On the other hand, local governments should clarify debt liabilities and share losses with other stakeholders. Local debt formation is not solely a result of government actions; financial institutions often contributed by overestimating local governments' repayment capacity or inadequately assessing risks. For debts that cannot be fully resolved, the debt chains should be thoroughly audited and liabilities clearly identified. A 'mutual offset' mechanism could be implemented to clear debts through reciprocal write-offs. Stakeholders should bear their respective responsibilities and share part of the debt losses with local governments. This approach not only strengthens risk awareness in future credit decisions but also optimises debt management and enhances financial market discipline.

Step 3: Preventing the generation of new debt

The Chinese Government should accelerate the reform announced by the Third Plenum of the Twentieth Party Congress to create an asset balance sheet for every level of government. This would help governments comprehensively understand economic conditions, develop informed policies and strengthen local debt management. This system should cover enterprises, banks and governments, incorporating all types of local government assets and liabilities to ensure transparency and dynamic updates. Specifically, commercial and public welfare debts should be unified within the budget and final accounts system to improve financial transparency. Standardised statistical and disclosure guidelines should ensure data comparability and consistency, with regular public disclosures creating a transparent debt management environment. Additionally, incentives and penalties should be implemented, linking the quality of debt disclosure to government performance evaluations and holding local governments accountable for incomplete or misleading information disclosures.

The asset balance sheet will strengthen the central government's monitoring of local government debt. A comprehensive, normalised monitoring framework for local government debt should be established, integrating implicit and statutory debts under unified supervision. Collaborative monitoring mechanisms should be improved, involving financial, banking, state-owned asset, auditing, legislative and disciplinary inspection bodies. Auditing agencies should enhance scrutiny of debt fund usage, while legislative bodies must reinforce monitoring throughout the debt lifecycle to ensure compliance and transparency. Violations leading to debt issues should face stringent penalties and responsible parties be held accountable. Highlighting and exposing cases can reinforce deterrence and serve as a warning.

References

Adelino, Manuel, Igor Cunha, and Miguel A. Ferreira. 2017. 'The Economic Effects of Public Financing: Evidence from Municipal Bond Ratings Recalibration.' *The Review of Financial Studies* 30, no. 9: 3223–68. doi.org/10.1093/rfs/hhx049.

Ang, Andrew, Jennie Bai, and Hao Zhou. 2018. *The Great Wall of Debt: Real Estate, Political Risk, and Chinese Local Government Financing Cost.* PBCSF-NIFR Research Paper No. 15-02. Washington, DC: McDonough School of Business, Georgetown University.

Aschauer, David Alan. 1989. 'Does Public Capital Crowd Out Private Capital?' *Journal of Monetary Economics* 24, no. 2: 171–88. doi.org/10.1016/0304-3932(89)90002-0.

Bai, Chong-En, Chang-Tai Hsieh, and Zheng Michael Song. 2016. *The Long Shadow of China's Fiscal Expansion. Brookings Papers on Economic Activity* 2: 129–81. doi.org/10.1353/eca. 2016.0027.

Bai, Chong-En, and Yingyi Qian. 2010. 'Infrastructure Development in China: The Cases of Electricity, Highways, and Railways.' *Journal of Comparative Economics* 38, no. 1: 34–51. doi.org/10.1016/j.jce.2009.10.003.

Banerjee, Abhijit, Esther Duflo, and Nancy Qian. 2020. 'On the Road: Access to Transportation Infrastructure and Economic Growth in China.' *Journal of Development Economics* 145: 102442. doi.org/10.1016/j.jdeveco.2020.102442.

Bank for International Settlements (BIS). n.d. 'About Debt Securities.' *BIS Statistics.* Basel: Bank for International Settlements. www.bis.org/statistics/dataportal/debt_sec.htm?m=201.

Chang, Jian, Lingxiu Yang, and Yiping Huang. 2013. 'How Big is the Chinese Government Debt?' *China Economic Journal* 6, nos 2–3: 152–71. doi.org/10.1080/17538963.2013.861118.

Chen, Kaiji, Jue Ren, and Tao Zha. 2018. 'The Nexus of Monetary Policy and Shadow Banking in China.' *American Economic Review* 108, no. 12: 3891–936. doi.org/10.1257/aer.20170133.

Chen, Kaiji, and Yi Wen. 2017. 'The Great Housing Boom of China.' *American Economic Journal: Macroeconomics* 9, no. 2: 73–114. doi.org/10.1257/mac.20140234.

Chen, Ting, and James Kai-sing Kung. 2019. 'Busting the "Princelings": The Campaign Against Corruption in China's Primary Land Market.' *The Quarterly Journal of Economics* 134, no. 1: 185–226. doi.org/10.1093/qje/qjy027.

Chen, Zhuo, Zhiguo He, and Chun Liu. 2020. 'The Financing of Local Government in China: Stimulus Loan Wanes and Shadow Banking Waxes.' *Journal of Financial Economics* 137, no. 1: 42–71. doi.org/10.1016/j.jfineco.2019.07.009.

China Electronic Local Government Bond Market Access (CELMA). n.d. *National Local Government Debt Limit.* Beijing: Government Debt Research and Evaluation Centre of the Ministry of Finance. www.celma.org.cn/ndsj/index.jhtml.

Cochrane, John H. 2011. 'Understanding Policy in the Great Recession: Some Unpleasant Fiscal Arithmetic.' *European Economic Review* 55, no. 1: 2–30. doi.org/10.1016/j.euroecorev.2010. 11.002.

Fan, Jianyong, Yu Liu, Qi Zhang, and Peng Zhao. 2022. 'Does Government Debt Impede Firm Innovation? Evidence from the Rise of LGFVs in China.' *Journal of Banking & Finance* 138: 106475. doi.org/10.1016/j.jbankfin.2022.106475.

Gao, Haoyu, Hong Ru, and Dragon Yongjun Tang. 2021. 'Subnational Debt of China: The Politics–Finance Nexus.' *Journal of Financial Economics* 141, no. 3: 881–95. doi.org/10.1016/j.jfineco.2021.05.028.

General Office of the Communist Party of China (CPC) Central Committee. 2006. 'Interim Provisions on Tenure of Leading Party and Government Cadres.' *Xinhua*, 7 August. www.gov.cn/jrzg/2006-08/06/content_355808.htm.

General Office of the State Council. 2018. 'My Country's Local Government Debt Risks Are Generally Controllable.' *People's Daily*, 9 January. www.gov.cn/xinwen/2018-01/09/content_5254545.htm.

Hu, Jiayin, Songrui Liu, Yang Yao, and Zhu Zong. 2022. *Government Deleveraging and the Reverse Crowding-in Effect: Evidence from Subnational Debt and Government Contractors.* National School of Development Working No. Paper 4296428. Beijing: Peking University. doi.org/10.2139/ssrn.4296428.

Huang, Yasheng. 2012. 'How Did China Take Off?' *Journal of Economic Perspectives* 26, no. 4: 147–70. doi.org/10.1257/jep.26.4.147.

Huang, Yi, Marco Pagano, and Ugo Panizza. 2020. 'Local Crowding-Out in China.' *The Journal of Finance* 75, no. 6: 2855–98. doi.org/10.1111/jofi.12966.

Huang, Zhonghua, and Xuejun Du. 2018. 'Holding the Market under the Stimulus Plan: Local Government Financing Vehicles' Land Purchasing Behavior in China.' *China Economic Review* 50: 85–100. doi.org/10.1016/j.chieco.2018.04.004.

International Monetary Fund (IMF). 2022. *Country Report No. 22/22: People's Republic of China.* Washington, DC: IMF. www.imf.org/zh/Publications/CR/Issues/2022/01/26/Peoples-Republic-of-China-Selected-Issues-512253.

IMF. 2024. *Fiscal Monitor: Putting a Lid on Public Debt.* Washington, DC: IMF. www.imf.org/en/Publications/FM/Issues/2024/10/23/fiscal-monitor-october-2024.

Kou, Chien-wen, and Wen-Hsuan Tsai. 2014. '"Sprinting with Small Steps" Towards Promotion: Solutions for the Age Dilemma in the CCP Cadre Appointment System.' *The China Journal* 71, no. 1: 153–71. doi.org/10.1086/674558.

Kroeber, Arthur R. 2016. *China's Economy: What Everyone Needs to Know.* Oxford: Oxford University Press.

Li, D.K., and H. Zhang. 2024. 'A Study on the Scale of Local Government Debt in China.' *Finance and Trade Economics* 45, no. 12: 5–21.

Li, Hongbin, and Li-An Zhou. 2005. 'Political Turnover and Economic Performance: The Incentive Role of Personnel Control in China.' *Journal of Public Economics* 89, nos 9–10: 1743–62. doi.org/10.1016/j.jpubeco.2004.06.009.

Liang, H., and L. Liu. 2019. 'The "Gray Rhino" in China's Balance Sheet: The Massive Debt of Local Financing Platforms.' *New Financial Review*.

Liu, Adam Y., Jean C. Oi, and Yi Zhang. 2022. 'China's Local Government Debt: The Grand Bargain.' *The China Journal* 87, no. 1: 40–71. doi.org/10.1086/717256.

Liu, Laura Xiaolei, Yuanzhen Lyu, and Fan Yu. 2017. 'Local Government Implicit Debt and the Pricing of LGFV Bonds.' *Journal of Financial Research* (Chinese) 12: 170–88, Available at SSRN: doi.org/10.2139/ssrn.2922946.

Lu, Yi, Jin Wang, and Lianming Zhu. 2019. 'Place-Based Policies, Creation, and Agglomeration Economies: Evidence from China's Economic Zone Program.' *American Economic Journal: Economic Policy* 11, no. 3: 325–60. doi.org/10.1257/pol.20160272.

Ludvigson, Sydney. 1996. 'The Macroeconomic Effects of Government Debt in a Stochastic Growth Model.' *Journal of Monetary Economics* 38, no. 1: 25–45. doi.org/10.1016/0304-3932(96)01271-8.

Mao, Z.H., H.X. Yuan, X.H. Liu, Q.F. Wang, and Y.H. Wang. 2018. 'Analysis of Current Local Government Debt Risks and the Transformation of Financing Platforms in China.' Fiscal Science, no. 5: 5–21.

Ministry of Finance (MoF). 2009. *Notice on Accelerating the Implementation of Central Government Investment Projects to Expand Domestic Demand and Local Matching Funds and Other Related Issues*. Beijing: State Council of the People's Republic of China. www.gov.cn/zwgk/2009-10/13/content_1437713.htm.

MoF. 2014. 'Notice on Measures for the Clearance and Screening of Local Government Outstanding Debt into Budget Management.' *State Council Gazette* [2015], no. 6 (23 October). Beijing: State Council of the People's Republic of China. www.gov.cn/gongbao/content/2015/content_2821641.htm.

National Audit Office of China (NAO). 2011. *Results of the National Local Government Debt Audit*. No. 35. Beijing: NAO. www.audit.gov.cn/n5/n25/c63566/content.html.

NAO. 2013. *Results of the National Government Debt Audit*. 30 December. Beijing: NAO. www.gov.cn/gzdt/att/att/site1/20131230/1c6f6506c7f8142b809901.pdf.

NAO. 2017. *Report of the State Council on the Rectification of Problems Found in the Audit of the Implementation of the Central Budget and Other Fiscal Revenues and Expenditures in 2016*. Beijing: NAO. www.audit.gov.cn/n5/n26/c118298/content.html.

NAO. 2023. *The State Council's Audit Report on the Execution of the Central Budget and Other Fiscal Revenues and Expenditures for 2022*. Beijing: National People's Congress of the People's Republic of China. www.npc.gov.cn/c2/c30834/202306/t20230628_430337.html.

National People's Congress (NPC) Finance and Economics Committee. 2024. 'Report of the National People's Congress Finance and Economics Committee on the Review of the State Council's Proposal to Increase the Local Government Debt Limit to Replace the Existing Hidden Debt.' *Xinhua*, 8 November. www.gov.cn/yaowen/liebiao/202411/content_6985597.htm.

Oi, Jean C., Jason M. Luo, and Yunxiao Xu. 2025. 'A Perfect Storm: Fiscal Discipline, COVID, and Local Government Debt in China.' *The China Journal* 93, no. 1: 1–25. doi.org/10.1086/734005.

People's Bank of China (PBOC). 2024. *China Monetary Policy Report Q4 2024*. Beijing: People's Bank of China. www.pbc.gov.cn/en/3688229/3688353/3688356/5188141/5621959/index.html.

PBOC, and China Banking Regulatory Commission (CBRC). 2009. *Guiding Opinions on Further Strengthening Credit Structure Adjustment and Promoting the Steady and Rapid Development of the National Economy*. State Council Gazette No. 17. Beijing: State Council of the People's Republic of China. www.gov.cn/gongbao/content/2009/content_1336375.htm.

Piketty, Thomas, Li Yang, and Gabriel Zucman. 2019. 'Capital Accumulation, Private Property, and Rising Inequality in China, 1978–2015.' *American Economic Review* 109, no. 7: 2469–96. doi.org/10.1257/aer.20170973.

Ru, Hong. 2018. 'Government Credit, A Double-Edged Sword: Evidence from the China Development Bank.' *The Journal of Finance* 73, no. 1: 275–316. doi.org/10.1111/jofi.12585.

Shen, Cheng. 2024. 'The Most Powerful Debt Reduction Measure in Recent Years Is About to Be Launched. What Do You Think?' *Xinhua*, 14 October. www.gov.cn/zhengce/202410/content_6980079.htm.

Standing Committee of the National People's Congress (NPC). 2024. 'The Debt Reduction "Combination Punch" Has Been Launched to Reduce the Burden and Increase Momentum for Local Development.' *People's Daily*, 9 November. www.gov.cn/zhengce/202411/content_6985762.htm.

State Council of the People's Republic of China (State Council). 2023. *Report of the State Council on Government Debt Management for the Year 2023*. Beijing: National People's Congress of the People's Republic of China.

Tian, Guoqiang. 2000. 'Property Rights and the Nature of Chinese Collective Enterprises.' *Journal of Comparative Economics* 28, no. 2: 247–68. doi.org/10.1006/jcec.2000.1658.

Tsui, Kai-yuen. 2005. 'Local Tax System, Intergovernmental Transfers and China's Local Fiscal Disparities.' *Journal of Comparative Economics* 33, no. 1: 173–96. doi.org/10.1016/j.jce.2004.11.003.

United Nations Conference on Trade and Development (UNCTAD). 2024. *A World of Debt Report 2024: A Growing Burden to Global Prosperity*. Geneva: UNCTAD. unctad.org/system/files/official-document/osgttinf2024d1_en.pdf.

Wang, Jin. 2013. 'The Economic Impact of Special Economic Zones: Evidence from Chinese Municipalities.' *Journal of Development Economics* 101: 133–47. doi.org/10.1016/j.jdeveco.2012.10.009.

Wong, Christine. 2025. 'Local Government Debt in China: The 2023 Bailout and Future Prospects.' *The China Journal* 93, no. 1: 26–50. doi.org/10.1086/733767.

Wu, Ho-Mou, and Shiliang Feng. 2014. 'A Study of China's Local Government Debt with Regional and Provincial Characteristics.' *China Economic Journal* 7, no. 3: 277–98. doi.org/10.1080/17538963.2014.961688.

Xinhua. 2016. 'Resolutely Block Illegal Debt Channels: Relevant Officials of the Ministry of Finance Responded to Hot Issues on Local Government Debt.' *Xinhua*, 4 November. www.gov.cn/xinwen/2016-11/04/content_5128634.htm.

Xu, Chenggang. 2011. 'The Fundamental Institutions of China's Reforms and Development.' *Journal of Economic Literature* 49, no. 4: 1076–151. doi.org/10.1257/jel.49.4.1076.

Xu, J., J. Mao, and X. Guan. 2020. 'Recognition of Implicit Local Public Debts: Views Based on the Accurate Definition of Local Government Financing Vehicles and Financial Potential.' *Management World*, no. 9: 37–59.

Xu, J., and X. Zhang. 2013. 'China's Sovereign Debt: Situation, Investment and Risk Analysis.' *South China Journal of Economics* 1: 14–34.

Yakita, Akira. 2008. 'Sustainability of Public Debt, Public Capital Formation, and Endogenous Growth in an Overlapping Generations Setting.' *Journal of Public Economics* 92, nos 3–4: 897–914. doi.org/10.1016/j.jpubeco.2007.05.007.

Yao, Yang, and Muyang Zhang. 2015. 'Subnational Leaders and Economic Growth: Evidence from Chinese Cities.' *Journal of Economic Growth* 20: 405–36. doi.org/10.1007/s10887-015-9116-1.

Zhang, B., S.N. Wang, J.J. Zhang, and H. Deng. 2018. 'Macroeconomic Policy Report for the First Quarter of 2018.' *China Finance 40 Forum Report*.

Zhang, Ping, Lin Sun, and Chuanyong Zhang. 2021. 'Understanding the Role of Homeownership in Wealth Inequality: Evidence from Urban China (1995–2018).' *China Economic Review* 69: 101657. doi.org/10.1016/j.chieco.2021.101657.

Zhang, Shidai, and Kellee S. Tsai. 2024. '"One System, Two Shadows": A Local Public Finance Perspective on China's Shadow Banking System.' *China Economic Review* 87: 102209. doi.org/10.1016/j.chieco.2024.102209.

Zhang, X.J., X. Chang, and L. Liu. 2018. 'Structural Deleveraging: Progress, Logic, and Prospects —China Deleveraging Annual Report 2017.' *Economic Dynamics*, no. 5: 3–25.

Zhu, Xiaodong. 2021. 'The Varying Shadow of China's Banking System.' *Journal of Comparative Economics* 49, no. 1: 135–46. doi.org/10.1016/j.jce.2020.07.006.

Appendix 9.1

Table A9.1: Top-10 government spending categories, central versus local government

Central government spending (top 10)	RMB billion	%	Local government spending (top 10)	RMB billion	%
Total	**1,876**	**100.0**	**Total**	**10,719**	**100.0**
National defence	648	34.5	Education	2,014	18.8
Science and technology	221	11.8	Social security and employment	1,200	11.2
Interest payment for debt	206	11.0	General public service	1,170	10.9
Public security	118	6.3	Agriculture, forestry and water conservation	1,147	10.7
Education	110	5.9	Urban and rural community affairs	906	8.5
General public service	100	5.3	Transportation	733	6.8
Transportation	86	4.6	Medical and health care	717	6.7
Grain, oil & materials reserve & management services	65	3.4	Public security	593	5.5
Social security and employment	59	3.1	Housing security	407	3.8
Agriculture, forestry and water conservation	50	2.7			

Sources: Ministry of Finance (www.mof.gov.cn/en/data/); Barclays Research (research-centre. barclays.co.uk/).

10

From growth engine to fiscal drag: Rethinking China's local government finance

Christine Wong

It is well known that China's decentralised fiscal system assigns most public service provision to local governments and that, in aggregate, subnational governments account for 85–86 per cent of national budget expenditures. What is less well known but even more notable is that subnational governments also undertake and finance some 80 per cent or more of the investments in infrastructure nationally.[1] Given their large weight in the performance of both these tasks, how local governments are faring not only dominates fiscal outcomes but also has major spillovers to the broader economy. The recent travails of local government finance have clearly hampered the central government's capacity for macro-management.

Local government finances are under severe stress from three sources: a long-term erosion of the aggregate fiscal position of government, the collapse of land revenues and a looming debt crisis. In this chapter, I argue that although these problems manifest at the local level, their root causes are structural and systemic and often stem from larger underlying problems of the wider economy. I will also explain the central role of land revenues in the fiscal system in funding capital spending and service provision (including debt servicing). Given the restructuring of the property sector and demographic trends that point to a long-term reduction in the level of land receipts, this will require substantial adjustment in China's public finances, not just a realignment of resources between the central and local governments. Resolving the problems of local finance will require structural and systemic reform rather than partial reforms focused on local government behaviours.

---

1   The exact figure cannot be ascertained because a significant portion is financed by the off-balance-sheet borrowing of local government financing vehicles.

Long-term erosion of the budget

In China, the government presents four 'budgets' in its annual report to the National People's Congress (NPC).[2] Among them, the general budget is the main and by far the largest account. It draws revenue from taxes and administrative fees to finance government operations and public services. New taxes introduced by reforms in 1994 brought a long fiscal expansion that saw revenue growing robustly from the mid-1990s through the first decade of the 2000s (see Wong and Bird 2008; Wong 2021). During the period 2000–12, when China's GDP grew at double-digit rates in the wake of its accession to the WTO, revenues were growing 20 per cent per annum, and their share in GDP climbed from 13.4 per cent to 22 per cent (Figure 10.1). As China's economic growth moderated from 2012, revenue growth slowed and the revenue–GDP ratio plateaued. After 2015 the trend turned decisively downward under the combined weight of a further slowdown in economic growth and tax reform and tax cuts that began in 2016, including replacing business tax with an expanded value-added tax, followed by a series of tax cuts to support the economy during the Covid-19 pandemic and, later, the housing market downturn. By 2024, the revenue–GDP ratio had fallen to 16 per cent, and it is projected to fall further, to 15.5 per cent, in 2025.

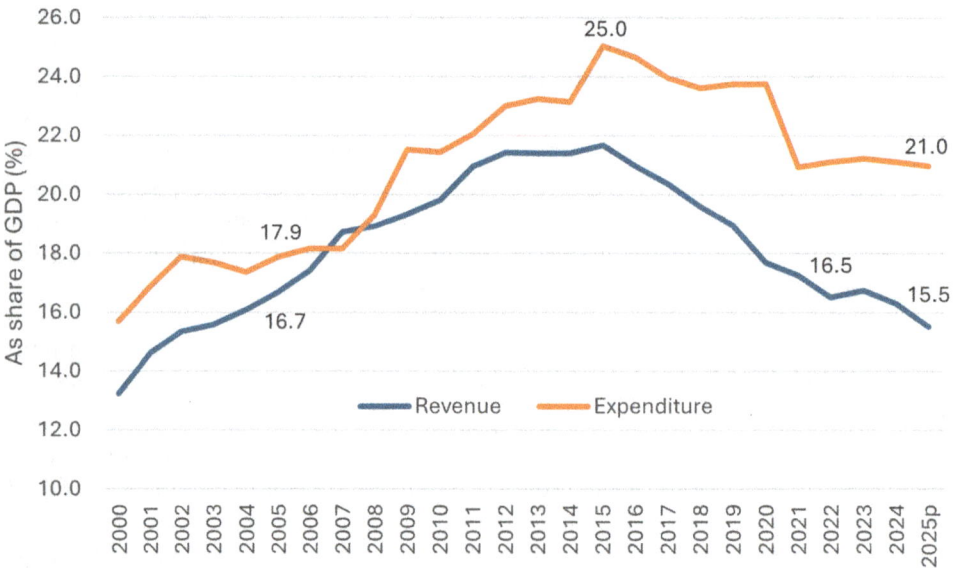

Figure 10.1: China's general budget revenues and expenditures, 2000–25
Note: The figures for 2025 are projected from the 2025 budget presented in March.
Source: National Bureau of Statistics of China (data.stats.gov.cn/english/).

---

2   The four budgets and how they evolved are discussed in detail in Wong (2024).

Table 10.1: Social spending programs introduced under the Hu Jintao–Wen Jiabao administration

Program	Launch date	Beneficiaries (no. of people)
Free Rural Compulsory Education: To eliminate out-of-pocket costs of basic education for rural families and raise the minimum standard of school spending.	2006	150 million at peak (2007)
New Rural Cooperative Medical Scheme: To provide cooperative insurance for rural families. Later merged with the urban program and renamed the Basic Residents Medical Insurance Scheme.	2003	835 million in 2010; 963 million in 2023
Rural minimum income guarantee (*Dibao*): To provide income support for the poor.	2007	53.9 million at peak (2013)
Rural pension scheme to cover all rural residents. In 2014, it was merged with the urban program to become the Basic Residents Pension Scheme.	2008	1.06 billion in 2024

Source: Author's summary.

To put this in perspective, from their peak at close to 22 per cent of GDP in 2015, revenues had fallen to 16.5 per cent by 2022—a level that was lower than the 16.7 per cent collected in 2005. This is significant because several large social spending programs introduced during the Hu Jintao–Wen Jiabao administration—such as free basic education and near-universal coverage of health, pension and income support—came online in the mid-2000s. Because of their mammoth size, these programs pushed budget expenditures up by about 7 per cent of GDP from 2005 to the peak in 2015 (Table 10.1). Since most of these programs have continued to grow, the current level of revenue collection is clearly insufficient to meet spending needs, resulting in growing deficits and spending cuts.

As they are on the frontline of service provision, local governments bear the brunt of this fiscal erosion and widening deficits. Under the current revenue-sharing arrangements, local governments 'earn' about 55 per cent of total revenues but account for 85 per cent of expenditures and depend on central government transfers to fund the gap.

In Figure 10.2, we show that the fiscal erosion and the need to fund transfers have in recent years squeezed the central government as well. In the years since 2022, the central government has turned over all its revenues to transfers.[3] Even so, many local governments are unable to meet their spending needs. In their review of the 2022 budget implementation, the Finance and Economics Committee of the NPC noted the 'relatively prominent' funding gap at the grassroots (county) level, and that many faced 'difficulty in meeting the "three guarantees"' of payroll, government operations and basic services (MoF 2023). In the same year, a survey of 521 counties and cities by the Chinese Academy of Fiscal Sciences, a Ministry of Finance inhouse think tank, found 'most' local governments admitting to 'leaving a considerable portion of policy mandates unmet' due to funding shortages (Cheng 2023a).

---

3   In 2020, the government introduced a RMB1-trillion special treasury ('Covid') bond to supplement the additional RMB1-trillion deficit in response to the pandemic. In 2022, the central government called up a special remittance of RMB1.65 trillion from central government institutions to meet the shortfall.

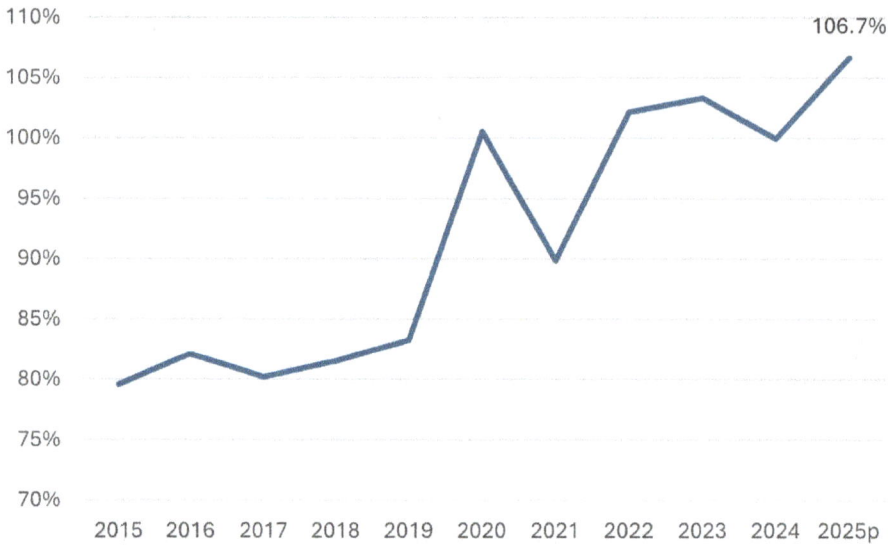

Figure 10.2: Transfers as a share of central government revenues, China
Note: Data for 2025 are projected from the 2025 budget presented in March.
Source: National Bureau of Statistics of China (data.stats.gov.cn/english/).

The growing dependence on land revenues

Land has long been a key source of financing for local governments (Wong 1997, 2013). From the 1990s, as China's rapid urbanisation created enormous need for infrastructure, local governments began to tap land as a source of finance, both directly and as collateral for borrowing off-budget through local government financing vehicles (LGFVs). Even after the 2015 reforms allowed local governments to issue bonds for capital spending, land has remained the main source of funds for investment as well as debt servicing for both the special project bonds (SPBs) and the LGFV debt that is not included in local government balance sheets (Wong 2023).

In addition, over the past decade land revenues have been increasingly used to finance recurrent expenditures in local government budgets. This began in 2015, when budget reforms promoted more unified management of fiscal resources and local governments were encouraged to draw on land revenues to supplement budgetary resources. In 2022, for example, local governments brought in RMB1.2 trillion 'from other budgets, reserve funds and carry-overs'.[4] Nearly all of it presumably came from land revenues, which make up more than 80 per cent of government fund revenues; social security funds are earmarked for pension and health expenditures and the state capital operating budget for state-owned enterprises. Then finance minister Liu Kun confirmed in early 2023 that 15 per cent of revenues from the local government fund budget have been transferred annually into the general budget (Cheng 2023b).

---

4   Report on the execution of the central and local budgets for 2022 and on the draft central and local budgets for 2023 at the first session of the Fourteenth National People's Congress, 5 March 2023 (State Council 2023).

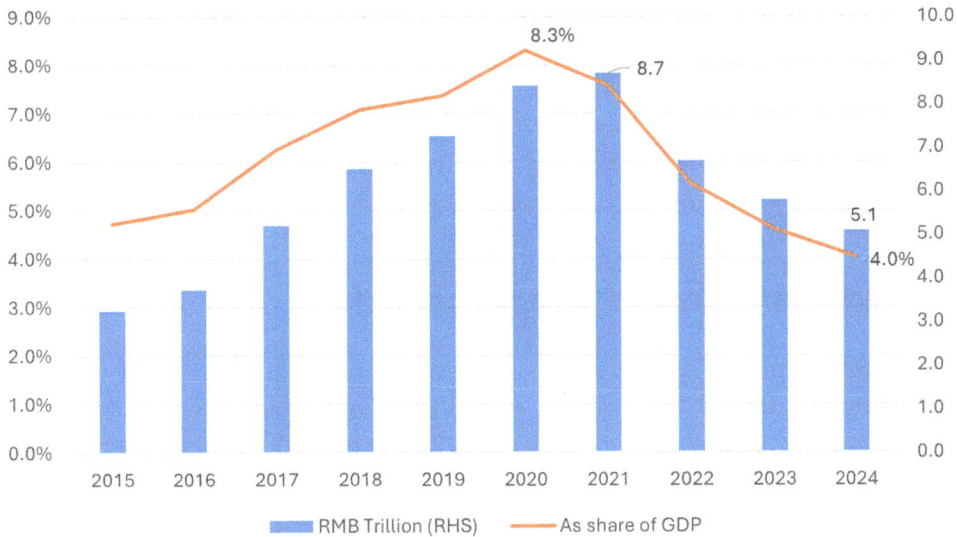

Figure 10.3: Local government gross receipt of land revenues

Source: Ministry of Finance data via CEIC (www.ceicdata.com/en).

Figure 10.4: Funds brought into local government general budgets

Source: Calculated from annual budget reports and CEIC data (www.ceicdata.com/en).

Indeed, as general revenues declined and fiscal pressure grew, it appeared that local governments stepped up efforts to raise land revenues, which grew from 5 per cent of GDP in 2016 to 8.3 per cent in 2020 (Figure 10.3), and local governments tapped these revenues through 'inter-budget transfers' to fund as much as 9 per cent of their budget expenditures—until the Covid-19 pandemic and housing sector meltdown disrupted this funding source (Figure 10.4) (Wong 2023).

In nominal terms, gross receipts from land peaked in 2021 at RMB8.7 trillion, compared with RMB11.1 trillion in local government general revenues, and made up 43 per cent of the combined fiscal resources under local government disposal.[5] Their collapse in recent years has sharply curtailed local governments resources, forcing the central government to step up transfers and precipitating a debt crisis as it cut sharply into local governments' debt-servicing capacity.

The looming local government debt crisis

Local government debt has long been a problem in China. Large-scale local government borrowing began in the 1990s, when rapid urbanisation ushered in by market reforms created huge demand for new infrastructure. As the same market forces were eroding the monopoly profits of SOEs that had been the mainstay of government revenues under the planned economy, government revenues plummeted, from more than 30 per cent of GDP in 1978 to 10 per cent in 1995–96. Amid this steep fiscal decline, capital spending was largely pushed out of the budget, the fiscal system was decentralised and local governments were left on their own to cope with the needs of urban growth (see Wong 2009). With neither budget resources nor the legal authority to borrow, local governments turned to tapping the rising value of land and borrowing off-budget through state-owned urban development corporations and other financial vehicles to finance infrastructure (Wong 2013).

The size and viability of local government debt came under central government scrutiny in the wake of the massive RMB4-trillion stimulus during the GFC. Under permissive policies, LGFVs proliferated and quickly became the most active users of stimulus funds, adding rapidly to their stock of debt (Wong 2011; Bai et al. 2016). To get a handle on the problem, central authorities ordered nationwide audits of local government debt and found that, by 2013, local governments had accumulated as much as RMB30 trillion in debt (or more than 40 per cent of GDP) when guaranteed and partially guaranteed debts were included (NAO 2013).

Reining in local government debt and off-budget borrowing was a top priority of the fiscal reforms in 2015. The central government's approach was two-pronged. First, it gave local governments the legal right to borrow through bond issuance under annual quotas approved by the NPC—known as 'opening the front door'. Second, it sought to stop the unregulated borrowing off-budget and bring the existing debt onto local government balance sheets and under unified budget management. Over the period 2015–18, RMB14.4 trillion in low-interest government bonds were offered to swap out high-cost LGFV debt, and LGFVs were banned from further borrowing on behalf of local governments, to 'close the back door'.[6]

---

5   General revenues and government fund revenues are the main fiscal resources under local government disposal. There are also revenues from the state capital operating budget and the social security fund budget, but these are earmarked for use in their relevant areas—for state-owned enterprises and pension and healthcare expenditures, respectively.
6   This is discussed in Wong (2018).

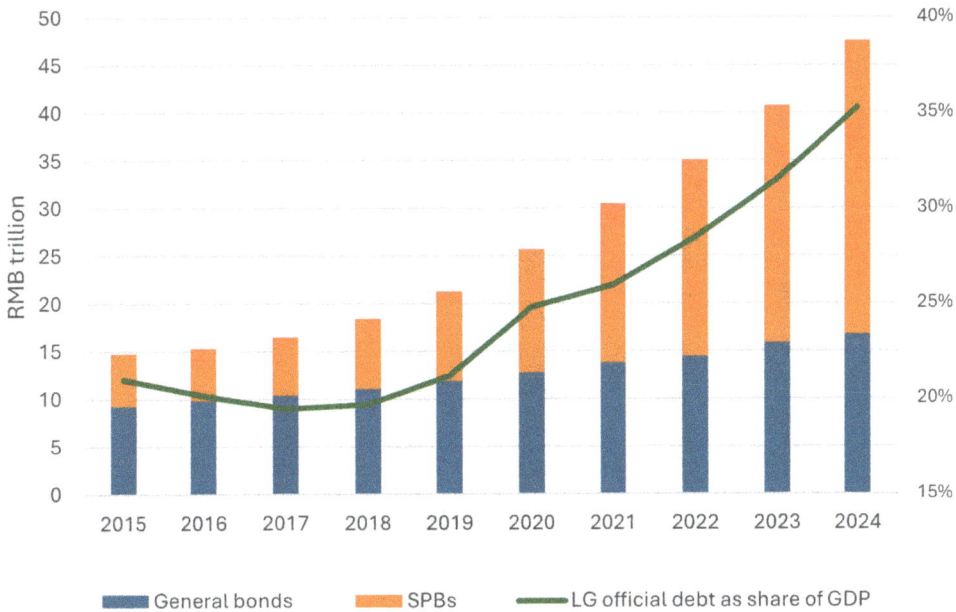

Figure 10.5: Local government outstanding bonds
Source: Ministry of Finance (www.mof.gov.cn/en/data/).

With the introduction of local government bonds, the official debt of local governments has grown rapidly, especially after 2018, with its share of GDP growing from 20 per cent to 35 per cent in 2024 (Figure 10.5).

Despite the reform, however, off-budget debt has grown rapidly. Estimates vary, partly due to differing coverage. Based on projections from LGFV bond issues, the International Monetary Fund (IMF) estimates that LGFV debt grew from RMB40 trillion in 2019 to RMB66 trillion in 2024 (Bloomberg 2024). Added to the official debt of RMB47.5 trillion, total ('augmented') local government debt reached RMB113.5 trillion by 2024 and was equal to 84 per cent of GDP (Figure 10.6). If state guidance funds were included, total augmented local government debt would rise to 92 per cent of GDP (IMF 2024). In addition, there are informal debts that may add as much as 5–10 per cent more, including widespread and mounting payment arrears to construction companies, suppliers, contractors, public employees, civil servants and even ordinary citizens.

The reason for the continued growth of off-budget debt after 2015 can be found in the funding structure for investment in infrastructure, which shows a persistent, huge gap between annual investments in infrastructure and local government borrowing capacity as approved by the NPC (Table 10.2). This gap is expected to be filled by land revenues, fiscal resources and private investments through public–private partnerships and the like. When these resources fall short, local governments turn to LGFVs and, with the collapse of land revenues, this shortfall has grown in recent years even in the face of growing SPB quotas.

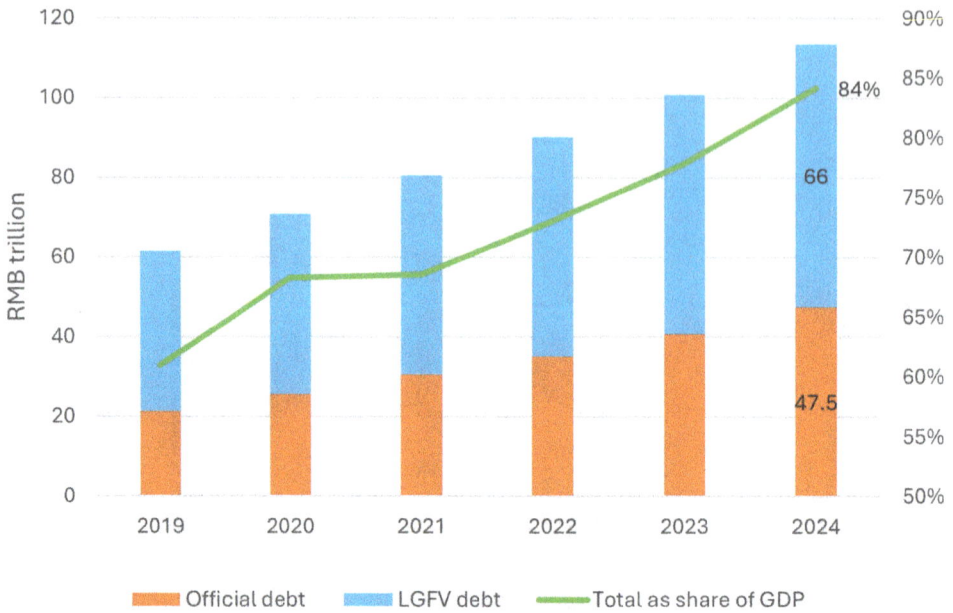

Figure 10.6: Estimated augmented local government debt

Sources: Ministry of Finance (www.mof.gov.cn/en/data/). Hidden debt estimate is from IMF (2024).

Table 10.2: Local government investment in infrastructure and funding gaps, 2016–21 (RMB billion)

Year	Local government investment in infrastructure*	Quotas for local government general bonds	Quotas for local government SPBs	Funding gap
2016	9,510	780	400	8,330
2017	11,200	830	800	9,570
2018	11,626	830	1,350	9,446
2019	12,068	930	2,150	8,988
2020	12,176	980	3,750	7,446
2021	12,225	820	3,650	7,755

Note: * Calculations based on an average 80 per cent share by local governments.

Sources: Ministry of Finance (www.mof.gov.cn/en/data/); Liu and Liu (2022); Wu (2022).

As to what lies behind the growth of infrastructure investment—aside from the incentives for local officials to show 'political achievements'—a big driver is the central government's preference for infrastructure investment for fiscal stimulus and assigning it mostly to local governments. Since the 1990s, local governments have accounted for 70–90 per cent of the total infrastructure investment (Wong 2013, 2014). During the pandemic, and after it to counter the economic drag posed by the housing market meltdown, policymakers called for an 'all-out push' on infrastructure investment and local governments were allocated quotas for SPBs of more than RMB20 trillion over the period 2019–24, adding 14 per cent of GDP to official local government debt. During the same period, an additional RMB26 trillion was added to LGFV debt, according to IMF estimates (Bloomberg 2024).

Squeezed by the slowing economy and weakening revenue receipts, the risk of default on local government debt rose sharply in 2022–23, exacerbated by a looming spike in debt maturities. The steep drop in land revenues hit LGFVs especially hard since a large proportion of their debt is tied to infrastructure and development projects, most of which depend on land finance. To avert the rising default risks, in late 2023, the central government rolled out a debt resolution program that offered refinancing bonds and concessions from state-owned banks to restructure the maturing loans of at-risk LGFVs (Wong 2025). A blanket guarantee of all maturing debt of LGFVs through 2025 helped to calm corporate bond markets, but the huge debt overhang and pressure to deleverage have continued to strain local finances and drag on economic growth. An additional package was approved by the NPC Standing Committee in November 2024, offering a total of RMB10 trillion in SPBs over the next three to five years to replace LGFV debt.[7]

To date, the approach remains the same as that applied in 2015. Holding firm to the belief that 'each family must carry its own child', the program offers no debt relief to local governments. The program aims to bring off-budget debt onto local government balance sheets under unified budget management and reduce the debt-servicing burden by offering low-cost government bonds, to buy time for local governments to gradually work down the debt and return to fiscal sustainability.

However, the debt resolution program is unlikely to achieve its objective of returning local governments to a sustainable level of debt. First, the much higher current level of debt on local government balance sheets does not leave much room for taking on off-balance-sheet debt.

This is shown in Figure 10.7, where the stock of official local government debt grew in 2024 to 172 per cent of fiscal capacity—calculated as the sum of local government general budget revenues, local government fund revenues and central government transfers.[8] This far surpasses the 'red line' of 120 per cent set by the Ministry of Finance for monitoring local government debt risks in 2016 (State Council 2016). Even under the more permissive 'traffic light' standards proposed in 2020, the debt ratio is currently near the high end of the 'yellow light' of 120–200 per cent (Li 2021).

Nor will the debt resolution program change the dynamics of the local government debt problem since it does not address the core issue of how to finance infrastructure without loading local governments with more debt. Until the central government stops using infrastructure for stimulus and assigning it to local governments, the long-term dynamics of the fiscal system continue to be towards debt financing and debt accumulation, both on and off budget.

---

7   This consists of RMB6 trillion in additional quotas of SPBs, to be used in three equal-sized increments over three years (2024–26), and the use of RMB800 billion annually from the allocated SPB quotas over five years (2024–28) (Cheng 2024).

8   Fiscal capacity calculations are based on figures projected from the first 10 months of 2024, except for transfers, for which the budgeted figure is used. Government fund revenues include mostly gross receipts from land sales, which constitute 75–90 per cent of the total, and significantly overstate the availability of disposable revenue.

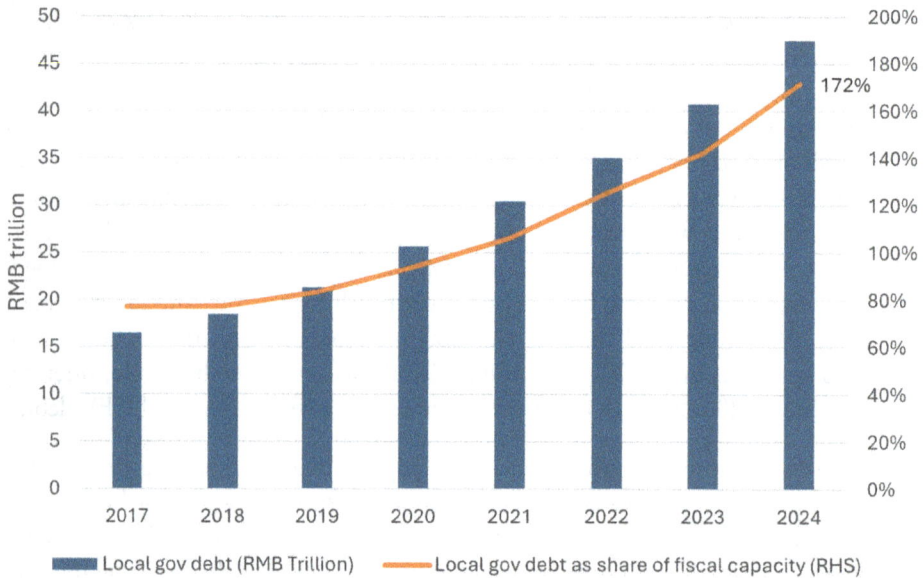

Figure 10.7: Sustainability metrics for official local government debt, 2017–24
Sources: Ministry of Finance data from budget reports (www.mof.gov.cn/en/data/); and CEIC (www.ceicdata.com/en).

The urgent need for reform

China is currently on an unsustainable fiscal path. The challenge extends beyond the misalignment of resources and responsibilities between the central and local governments. There is simply not enough revenue to meet expenditure needs. Addressing this fiscal gap requires more than just redistributing shares of revenues and expenditures; it requires a comprehensive overhaul of the system.

The government faces a dual crisis: declining tax and land revenues, compounded by a hardline debt resolution campaign since 2018 that has severely constrained local government finances. The resulting fiscal squeeze has triggered a host of dysfunctional behaviours—from service cuts, delayed salaries and pay cuts for public employees to surging fines and fees, hasty liquidation of state assets and alarming reports of 'profit-seeking law enforcement' targeting enterprises, particularly affluent, non-local firms. These practices are eroding investor and consumer confidence, ultimately threatening China's economic growth.

With China entering its third year of deflationary pressures, a large fiscal stimulus is urgently needed to fend off the risk of falling into a Japan-style deflationary spiral. Injecting funds to support the housing sector recovery should be a priority given the outsized role housing plays in GDP, wealth holdings and consumer confidence. Direct subsidies to consumers would also add a quick boost to consumption and build confidence, in addition to providing much-needed assistance to hard-hit families at the bottom of the income distribution.

For the longer term, it is imperative for authorities to formulate a plan for reforming the structure of the fiscal system, starting with tax reform to rebuild its revenue mechanism. After four decades of economic and demographic transformation, a thorough reassessment and restructuring of public expenditures are also essential to ensure fiscal sustainability.

References

Bai, Chong-En, Chang-Tai Hsieh, and Zheng Michael Song. 2016. *The Long Shadow of a Fiscal Expansion*. NBER Working Paper No. 22801. Cambridge: National Bureau of Economic Research. doi.org/10.3386/w22801.

Bloomberg. 2024. 'China Unveils $1.4 Trillion Debt Swap, Saves Stimulus for Trump.' *Bloomberg News*, 8 November. www.bloomberg.com/news/articles/2024-11-08/china-unveils-839-billion-debt-swap-to-rescue-local-governments.

Cheng, Siwei. 2023a. 'What Are the Risks of Local Financial Operations? Research and Calculation by the Academy of Finance and Sciences.' *Caixin*, 28 February. economy.caixin.com/2023-02-28/102002952.html.

Cheng, Siwei. 2023b. 'Liu Kun: Local Finances Will Gradually Improve, and the Impact of the Decline in Land Revenues on General Expenditures Is Controllable.' *Caixin*, 1 March. economy.caixin.com/2023-03-01/102003242.html.

Cheng, Siwei. 2024. 'The Details of the Local Debt Reduction Policy Are Clear, and the Total Debt Reduction Resources Will Increase By 10 trillion Yuan in Five Years.' *Caixin*, 8 November. economy.caixin.com/2024-11-08/102254984.html.

International Monetary Fund (IMF). 2024. *People's Republic of China: 2024 Article IV Consultation —Press Release; Staff Report; and Statement by the Executive Director for the People's Republic of China*. IMF Staff Country Reports No. 258. Tokyo: IMF Asia and Pacific Department. doi.org/10.5089/9798400284281.002.

Li, Xunlei. 2021. 'How Big is the Risk of Local Debt in My Country?' *Panorama Network*, 9 April. weyt.p5w.net/article/2566301.

Liu, Zhi, and Xiuying Liu. 2022. 'Is China's Infrastructure Development Experience Unique?' *Journal of Chinese Economic and Business Studies* 21, no. 3: 323–40. doi.org/10.1080/1476 5284.2022.2040074.

Ministry of Finance of the People's Republic of China (MoF). 2023. 'Report of the Financial and Economic Committee of the 14th National People's Congress Regarding the Implementation of the Central and Local Budgets in 2022 and the Draft Central and Local Budgets for 2023.' *Xinhua*, 9 March. www.mof.gov.cn/zhengwuxinxi/caizhengxinwen/202303/t2023 0309_3871498.

National Audit Office of the People's Republic of China (NAO). 2013. *National Government Debt Audit Results*. Announcement No. 32, 30 December. Beijing: NAO. www.audit.gov.cn/n5/n25/c63642/content.html.

State Council of the People's Republic of China (State Council). 2016. *Plan for Emergency Response to Local Government Debt Risks*. Office of the State Council Letter No. 88, 27 October. Beijing: Office of the State Council of the People's Republic of China.

State Council. 2023. 'Full Text: Report on China's Central and Local Budgets.' *Xinhua*, 16 March. english.www.gov.cn/news/topnews/202303/16/content_WS64124cccc6d0f528699db536.html.

Wong, Christine. 2009. 'Rebuilding Government for the 21st Century: Can China Incrementally Reform the Public Sector?' *China Quarterly* 200: 929–52. doi.org/10.1017/S030574100999 0567.

Wong, Christine. 2011. 'The Fiscal Stimulus Program and Public Governance Issues in China.' *OECD Journal on Budgeting* 2011/3: 53–73. doi.org/10.1787/budget-11-5kg3nhljqrjl.

Wong, Christine. 2013. 'Paying for Urbanization in China: Challenges of Municipal Finance in the Twenty-First Century.' In *Financing Metropolitan Governments in Developing Countries*, edited by Roy W. Bahl, Johannes F. Linn, and Deborah L. Wetzel, 273–308. Cambridge: Lincoln Institute of Land Policy. www.lincolninst.edu/sites/default/files/pubfiles/financing-metropolitan-governments-developing-full_0.pdf.

Wong, Christine. 2014. *China: PIM under Reform and Decentralization*. Washington, DC: World Bank. hdl.handle.net/10986/21045.

Wong, Christine. 2018. 'An Update on Fiscal Reform.' In *China's Forty Years of Reform and Development: 1978–2018*, edited by Ross Garnaut, Ligang Song, and Cai Fang, 271–90. Canberra: ANU Press. press-files.anu.edu.au/downloads/press/n4267/pdf/ch15.pdf.

Wong, C. (2021, September 7). 'What's Up with China's Fiscal Policy? The Puzzle of Recent Budget Data.' *EAI Commentary* no. 34, 7 September. Singapore: East Asian Institute, National University of Singapore. research.nus.edu.sg/eai/wp-content/uploads/sites/2/2021/09/EAIC-34-20210907.pdf.

Wong, Christine. 2023. 'Why There Is No Fiscal Stimulus in China in 2023.' *EAI Commentary*, no. 66, 30 August. Singapore: East Asian Institute, National University of Singapore. research.nus.edu.sg/eai/wp-content/uploads/sites/2/2023/08/EAIC-66-20230830.pdf.

Wong, Christine. 2024. 'The State of Public Finance in China: Why Tax and Intergovernmental Reform are Urgently Needed.' In *China: Regaining Growth and Momentum After the Pandemic*, edited by Ligang Song and Yixiao Zhou, 67–83. Canberra: ANU Press. doi.org/10.22459/CRGMP.2024.04.

Wong, Christine. 2025. 'Local Government Debt in China: The 2023 Bailout and Future Prospects.' *The China Journal*, no. 93. doi.org/10.1086/733767.

Wong, Christine. and Rirchard Bird. 2008. China's fiscal system: A work in progress. In *China's Great Economic Transformation: Origins, Mechanism, and Consequences of the Post-Reform Economic Boom*, edited by L. Brandt and T. G. Rawski, 429–66. Cambridge:. Cambridge University Press.

Wu, Yaping. 2022. *Infrastructure Investment and Financing: Theory, Practice and Innovation*. Beijing: Economic Management Press.

11

National saving and macroeconomic imbalances in China

Weifeng Larry Liu, Ligang Song and Yixiao Zhou

China has experienced rapid economic growth since its economic reforms began in 1978. This strong growth has been accompanied by high national saving rates and persistent current account surpluses. These patterns were most pronounced during the 2000s after China's accession to the World Trade Organization (WTO) in 2001. China's current account surpluses have significantly contributed to global structural imbalances, drawing considerable international attention, especially after the Global Financial Crisis (GFC) in 2008 (McKay and Song 2010). The patterns in national saving and current account balances reflect internal and external imbalances in China's macroeconomic landscape.

Whether high saving rates and large current account imbalances are problems depends critically on their underlying causes (Blanchard and Milesi-Ferretti 2009). If the patterns arise from optimal intertemporal choices such as those driven by demographic changes or temporary productivity gains, they can be desirable as mechanisms for smoothing consumption, financing investment and sharing risks. However, if they result from domestic distortions such as market friction, financial repression or fiscal misbehaviour, they may reflect deeper structural problems that warrant policy intervention.

China's persistently high saving rates and current account surpluses can be traced to longstanding structural issues in its economy. High saving rates are symptomatic of low consumption, suggesting that rapid economic growth has not fully translated to consumption-based welfare improvement. Low consumption, combined with substantial trade surpluses, suggests that China's economic growth has heavily relied on domestic investment and external demand. This growth strategy is not sustainable in the long run when investment efficiency wanes and external demand declines. Also, as China moves towards high-income status, its consumption remains markedly low relative to GDP compared with its high-income peers. At its current scale, China can no longer rely primarily on investment and external demand. Instead, it must shift towards domestic consumption as the main engine of economic growth. The structural problems have also exacerbated the economic challenges China has faced in the aftermath of the Covid-19 pandemic.

This chapter presents a long-term retrospective on the dynamics of China's macroeconomic imbalances over the past five decades and explores the prospects for rebalancing the economy in the current context of population ageing, the energy transition and geopolitical fragmentation. While there are numerous studies of China's economic growth, much less attention has been paid to its macroeconomic imbalances. Yang (2012) provides an excellent account of China's saving and external balances from the 1980s to 2010, arguing that the high saving rates and current account surpluses are primarily attributable to a set of policies, institutions and structural distortions. This chapter extends his analysis to the present, with a particular focus on the period since 2000, when China's accession to the WTO marked a significant step in its integration into the global economy.

Many factors can shape the dynamics of saving, investment and current account balances. This chapter focuses on long-term structural factors, including demographic change, institutional frameworks (such as social welfare systems and state-owned enterprises), market regulations (related to production factors) and key policies (such as exchange rate, trade and fiscal policies). The analysis abstracts from short-term cyclical factors such as commodity price shocks, monetary policy shifts and changes in inflation, with the aim of providing a deeper understanding of the enduring forces at play in shaping macroeconomic balances.

China's economic development since 1978 has occurred in several episodes, each characterised by distinctive economic reforms and transformations. For narrative convenience, we divide the entire period into six episodes: initial reform and opening-up (1978–91), accelerated market reforms (1992–2000), accelerated global integration (2001–08), economic rebalancing (2009–19) and the post-Covid era (including the Covid period beginning in 2020).

The remainder of this chapter is organised as follows. Section two outlines key macroeconomic accounting relations for an open economy and presents data on major macroeconomic variables, including saving, investment and current account balances. Section three examines saving, decomposed into household, corporate and government sectors, while section four focuses on investment with a similar sectoral breakdown. Section five examines the role of institutions, regulation and policy. Section six explores prospects in the context of population ageing, the energy transition and geopolitical fragmentation, while section seven concludes with policy suggestions for economic rebalancing.

Macroeconomic accounting and data

This section starts with macroeconomic accounting. The balance of payments comprises the current account, the financial account and the capital account. The current account includes the trade balance (goods and services) and the income balance. The financial account covers direct investment, portfolio investment, other investment and reserve assets. According to the balance of payments identity, the current account balance must equal net international capital flows.

From the perspective of national income accounting, the current account reflects the difference between national saving and investment. National saving can be further divided into household saving, corporate saving and government saving. Similarly, investment can be divided into three sectors: household, corporate and government. Thus, the current account is jointly driven by the saving–investment gaps in the household, corporate and public sectors.

Therefore, the current account can be analysed from three complementary perspectives: the goods market, the asset market and the national income identity. It is important to stress that they are accounting identities, and do not imply causality by themselves. A current account imbalance can originate from a trade imbalance, which then necessitates a corresponding saving–investment gap. Alternatively, it can be driven by a domestic saving–investment imbalance, which subsequently manifests as a trade imbalance. Trade flows and capital flows are two sides of the same coin, and determining which side is driving the imbalance requires a deeper understanding of economic fundamentals and policy frameworks. For example, if a country allocates a portion of its domestic saving to purchase safe assets from abroad such as US Treasury bonds, it must run a current account surplus to finance those capital outflows. Conversely, if a country exports more than it imports, it accumulates a current account surplus and builds up foreign asset holdings.

Figure 11.1a presents the current account balance since 1980. From 1980 to the mid-1990s, the current account fluctuated significantly, reflecting a period of economic opening, trade liberalisation and investment-driven growth. In the late 1990s and early 2000s, the current account shifted into consistent surpluses, gradually increasing in line with China's integration into the global economy and its export-oriented development strategy. The surplus peaked just before the GFC in 2008, reaching nearly 10 per cent of GDP. This period coincided with China's accession to the WTO and strong global demand for Chinese goods. After 2008, the surplus began to decline sharply, falling below 2 per cent of GDP by the mid-2010s. The current account has since stabilised at moderate surplus levels—generally between 1 and 2 per cent of GDP—suggesting a more balanced external position amid a transition towards consumption-driven growth and changing global trade dynamics.

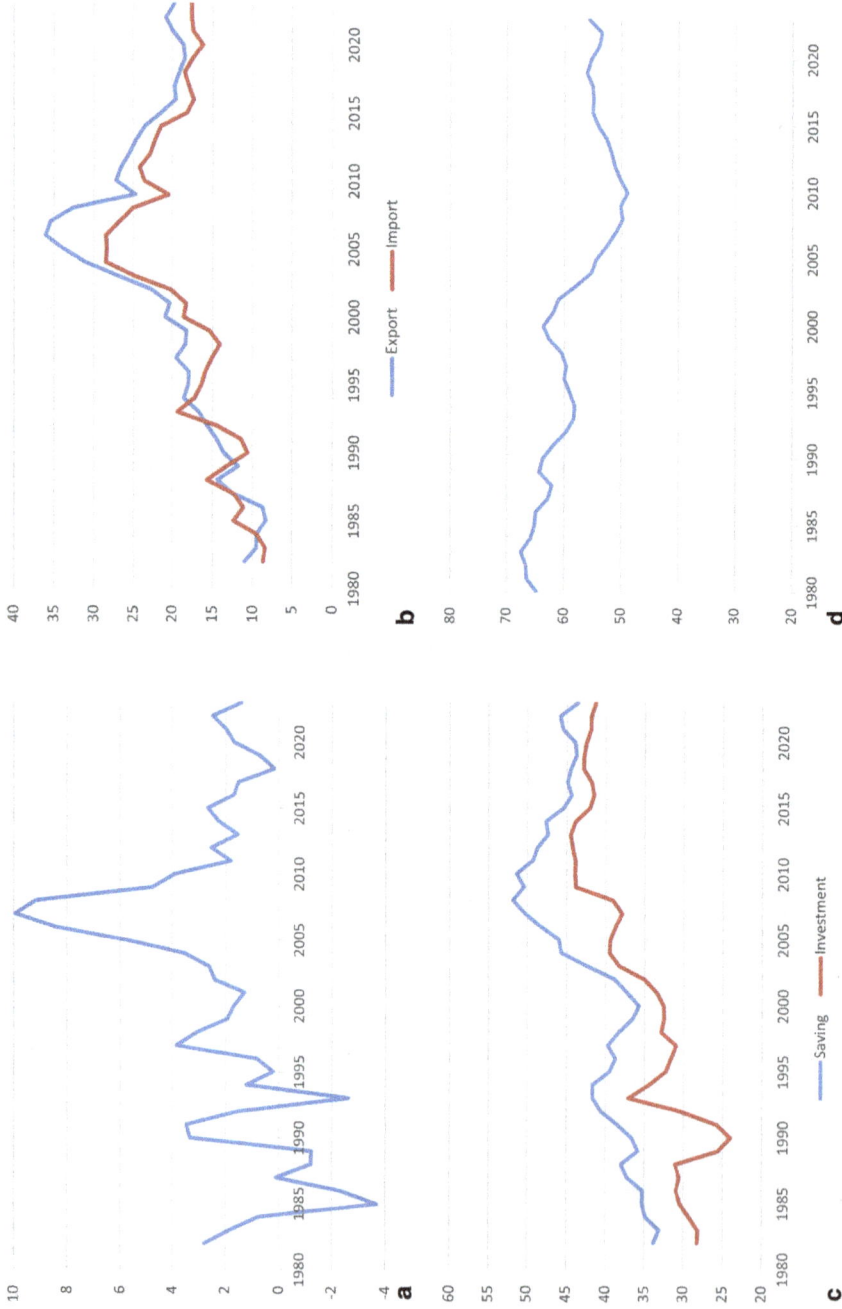

Figure 11.1: China's current account and related variables, 1980–2023 (% of GDP): a. Current account balance; b. Exports and imports; c. Savings and investment; d. Consumption

Source: World Bank, *World Development Indicators* (databank.worldbank.org/source/world-development-indicators).

The current account pattern mirrors both the difference between exports and imports (Figure 11.1b) and the gap between saving and investment (Figure 11.1c). Saving as a share of GDP rose steadily, moving from just over 30 per cent in the early 1980s to about 40 per cent by the early 2000s. It then surged rapidly, peaking above 50 per cent of GDP around 2010, followed by a gradual decline in the 2010s. In recent years, saving has stabilised at about 45 per cent of GDP—still high by global standards. On the other hand, investment hovered about 30 per cent of GDP before 2000, with volatility during the early 1990s reform period. A sharp increase in investment began in the early 2000s, peaking around 50 per cent of GDP in 2010. After 2010, investment gradually declined, remaining slightly above 40 per cent of GDP in recent years. In addition, the pattern of consumption mirrors that of saving (Figure 11.1d). Consumption as a share of GDP followed a declining tread over the first several decades, falling to 50 per cent before the GFC, and then slowly increased throughout the 2010s.

Saving

China has maintained one of the highest saving rates in the world over the past several decades, with the national saving rate exceeding 50 per cent of GDP in the 2000s. In contrast, the United States has experienced a relatively stable saving rate of about 20 per cent over the past several decades. China's exceptionally high saving rate has been recognised as the major driver of its current account surpluses. To understand the drivers of China's saving rate, it is essential to distinguish between household, corporate and government sectors, as their saving behaviours differ significantly. In particular, the prominent role of state-owned enterprises (SOEs) and the strong economic influence of the government set China apart from many other economies. China's high saving rate has been driven not only by household saving, but also by substantial contributions from the corporate and government sectors (Ma and Wang 2010). Figure 11.2a presents saving by households, corporations and the government, while Figure 11.2b shows the breakdown of consumption across households and the government.

Household saving

Household saving is defined as the difference between disposable income and household consumption, so it is closely linked to household consumption behaviour. There is a broad consensus that individuals tend to smooth consumption over time. The degree of consumption smoothing depends on factors including income expectations and uncertainty, borrowing constraints, the availability of saving vehicles and so on. At the national level, aggregate consumption—the sum of consumption across individuals—further depends on age structure and income inequality, as the propensity to consume varies across age groups and income levels.

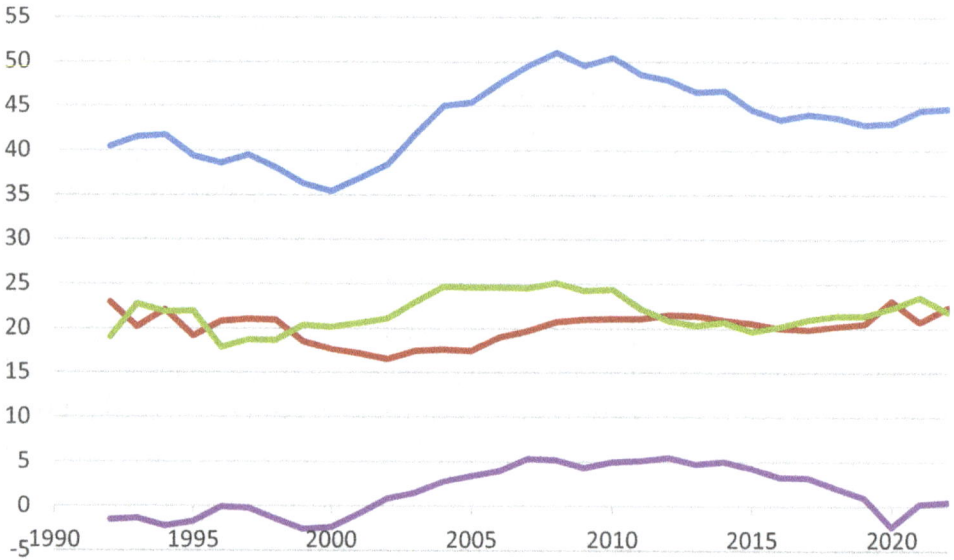

a

Legend: Total — Household — Corporate — Government

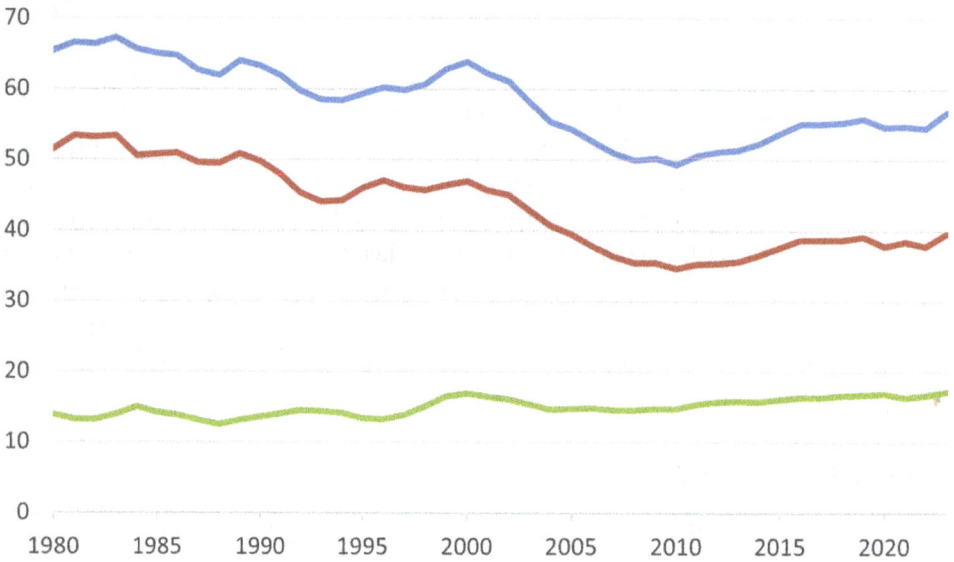

b

Legend: Total — Household — Government

Figure 11.2: China's saving and consumption by sector (% of GDP): a. Saving; b. Consumption

Sources: Authors' compilation from Chinese *Statistical Yearbooks*, various years (data.stats.gov.cn/english/).

Many studies examine household saving from different perspectives, including demographic change, underdeveloped social welfare systems, the transition from public to private provision of essential services (such as housing, health care and education), income inequality and precautionary motives related to income uncertainty. These explanations are not mutually exclusive but rather complementary.

Demographic change

Demographic change has been an important driver of China's household saving rates. Declining fertility, particularly driven by the One-Child Policy, along with increasing longevity have resulted in China's population ageing—a long-term transition that is expected to intensify over the coming decades. According to the lifecycle hypothesis, individuals tend to smooth consumption over their lifetime by saving during their prime to support their consumption in retirement when their income declines. As the population ages, aggregate saving may increase as more prime-age workers save for retirement. Increasing longevity requires individuals to accumulate more savings to support longer retirements, while declining fertility allows households to save more by reducing the financial burden of raising children. Also, income growth across generations can influence aggregate saving as a young individual would save more than an old individual spends because young generations tend to have higher lifetime incomes.

Some empirical studies show that demographic factors significantly contributed to China's high household saving rates. For example, Modigliani and Cao (2004) argue that China's demographic structure, of which the working-age population makes up a large share, significantly boosted aggregate saving. Bonham and Wiemer (2013) highlight that demographic change, including population ageing and declining dependency ratios, is key to shaping China's household saving. Curtis et al. (2015) show that demographic structure with a rising share of middle-aged savers can explain much of the increase in China's saving rate. Similarly, Ge et al. (2018) suggest that demographic change accounts for a substantial portion of the rise in household saving. Choukhmane et al. (2023) find that the One-Child Policy reduced expected old-age support from children and thus increased household saving, accounting for at least 30 per cent of the rise in China's household saving rate. Zhang et al. (2018) suggest that rising household saving resulted from a combination of factors, including demographic change and the transformation of social welfare and job security. He et al. (2019) find that demographic change and pension reforms have contributed significantly to the rise in China's household saving rate.

Social welfare

Social welfare can affect household saving through several channels.[1] A strong welfare system reduces household risk and shapes expectations, thus reducing the incentive for precautionary saving. Social welfare also raises disposable income through transfers,

---

[1] Various terms are used in the literature to refer to state-provided support systems, including social welfare, social safety net, social security, social insurance, social protection, etcetera.

subsidies and public services, especially benefiting lower-income households who have a higher propensity to consume. A comprehensive pension system further reduces the need for retirement saving. In addition, a social welfare system redistributes income, increasing aggregate consumption by shifting resources to low-income groups.

China's social welfare system has undergone significant changes over the past several decades. The 1990s marked a turning point, as the government dismantled the system of lifetime employment and SOE-provided welfare. As SOEs restructured or downsized, millions of urban workers lost access to job-based welfare benefits such as housing, health care and pensions. In the 2000s, China began to significantly expand its social welfare programs, including the basic pension system for both urban and rural areas, the New Rural Cooperative Medical Scheme for health insurance and various unemployment insurance schemes. Despite these improvements, China's social welfare system remains constrained in both coverage and generosity, particularly for rural areas and informal sectors. This underdeveloped welfare system has been a key driver of households' high saving and low consumption.

Also, the expansion of social welfare benefits has been accompanied by a rapid rise in the cost of living. The increasing cost of essential services, such as health care, education and aged care, have continued to drive precautionary saving behaviour. On the other hand, housing has been moved away from the social welfare system, making homeownership a major financial burden, further increasing household saving incentives, as discussed later.

Some empirical studies highlight the role of income uncertainty and social welfare in shaping household saving. Blanchard and Giavazzi (2005) argue that China's high household saving rate stems from weak social welfare, limited public services and underdeveloped financial markets, which drive precautionary saving. Demographic shifts further reduce reliance on family support, reinforcing the need for households to save for future uncertainty. Feng et al. (2011) find that weaker pension coverage and reduced expected benefits after the SOE reforms in the late 1990s led to significantly higher household saving rates. Chamon et al. (2013) attribute China's high household saving rates primarily to income uncertainty and inadequate social welfare. Ma and Yang (2013) also identify inadequate social welfare and limited pension and healthcare coverage, among other factors, as key drivers of household saving. He et al. (2018) find that the increased unemployment risk from the SOE reforms significantly increased precautionary saving among affected households, accounting for 40 per cent of their wealth accumulation between 1995 and 2002. İmrohoroğlu and Zhao (2018) suggest that the combination of increased long-term care risks for the elderly and the weakening of traditional family insurance due to the One-Child Policy accounts for half of the rise in China's saving rate between 1980 and 2010.

Housing reform and prices

China's housing reform in the late 1990s marked a transition from a welfare-based to a market-based system, leading to significant housing price surges in the subsequent two decades. Figure 11.3 shows housing price trends between 2000 and 2021 across different city tiers. This reform had profound impacts on consumption and saving. The expansion of homeownership was a key driver of household wealth accumulation. Households needed to save significantly for down payments and prioritised saving for future home purchases over non-housing consumption. Parents often saved aggressively to help children afford homes. The high cost of homeownership reduced expenditure on non-housing consumption, slowing China's transition to a consumption-driven economy.

Several empirical studies explore the impacts of China's rising house prices on household saving. Chen and Yang (2013) show that housing price increases drove households to save more, accounting for 45 per cent of the overall increase in the household saving rate between 2002 and 2007. Wang and Wen (2012) find that rising housing prices increased the saving rate by 2–4 percentage points. Chen et al. (2020) find that after China's 1998 urban housing reform, households significantly increased their saving rates to meet housing purchase needs, especially young and middle-aged households.

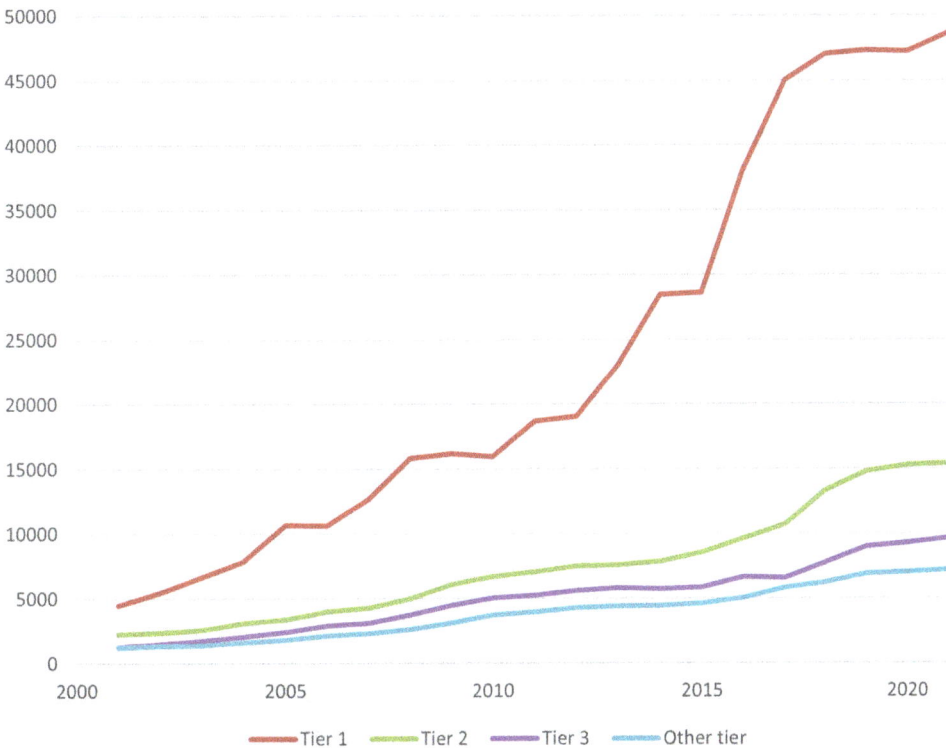

Figure 11.3: China's housing prices, 2000–21

Note: The National Bureau of Statistics of China classifies cities into four tiers: there are four first-tier cities (Beijing, Shanghai, Guangzhou and Shenzhen), 31 second-tier cities, 35 third-tier cities, and all other cities fall into the fourth tier.

Source: Qiu and Liu (2024).

The recent housing market crisis has strong implications for household saving. Housing prices have started to decline, and young generations have weaker preferences for homeownership compared with previous generations. These changes helped reduce the incentives for young generations to save. Local governments face revenue challenges as land sales slow, affecting public spending and saving.

Income inequality

Income inequality has widened significantly over the past several decades. While economic growth has lifted millions out of poverty, income disparities between urban and rural areas, between social classes and across regions have persisted (Zhou and Song 2016). The Gini index rose from 0.3 in the 1980s to 0.45 in 2010 and then gradually declined. The top 10 per cent income share rose from 27 per cent to 41 per cent during the period 1978–2015 (Piketty et al. 2019). This inequality influenced national saving because saving behaviour differs across income levels.

High-income households save excessively for investment rather than increased consumption. On the other hand, low-income households have limited access to health care, pensions and education benefits and must self-insure against risks through saving. Job instability, especially in the private sector, also leads to high precautionary saving to prepare for periods of unemployment. The growing middle class faces strong homeownership pressure, leading to high saving for deposits and mortgages. Rising education costs require substantial saving.

Several empirical studies examined the relationship between income inequality and household saving in China. Jin et al. (2011) show that wealth disparities drive precautionary saving behaviour among lower-income groups based on urban household survey data from 1997 to 2006. Chu and Wen (2017) also suggest that income inequality significantly shapes saving behaviour at both individual and aggregate levels using both household survey data and macro data.

Corporate saving

Corporate saving—defined as the value added for financial and non-financial companies minus labour compensation, production taxes, net asset payments and net transfer payments—captures firms' retained earnings. In China, understanding corporate saving requires distinguishing between SOEs and private firms, as their behaviours differ significantly.

Zhou (2009) argues that China's high saving rate was closely linked to its unique pattern of income distribution across sectors, where a large share of national income accrued to the corporate sector and the government, both of which tended to have higher saving propensity than households. For example, household saving remained stable at about 20 per cent of GDP over the period 1992–2007. In contrast, corporate saving surged in the 2000s following China's accession to the WTO and reached 23 per cent of GDP by 2007, roughly doubling the level in 1992. SOEs—particularly in energy, finance and heavy industry—benefited from state support and monopolistic advantages and accumulated substantial retained earnings, reflecting a combination of high profitability and low dividend payouts.

The high profitability of SOEs was attributed to productivity improvements and market distortions of production factors. SOE reforms in the 1990s increased their competitiveness (Garnaut et al. 2006). Accession to the WTO removed trade barriers and introduced better technologies and management. Meanwhile, SOEs benefited from access to cheap production factors due to market distortions arising from China's asymmetric market liberalisation (Huang and Tao 2010). The relaxation of worker mobility restrictions in the 1990s resulted in massive flows of migrants from rural to urban areas, creating a large pool of low-cost labour. SOEs enjoyed cheap credit due to low interest rates set by state-owned banks, which prioritised lending to SOEs over private firms. Since land is state-owned in China, SOEs could acquire land at artificially low prices—for example, in special economic zones. Also, price controls on coal and electricity kept energy costs low, further reducing production costs for SOEs, particularly those in heavy industry. In short, the combination of productivity-enhancing and cost-constraining policies increased profits, especially for SOEs.

On the other hand, SOEs were subject to weak dividend distribution requirements, which allowed them to retain a large share of earnings instead of returning them to the state or funding public spending. The excessive corporate saving of SOEs in the 2000s was thus a structural feature of the economy, rooted in institutional and governance factors rather than standard firm-level behaviour (Bayoumi et al. 2010). Corporate saving contributed to macroeconomic imbalances and highlighted the need for institutional reforms of SOEs, including taxation, social security and corporate governance, to support economic rebalancing (Ma and Wang 2010).

Private firms also accumulated savings, but for different reasons. Without equal access to credit, given the dominance of state-owned banks, private firms had to rely more on internal saving to finance investment or borrowing from the informal financial sector, paying much higher interest rates than in the formal financial sector, pushing up the economy-wide opportunity costs (Song 2005).

After the GFC, corporate saving began to moderate for several reasons. The global economic slowdown reduced China's export growth, which was a major driver of corporate profits and saving. The demographic dividend was vanishing and wages increased quickly (Cai and Lu 2013). Land prices, particularly in urban areas, increased rapidly during the housing boom. This increased costs for firms, especially in industries reliant on land for production or development, constraining their ability to accumulate savings.

Government saving

In China, government income comes from multiple sources: value added from SOE production, returns on state-owned assets, taxes on production and income and net revenues from social insurance funds. In the 1990s, government saving stayed around 5 per cent of GDP. However, it declined sharply in 2000 due to large-scale SOE reforms, which reduced profitability and the fiscal contributions from the state sector.

In the 2000s, after China's accession to the WTO, rapid economic growth significantly increased government revenues, particularly through increased corporate tax income. SOEs in key sectors such as energy, banking and heavy industry regained profitability, contributing to a rebound in public saving. As a result, government saving returned to approximately 5 per cent of GDP during this period. Despite increased government revenues, government saving remained relatively stable because the government increased spending on public infrastructure—in particular, a massive fiscal stimulus package in response to the GFC reduced government saving.

In the 2010s, public saving as a share of national saving continued to decline. Slower economic growth led to a deceleration of fiscal revenue growth. Also, government spending increased significantly, driven by increases in investment in infrastructure and rising social welfare spending, including pensions, health care and poverty alleviation programs. The combination of slower revenue growth and expanding social spending placed downward pressure on government saving, raising a big issue about governments' fiscal sustainability (especially at the local level), which continues today.

Investment

China's investment landscape has significantly transitioned from a state-dominated model in the early reform period to a mixed system in which private enterprises play an increasingly important role. China's investment as a share of GDP has been high compared with many other economies including major East Asian economies such as Japan, South Korea and Singapore, reinforcing the earlier point that it is the high saving rather than low investment that has primarily driven China's current account surpluses. The topic of investment has received far less attention than saving in the literature. But the role of investment is also important, especially given China's unique economic system, in which SOEs and the government at both the central and local levels play prominent roles in shaping investment dynamics. China's investment-to-GDP ratio was consistently lower than its saving-to-GDP ratio, leading to a pattern of excessive saving, which underscores the continued current account surpluses in China vis-a-vis the rest of the world. Figure 11.4 presents China's investment by sector: private investment, government and SOE investment and foreign direct investment (FDI).

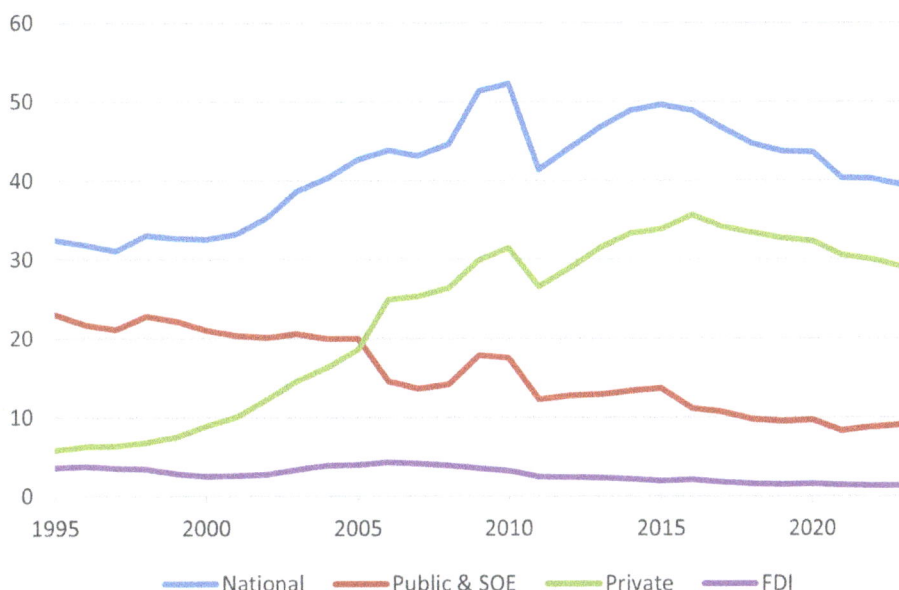

Figure 11.4: China's investment by sector (% of GDP)

Source: Authors' compilation from Chinese *Statistical Yearbooks*, various years
(data.stats.gov.cn/english/).

Private investment

During the 1980s and 1990s, the Chinese Government gradually allowed private firms and FDI to play a limited role in investing in the economy. Special economic zones were established to attract foreign capital and private investment. In the 2000s, following China's accession to the WTO, private investment (both domestic and foreign) surged, particularly in the real estate, manufacturing, technology and services sectors. Real estate investment became a dominant form of capital allocation for both households and businesses throughout the 2000s and 2010s, until the housing market crisis of 2021–22. The rapid increase in housing prices over the two decades fuelled further investment in housing and spurred local government investment in public infrastructure. Figure 11.5 presents housing investment in absolute terms and as a share of economy-wide investment.

In the 2010s, the private sector continued to drive investment in the manufacturing, technological and service industries. In particular, private investment in green technologies has grown rapidly due to a combination of policy support and market demand. Private companies were increasingly leading in solar energy, electric vehicles and battery storage. Meanwhile, as production costs, especially wages and housing prices, increased continuously, many low-end manufacturing industries began relocating to other developing countries, such as Vietnam and Bangladesh. The recent housing crisis after the Covid pandemic marked the end of the two-decade-long housing boom. With already high homeownership rates and unaffordable housing prices, housing demand declined sharply and private investment dramatically shifted away from real estate.

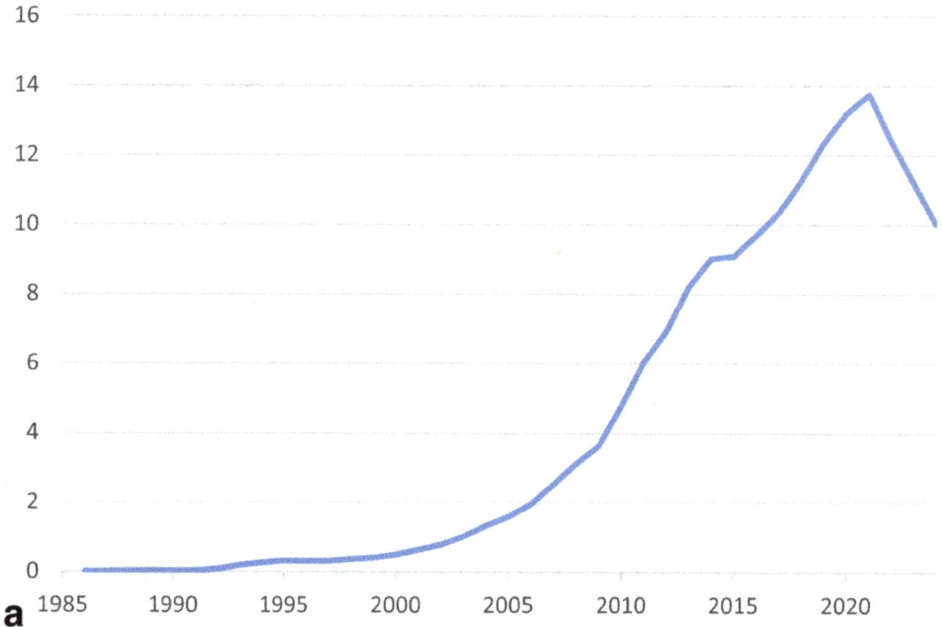

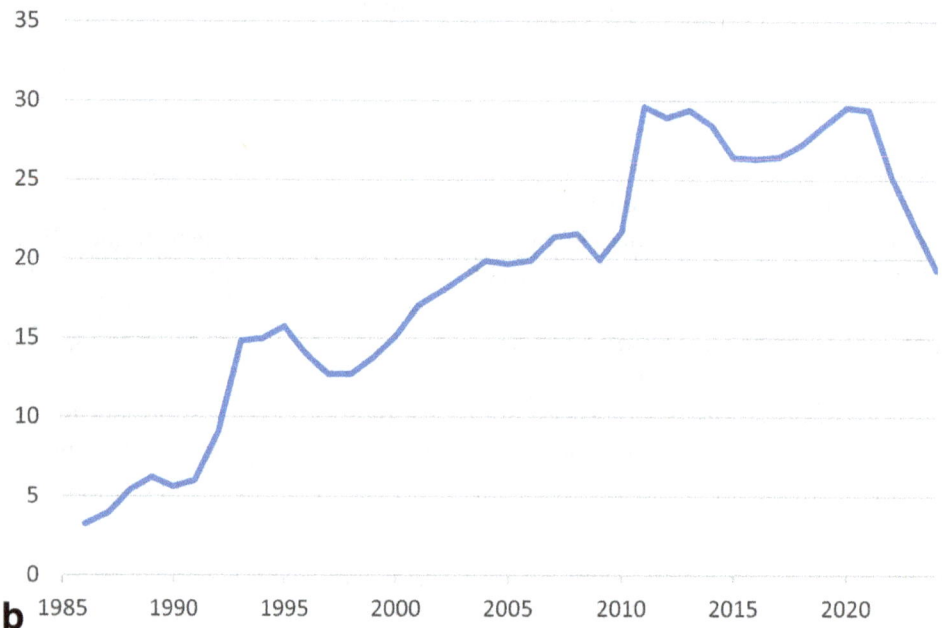

Figure 11.5: China's housing investment: a. Investment level (RMB trillion); b. Investment share (%)

Source: Authors' compilation from Chinese *Statistical Yearbooks*, various years (data.stats.gov.cn/english/).

Private investment was generally constrained by financing capacity due to the fragmentation of the financial markets (formal versus informal financial sectors) and non–market-based interest rate determination—a typical feature of financial repression (Song 2005), as the state-owned financial system was not efficient in funnelling vast savings into private firms. Despite high national saving rates, much of the capital was allocated to SOEs and public infrastructure projects, while private firms faced significant barriers to accessing credit although they were generally more productive. This financial repression limited the scope for private investment and innovation, hampering the development of domestic entrepreneurship (Son and Song 2015). As a result, excess domestic saving was not fully absorbed by the domestic economy and instead flowed abroad. These capital outflows contributed to persistent current account surpluses and the rapid accumulation of foreign exchange reserves. A significant part of these reserves was invested in low-yielding safe assets, particularly US Government bonds, which was not an optimal use of national saving.

SOE investment

SOEs have been a central topic in China's economic transition and policy debate as ownership reform constitutes a key part of the transition to a market-oriented economy. The evolution of SOE investment was largely shaped by a combination of policy directives, financial advantages and structural reforms.

The share of SOE investment declined during the 1980s and 1990s due to economic reforms and privatisation. Many inefficient SOEs were either closed or restructured (Garnaut et al. 2005). In the 2000s, SOEs regained momentum in heavy industries such as metals, materials, machinery, automobiles and chemical products as well as infrastructure, energy and finance, as promoted by the government. The investment boom after China's accession to the WTO was mostly driven by SOEs and FDI. As mentioned earlier, SOEs benefited from preferential access to capital through state-owned commercial banks and often operated under soft budget constraints, especially at the local level, where SOEs became tools for regional development under the auspices of local governments.

In the 2010s, SOE investment remained stable and maintained its dominant role in strategic sectors. SOEs increasingly aligned investment with national initiatives. In 2015, the government launched 'Made in China 2025' to boost investment in semiconductors, automation, advanced manufacturing and green technologies, with a focus on industrial upgrading and innovation. Meanwhile, the government introduced its 'Supply-Side Structural Reform' policy to reduce overcapacity in heavy industry. Investment in these areas was a combination of government, SOE and private sector investment, with SOEs central to the strategy and the private sector also playing an essential role, supported by government policies and financial incentives. Moreover, investment was also guided by ongoing urbanisation and environmental targets such as air pollution and climate policies, with SOEs taking the lead in areas such as high-speed rail expansion, space technology, the green energy transition and smart city infrastructure. China has now become the world's largest investor in renewable energy products. The government also encouraged SOEs to become more efficient and competitive through mixed-ownership reforms and overseas expansion through the Belt and Road Initiative and other overseas investments that generated some positive impact on domestic structural reform as SOEs investing abroad must comply with the rule-based market environment for their operation (Song et al. 2011). At the local level, governments increasingly relied on SOEs to carry out development projects and stimulate regional growth. The developments over the past decade have reinforced the enduring and complex role of SOEs in the economy.

Government investment

The government shifted away from direct control over investment to guiding investment through policies and incentives during the 1980s and 1990s. In the 2000s, the government heavily invested in infrastructure such as highways, railways including high-speed trains and urban development—all of which were key drivers of China's rapid economic growth. In response to the GFC, the government launched a massive stimulus package to boost investment, targeting infrastructure development.

In the 2010s, the government gradually shifted investment priority from large-scale infrastructure projects to strategic industries. Guided by initiatives such as 'Made in China 2025' and 'Supply-Side Structural Reform', public investment increasingly targeted technological innovation, advanced manufacturing, green energy and high-tech industries. This transition was also driven by growing concerns about diminishing returns from infrastructure investment, industrial overcapacity, environmental degradation and debt sustainability. Over the past decade, the government primarily acted as a facilitator rather than a direct investor, encouraging SOEs and private firms to invest in strategic industries through a range of policy tools such as subsidies, tax incentives, research and development grants, preferential access to credit and public–private partnerships.

Institutions, regulation and policy

This section examines how China's institutions, regulations and policies have shaped the patterns of saving, investment and external balances. We begin by extending the details of our earlier discussions of social welfare systems and factor market distortions, and then turn to other key policy areas, including exchange rate policy, trade policy, fiscal policy and financial regulations. Policies and regulations in these areas were not necessarily intended to promote external surpluses but were aimed at broader economic objectives such as sustained growth, full employment and financial stability. However, large trade surpluses emerged as a by-product of these policies (Song 1996; Goldstein and Lardy 2006; Fan 2008; Corden 2009; Huang and Tao 2010).

Social welfare system

As discussed earlier, social welfare plays an important role in shaping household consumption and saving. China dismantled its system of lifetime employment and job-based welfare in the 1990s as part of its broader transition from a planned to a market economy and began expanding a modern social welfare system in the 2000s. Figure 11.6 presents China's spending on social security with and without health care from 1990 to 2023. Spending on social security not including health care increased from 0.3 per cent to 3 per cent of GDP between 1990 and 2023 and, in absolute terms, rose from RMB5.5 billion to RMB3.9 trillion over the period. Meanwhile, healthcare spending grew from 0.73 per cent to 1.73 per cent of GDP from 2007 to 2023, bringing total spending on social security to 5 per cent of GDP in 2023.

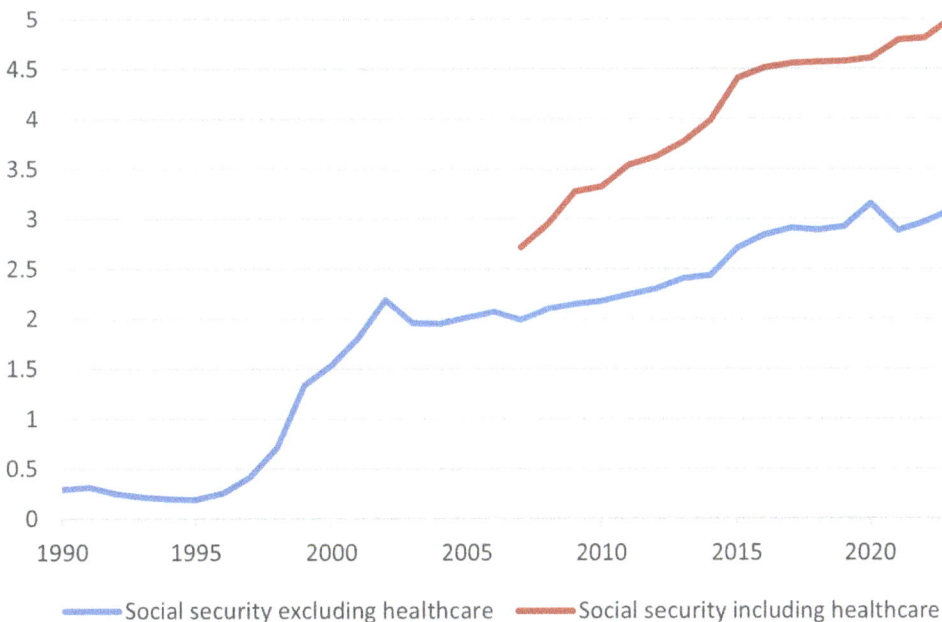

Social security excluding healthcare　　Social security including healthcare

Figure 11.6: China's social security spending (% of GDP)

Notes: Social security spending includes spending on pensions and social welfare, social security subsidies and expenditure for retirees by the government and other institutions. Healthcare spending was not separately reported in China's statistical yearbooks until 2007.

Source: Authors' compilation from Chinese *Statistical Yearbooks*, various years (data.stats.gov.cn/english/).

Despite significant improvements, China's social welfare system remains incomplete and faces significant challenges. Total social spending as a share of GDP was far below the Organisation for Economic Co-operation and Development average of 21 per cent in 2023 (OECD Statistics: www.oecd.org/en/data.html). Another major issue is the significant gap in welfare provision between urban and rural areas, which is maintained by the household registration (*hukou*) system. Migrant workers contribute to urban economic growth but have limited access to urban social welfare benefits, which discourages their consumption in cities (Song et al. 2010). Pension coverage is not yet universal and benefit levels remain relatively low, particularly in informal sectors and rural areas (Cai and Cheng 2014; Fang and Feng 2018). Many rural residents also lack access to adequate health insurance. The benefits of health insurance are often limited in scope and depth. Out-of-pocket health expenses remain high for rural residents, particularly for serious illnesses or treatments requiring travel to urban hospitals. These issues contribute to high financial vulnerability and reinforce precautionary saving behaviours by households.

Factor market regulation

Huang and Tao (2010) argue that China's current account surplus arose from the asymmetric liberalisation of product and factor markets. While China has significantly liberalised its product markets, with most goods and services priced by market forces, the government has retained strong regulations and distortions in factor markets, including labour, capital, land and energy. These distortions effectively repressed factor costs, particularly through low land prices and inadequate social welfare contributions for workers, which in turn reduced production costs, boosted corporate profits and saving and contributed to trade surpluses resulting from improved international competitiveness.

Despite some gradual reforms in the 2010s, factor markets in China remain significantly more regulated and distorted than product markets—a structural problem that still holds today. Labour mobility continues to be constrained by the *hukou* system, which limits migrant workers' access to public services and social protection in urban areas. Social welfare contributions are not uniformly enforced, particularly for informal and migrant workers. Capital markets, although increasingly open to foreign participation, are still influenced by state-led financial intermediation, with SOEs often enjoying preferential access to credit and financing at below-market rates. Land markets remain under government control, with urban land allocated administratively and rural land locked in collective ownership structures that prevent free transfers and market pricing. While energy pricing has become more market-based in recent years, subsidies and regulated prices continue to affect energy-intensive sectors. These enduring distortions continue to suppress factor costs, bolster corporate saving and indirectly support China's external surpluses.

Financial market regulation

As mentioned earlier, China's financial system is highly regulated and overwhelmingly dominated by state-owned institutions, which are inefficient in channelling saving into private investment. Generally, firms finance investment through external debt or equity financing. In terms of debt financing, firms can borrow money from banks or issue corporate bonds. But China's banking system is mostly state-owned and primarily lends funds to SOEs and government-backed projects. SOEs dominate the corporate bond market, while private firms face higher borrowing costs. In terms of equity financing, China's initial public offering (IPO) market is highly regulated, with government control over listings. As a result, it is difficult for small and medium-sized private firms to access funding, although China has made important progress in establishing platforms at both the Shanghai and the Shenzhen stock exchanges to attract investment for small and medium-sized enterprises and high-tech start-ups.

Also, China has not fully liberalised the capital account and its financial system has strict controls on international capital flows. This regulation limits the private sector (households and private firms) from investing abroad. Saving accumulates within the domestic financial system, mainly in the state-dominated banking system. Excess saving must be invested elsewhere—often abroad—primarily through government channels such as foreign exchange reserves, sovereign wealth funds and state-led overseas investment, although overseas investments by private firms have been steadily increasing over time. Figure 11.7 presents China's foreign reserves since 1980. Reserves grew modestly until the early 2000s, and then surged rapidly after China's accession to the WTO, peaking at nearly US$4 trillion in 2014. A sharp decline followed, after which reserves stabilised slightly above US$3 trillion.

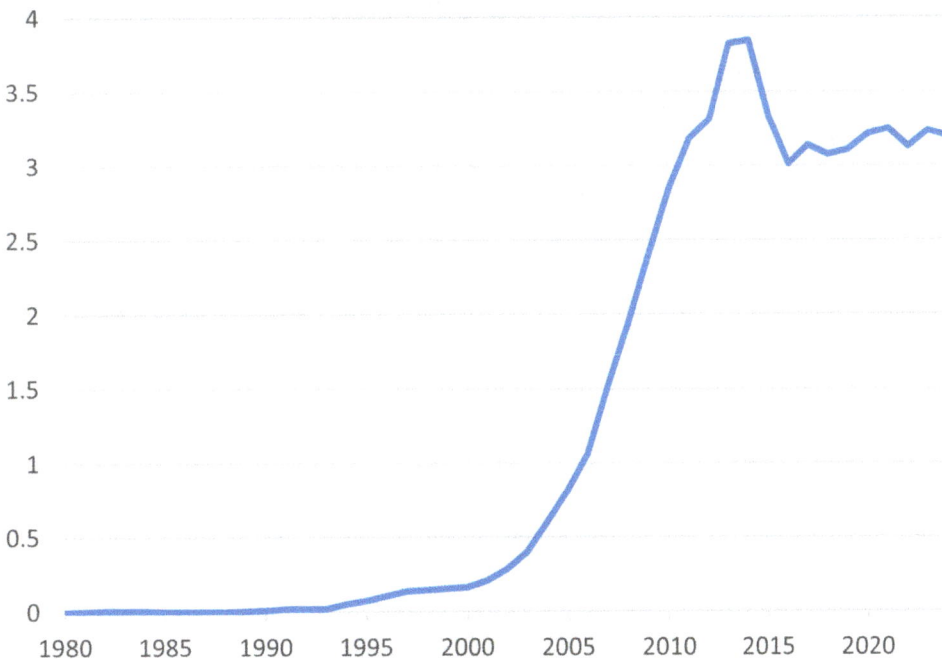

Figure 11.7: China's foreign reserves (US$ trillion)
Source: State Administration of Foreign Exchange of China (www.safe.gov.cn/en/).

Exchange rate policy

A country's exchange rate policy directly affects its current account balance. An undervalued fixed exchange rate increases export competitiveness and supports current account surpluses, while an overvalued currency can lead to deficits. China's exchange rate policy has been a focal point of public debate about its external imbalances. China maintained a dual (official and market) exchange rate system until the government unified the two rates in 1994 and then pegged the renminbi to the US dollar from 1994 to 2005 (Figure 11.8). The fixed exchange rate over the decade helped support trade surpluses, especially after China's accession to the WTO. Increasing trade surpluses led to concerns over the renminbi's undervaluation. The G7 and the International Monetary Fund (IMF) in 2005 called for revaluation of the renminbi and urged China to allow greater exchange rate flexibility. In response, China moved towards a more flexible exchange rate system, allowing the currency to appreciate gradually against the US dollar (or a basket of major currencies). This shift aimed to address trade imbalances and respond to international pressure. After the GFC, China allowed greater flexibility in the exchange rate, with more market forces in the exchange rate markets. China has also moved towards internationalising the renminbi, which became part of the IMF's special drawing rights basket in 2016. The government has also implemented reforms to open its financial markets (including both equity and bond markets) to foreign investors.

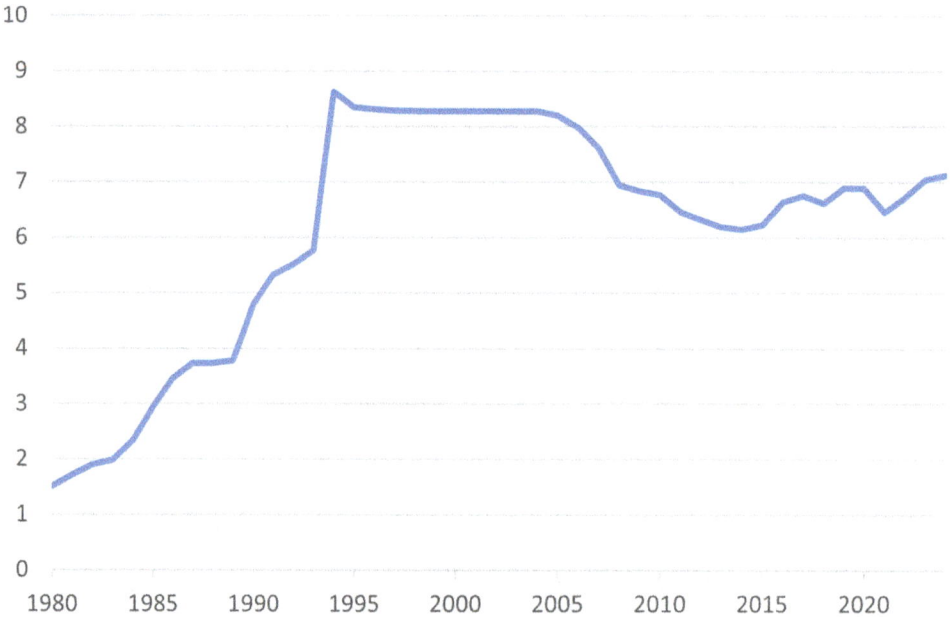

Figure 11.8: China's nominal exchange rate (RMB per USD)
Source: State Administration of Foreign Exchange of China (www.safe.gov.cn/en/).

China's exchange rate policies are often aligned with broader economic objectives. While promoting exports is frequently highlighted in public debates, the policies also aim to balance multiple goals, including maintaining short-term financial stability, managing inflation and capital flows, supporting renminbi internationalisation and reducing trade tensions.

Some studies argue that it is not realistic to expect a more flexible exchange rate regime to automatically correct the current account imbalance. Yu (2007) cautions that exchange rate reform alone is unlikely to lead to significant reductions in China's current account surplus, arguing that structural factors play a more decisive role than currency appreciation. Chinn and Wei (2013), using a panel dataset of 170 countries over the period 1971–2005, find no strong, robust or monotonic relationship between exchange rate regime flexibility and current account reversion. Woo (2006) argues that trade surpluses are more effectively addressed by the establishment of an efficient financial intermediation mechanism than by currency appreciation. Similarly, Goldstein and Lardy (2008) argue that exchange rate policy alone is unlikely to be effective in solving external imbalances. Similarly, Ma et al. (2013, 2016) argue that while appreciation of the renminbi can help rebalance the economy by reducing net exports, exchange rate adjustment alone is insufficient. Rather, sustainable rebalancing requires structural reforms such as improving the social safety net, reforming the financial system and encouraging household consumption to address the underlying causes of China's high saving rate.

Trade policy

China has pursued export promotion policies since its economic reform. Before its accession to the WTO in 2001, China practised a combination of export-promoting and import-restricting policies through tariffs, quotas and import licences (Yang 2012). After China's accession to the WTO, trade barriers fell dramatically, increasing the profitability of firms. A large fraction of profits was either saved in the corporate sector or collected by the government as revenue.

China's export policy has evolved through three stages, reflecting its economic development and integration into the global economy. Before 2000, China promoted exports by establishing special economic zones, reducing tariffs and trade barriers, attracting FDI and joint ventures, providing export tax rebates, low-interest loans, subsidies, preferential land deals and infrastructure support, alongside exchange rate polices. China's accession to the WTO further liberalised its trade policies by systematically reducing trade barriers, leading to significant increases in trade surpluses during the 2000s. After the GFC, China gradually moved away from low-cost manufacturing and rebalanced its economy from international trade to domestic consumption and technological innovation. The Belt and Road Initiative, launched in 2013, expanded China's trade network and infrastructure investments in developing countries. More recently, initiatives such as 'Made in China 2025' enhanced exports in high-tech industries like semiconductors, automation and green technology products (solar panels and electric vehicles).

International trade has shifted from traditional simple exchanges of final consumption goods to more complex global value chains (GVCs), in which production is fragmented across countries, thanks to technological advancements and trade liberalisation, among other factors. China plays a central role in GVCs. A significant fraction of China's trade surplus over the past two decades came from processing trade, in which intermediate goods are imported, assembled and then exported as finished products. This export model inflated China's gross trade surpluses, while much of the value added originated from other countries. China's move up the value chain increased the trade surplus and strengthened its net export position.

Rising geopolitical tensions and supply chain disruptions during Covid-19, together with rising labour costs, have diverted some low-cost manufacturing firms away from China, resulting in a slowdown in exports of low-cost manufacturing products. But China's exports of high-value manufacturing goods have increased in recent years. China's manufacturing exports have continued to account for about one-third of the world total.

Fiscal policy

Fiscal policy comprises two components: revenue and expenditure. In China, tax and land-related revenues have been the main sources of government income, with local governments heavily reliant on land sales and fiscal transfers from the central government. Government expenditure includes public investment, social welfare and transfer payments, debt servicing and policy instruments such as subsidies, grants and incentive programs.

The current account balance is the sum of the private sector balance and the public sector balance. The longstanding twin-deficit hypothesis suggests that a fiscal deficit is often associated with a trade deficit. If the government borrows more, it will increase domestic demand, which can increase imports and worsen the trade balance. If the deficit is financed by international capital flows, it can appreciate the domestic currency, further widening the trade deficit. This relationship is not always one-to-one but depends on the dynamic changes from exchange rate policy, monetary policy and international capital flows.

Fiscal policy can influence national saving and current account balances through expenditure. Government investment, as discussed above, is a key component of fiscal policy. An increase in public investment directly reduces public saving and thus national saving, which in turn reduces current account surpluses. Higher spending on social welfare can boost domestic consumption and thus reduce household saving and further narrow trade surpluses. In addition, fiscal instruments such as subsidies and tax incentives (including those targeting international trade and broader distortions in factor markets) change the relative prices of exports and imports, thereby influencing the current account balance.

On the revenue side, the tax system plays an important role in shaping income distribution, which in turn influences income inequality, aggregate consumption and national saving. A more progressive tax system can reduce income inequality by redistributing income from high-income to low-income households, who tend to have a higher marginal propensity to consume. This redistribution boosts overall consumption and may reduce household saving rates. At the same time, corporate tax policies affect firms' retained earnings and investment decisions, thereby influencing corporate saving. The structure of both personal and corporate taxes has important implications for the composition and level of national saving, as well as for broader macroeconomic imbalances.

Prospects

China's national saving and current account balances are expected to continue shrinking over the coming decades. The decline in the saving rate reflects structural shifts in the economy, including population ageing, evolving consumption patterns, less public investment and rising public debt. As the saving rate declines, the current account surplus will likely narrow further.

A major factor contributing to the projected decline in China's national saving is its rapid population ageing. According to the United Nations' 2024 *World Population Prospects* (population.un.org/wpp/), the elderly dependency ratio in China is expected to increase dramatically, from about 15 per cent today to nearly 50 per cent by 2050. This demographic shift implies that a shrinking share of the working-age population will support a growing elderly population. As more people enter retirement, they will begin to decumulate their savings to finance consumption, leading to a structural decline in aggregate saving. Also, the ageing population will place increased pressure on public pension and healthcare systems, raising the fiscal burden on the government.

China has been expanding its social welfare system. If the government continues to move towards levels of social welfare like those in advanced economies, rising expenditures will place upward pressure on fiscal balances and thus reduce government saving. On the other hand, as social welfare improves, households may reduce precautionary saving, leading to a decline in household saving rates as well. Such structural increases in public spending may also necessitate tax reforms or reallocation of fiscal resources, given that China has already accumulated significant public debt, particularly at the local level (Wong 2024).

Shifts in the consumption–saving preferences of young generations are also expected to reduce household saving rates. Unlike previous generations, young Chinese are more inclined to spend than save, driven by improved access to credit and changing lifestyles, and less concern about old-age security due to expectations of social safety nets. This behavioural shift will reinforce the downward trend in national saving.

The housing market, which has long been a cornerstone of household wealth and saving behaviour in China, has been experiencing dramatic structural change in recent years. With homeownership already near saturation and housing prices peaking around the 2021–22 housing market crisis, the role of real estate as a store of value and a saving vehicle is diminishing. Future declines in property prices would further weaken household saving incentives. In addition, government revenues, particularly at the local level, have been heavily reliant on land sales and housing-related taxes. As the real estate sector contracts, this revenue stream is shrinking, thereby reducing government saving.

China's 'dual circulation' strategy, introduced in 2020, emphasises strengthening domestic consumption (internal circulation) through building an integrated domestic market while maintaining openness to global markets (external circulation). This strategic pivot aims to rebalance growth drivers by reducing reliance on external demand and investment-led growth. Over time, this shift could reduce the trade surplus and change the composition of exports and imports.

Ongoing geopolitical tensions, particularly with the United States, may have long-term implications for China's trade relationships and supply chain integration. These disruptions may lead to a more fragmented global trading environment, further constraining China's external balances and reducing its reliance on export-led growth.

Nevertheless, China remains well positioned to expand in high-value export sectors such as green technologies. As the world transitions to low-carbon energy systems, demand for Chinese electric vehicles, solar panels and battery technologies is expected to grow. This could partially offset the decline in traditional manufacturing exports and support external balances, even amid broader headwinds from rising trade protectionism.

Conclusion

China's persistently high national saving rates and large current account surpluses have long reflected underlying structural imbalances in the economy, particularly the low levels of domestic consumption as a share of GDP. While China's rapid economic growth over the past few decades has lifted millions out of poverty and transformed the country into a global economic powerhouse, this growth has not been fully translated into consumption-based welfare improvements for households. Instead, a disproportionate share of income has been directed towards investment and net exports, rather than household spending.

The long-term macroeconomic imbalances stem from deep-rooted institutional, policy and structural distortions. Factors such as incomplete social welfare systems, constrained labour mobility, distorted factor markets and underdeveloped financial markets have suppressed consumption and encouraged excessive precautionary saving. Traditional macroeconomic policy tools have become increasingly less effective and less sustainable as mechanisms for promoting balanced growth.

Going forward, more sustainable and inclusive long-term growth in China must be driven mainly by domestic consumption and should aim to enhance household welfare more directly. This transition will require a strategic rebalancing of the economy from investment-led and export-led growth to consumption-led development. Such a shift is not only necessary for improving the quality of growth and the welfare of people, but also for ensuring economic resilience with an ageing population and growing global uncertainties.

Achieving this transformation will depend critically on broad-based policy and institutional reforms. These should focus on reducing structural distortions, better integrating domestic markets and fostering a more equitable distribution of income. Strengthening social welfare systems, including pensions, health care and unemployment insurance, can reduce the need for precautionary saving, while policies that boost household income, particularly for rural and low-income populations, and urbanise migrant workers will directly support consumption. Together, these reforms are essential for building a more balanced, welfare-enhancing and sustainable economic model for China's future.

Acknowledgements

Weifeng Larry Liu acknowledges financial support from the Australian Research Council Centre of Excellence in Population Ageing Research (CE170100005) and the Australian Centre on China in the World.

References

Bayoumi, Tamim, Hui Tong, and Shang-Jin Wei. 2010. *The Chinese Corporate Savings Puzzle: A Firm-Level Cross-Country Perspective*. NBER Working Paper No. 16432. Cambridge: National Bureau of Economic Research. doi.org/10.3386/w16432.

Blanchard, Olivier, and Francesco Giavazzi. 2005. 'Rebalancing Growth in China: A Three-Handed Approach.' *China & World Economy* 14, no. 4: 1–20. doi.org/10.1111/j.1749-124X.2006.00027.x.

Blanchard, Olivier, and Gian Maria Milesi-Ferretti. 2009. *Global Imbalances: In Midstream?* IMF Staff Position Note SPN/09/29. Washington, DC: International Monetary Fund. Available at SSRN: doi.org/10.2139/ssrn.1525542.

Bonham, Carl, and Calla Wiemer. 2013. 'Chinese Saving Dynamics: The Impact of GDP Growth and the Dependent Share.' *Oxford Economic Papers* 65, no. 1: 173–96. doi.org/10.1093/oep/gps020.

Cai, Fang, and Lu Yang. 2013. 'The End of China's Demographic Dividend: The Perspective of Potential GDP Growth.' In *China: A New Model for Growth and Development*, edited by Ross Garnaut, Cai Fang, and Ligang Song, 55–73. Canberra: ANU E Press. doi.org/10.22459/CNMGD.07.2013.04.

Cai, Yong, and Yuan Cheng. 2014. 'Pension Reform in China: Challenges and Opportunities.' *Journal of Economic Surveys* 28, no. 4: 636–51. doi.org/10.1111/joes.12082.

Chamon, Marcos, Kai Liu, and Eswar Prasad. 2013. 'Income Uncertainty and Household Savings in China.' *Journal of Development Economics* 105: 33–47. doi.org/10.1016/j.jdeveco.2013.07.014.

Chen, Binkai, Xi Yang, and Ninghua Zhong. 2020. 'Housing Demand and Household Saving Rates in China: Evidence from a Housing Reform.' *Journal of Housing Economics* 49, no. 3: 628–44. doi.org/10.1016/j.jhe.2020.101693.

Chen, B.K., and W. Yang. 2013. 'Land Supply, Housing Price and China Urban Residents Savings.' *Economic Research Journal* 1: 97–107.

Chinn, Menzie D., and Shang-Jin Wei. 2013. 'A Faith-Based Initiative Meets the Evidence: Does a Flexible Exchange Rate Regime Really Facilitate Current Account Adjustment?' *The Review of Economics and Statistics* 95, no. 1: 168–84. doi.org/10.1162/REST_a_00244.

Choukhmane, Taha, Nicolas Coeurdacier, and Keyu Jin. 2023. 'The One-Child Policy and Household Saving.' *Journal of the European Economic Association* 21, no. 3: 987–1022. doi.org/10.1093/jeea/jvad001.

Chu, Tianshu, and Qiang Wen. 2017. 'Can Income Inequality Explain China's Saving Puzzle?' *International Review of Economics & Finance* 52: 222–35. doi.org/10.1016/j.iref.2017.01.010.

Corden, W. Max. 2009. 'China's Exchange Rate Policy, Its Current Account Surplus, and the Global Imbalances.' *American Economic Review* 119: 176–85. fbe.unimelb.edu.au/__data/assets/pdf_file/0009/775278/Chinas_Current_Account_Surplus.pdf.

Curtis, Chadwick C., Steven Lugauer, and Nelson C. Mark. 2015. 'Demographic Patterns and Household Saving in China.' *American Economic Journal: Macroeconomics* 7, no. 2: 58–94. doi.org/10.1257/mac.20130105.

Fang, Hanming, and Jin Feng. 2018. *The Chinese Pension System*. NBER Working Paper No. 25088. Cambridge: National Bureau of Economic Research. www.nber.org/system/files/working_papers/w25088/w25088.pdf.

Feng, Jin, Lixin He, and Hiroshi Sato. 2011. 'Public Pension and Household Saving: Evidence from Urban China.' *Journal of Comparative Economics* 39, no. 4: 470–85. doi.org/10.1016/j.jce.2011.01.002.

Garnaut, Ross, Ligang Song, Stoyan Tenev, and Yang Yao. 2005. *China's Ownership Transformation: Process, Outcomes, Prospects*. Washington, DC: International Finance Corporation and The World Bank.

Garnaut, Ross, Ligang Song, and Yang Yao. 2006. 'Impact and Significance of SOE Restructuring in China.' *The China Journal* 55: 35–66. doi.org/10.2307/20066119.

Ge, Suqin, Dennis Tao Yang, and Junsen Zhang. 2018. 'Population Policies, Demographic Structural Changes, and the Chinese Household Saving Puzzle.' *European Economic Review* 101: 181–209. doi.org/10.1016/j.euroecorev.2017.09.008.

Goldstein, Morris, and Nicholas Lardy. 2006. 'China's Exchange Rate Policy Dilemma.' *American Economic Review* 96, no. 2: 422–26. doi.org/10.1257/000282806777212512.

Goldstein, Morris, and Nicholas Lardy, eds. 2008. *Debating China's Exchange Rate Policy*. Washington, DC: Peterson Institute for International Economics.

He, Hui, Feng Huang, Zheng Liu, and Dongming Zhu. 2018. 'Breaking the "Iron Rice Bowl": Evidence of Precautionary Savings from the Chinese State-Owned Enterprises Reform.' *Journal of Monetary Economics* 94: 94–113. doi.org/10.1016/j.jmoneco.2017.12.002.

He, Hui, Lei Ning, and Dongming Zhu. 2019. *The Impact of Rapid Aging and Pension Reform on Savings and the Labor Supply*. IMF Working Paper No. WP/19/61. Washington, DC: International Monetary Fund. doi.org/10.5089/9781498302890.001.

Huang, Yiping, and Kunyu Tao. 2010. 'Factor Market Distortion and the Current Account Surplus in China.' *Asian Economic Papers* 9, no. 3: 1–36. doi.org/10.1162/ASEP_a_00020.

İmrohoroğlu, Ayşe, and Kai Zhao. 2018. 'The Chinese Saving Rate: Long-Term Care Risks, Family Insurance, and Demographics.' *Journal of Monetary Economics* 96: 33–52. doi.org/10.1016/j.jmoneco.2018.03.001.

Jin, Ye, Hongbin Li, and Binzhen Wu. 2011. 'Income Inequality, Consumption, and Social-Status Seeking.' *Journal of Comparative Economics* 39, no. 2: 191–204. doi.org/10.1016/j.jce.2010.12.004.

Ma, Guonan, Robert McCauley, and Lillie Lam. 2013. 'The Roles of Saving, Investment and the Renminbi in Rebalancing the Chinese Economy.' *Review of International Economics* 21, no. 1: 72–84. doi.org/10.1111/roie.12021.

Ma, Guonan, Ivan Roberts, and Gerard Kelly. 2016. 'A Rebalancing Chinese Economy: Challenges and International Implications.' In *Structural Change in China: Implications for Australia and the World. Conference Proceedings*, edited by Iris Day and John Simon, 19–68. Sydney: Reserve Bank of Australia. www.rba.gov.au/publications/confs/2016/ma-roberts-kelly.html.

Ma, Guonan, and Yi Wang. 2010. *China's High Saving Rate: Myth and Reality*. BIS Working Paper No. 312. Basel: Bank for International Settlements. doi.org/10.2139/ssrn.1631797.

Ma, Guonan, and Dennis T. Yang. 2013. *China's High Saving Puzzle*. IZA Discussion Paper No. 7223. Bonn: Institute of Labor Economics. Available at SSRN: papers.ssrn.com/sol3/papers.cfm?abstract_id=2223150.

McKay, Huw, and Ligang Song. 2010. 'China as a Global Manufacturing Powerhouse: Strategic Considerations and Structural Adjustment.' *China and World Economy* 18, no. 1: 1–32. doi.org/10.1111/j.1749-124X.2010.01178.x.

Modigliani, Franco, and Shi Larry Cao. 2004. 'The Chinese Saving Puzzle and the Life-Cycle Hypothesis.' *Journal of Economic Literature* 42, no. 1: 145–70. doi.org/10.1257/00220510 4773558074.

Piketty, Thomas, Li Yang, and Gabriel Zucman. 2019. 'Capital Accumulation, Private Property, and Rising Inequality in China, 1978–2015.' *American Economic Review* 109, no. 4: 1234–71. doi.org/10.1257/aer.20170973.

Qiu, Tunye, and Weifeng Liu. 2024. 'Housing Prices and Marriage Delay: Evidence from China.' 11 October. Available at SSRN: doi.org/10.2139/ssrn.4985646.

Son, Ngoc Chu, and Ligang Song. 2015. 'Promoting Private Entrepreneurship for Deepening Market Reform in China: A Resource Allocation Perspective.' *China and World Economy* 23, no. 1: 47–77. doi.org/10.1111/cwe.12099.

Song, Ligang. 1996. 'Institutional Change, Trade Composition, and Export Supply Potential in China.' In *Inflation and Growth in China*, edited by Manuel Guitian and Robert Mundell, 190–225. Washington, DC: International Monetary Fund. www.elibrary.imf.org/display/book/9781557755421/ch014.xml.

Song, Ligang. 2005. 'Interest Rate Liberalisation in China and the Implications for Non-State Banking.' In *Financial Sector Reform in China*, edited by Yasheng Huang, Anthony Saich, and Edward Steinfeld, 111–30. Cambridge: Harvard University Asia Center. doi.org/10.1163/9781684171224_008.

Song, Ligang, Wu Jiang, and Yongsheng Zhang. 2010. 'Urbanisation of Migrant Workers and Expansion of Domestic Demand.' *Social Sciences in China* 31, no. 3: 194–216. doi.org/10.1080/02529203.2010.503080.

Song, Ligang, Jidong Yang, and Yongsheng Zhang. 2011. 'State-Owned Enterprises' Outward Investment and the Structural Reform in China.' *China and World Economy* 19, no. 4: 38–53. doi.org/10.1111/j.1749-124X.2011.01249.x.

Wang, Xin, and Yi Wen. 2012. 'Housing Prices and the High Chinese Saving Rate Puzzle.' *China Economic Review* 23, no. 2: 265–83. doi.org/10.1016/j.chieco.2011.11.003.

Wong, Christine. 2024. 'The State of Public Finance in China: Why Tax and Intergovernmental Reform Are Urgently Needed.' In *China: Regaining Growth Momentum after the Pandemic*, edited by Ligang Song and Yixiao Zhou, 67–83. Canberra: ANU Press. doi.org/10.22459/CRGMP.2024.04.

Woo, Wing Thye. 2006. 'The Structural Nature of Internal and External Imbalances in China.' *Journal of Chinese Economic and Business Studies* 4, no. 1: 1–19. doi.org/10.1080/1476528 0600551208.

Yang, Dennis Tao. 2012. 'Aggregate Savings and External Imbalances in China.' *Journal of Economic Perspectives* 26, no. 4: 125–46. doi.org/10.1257/jep.26.4.125.

Yu, Yongding. 2007. 'Global Imbalances: China's Perspective.' Paper prepared for Global Imbalances conference, Institute of International Education, Washington, DC, 8 February. www.piie.com/sites/default/files/publications/pb/pb07-4/yu.pdf.

Zhang, Longmei, R. Brooks, Ding Ding, Haiyan Ding, Hui He, Jing Lu, and Rui Mano. 2018. *China's High Savings: Drivers, Prospects, and Policies.* IMF Working Paper No. WP/18/277. Washington, DC: International Monetary Fund. doi.org/10.5089/9781484388778.001.

Zhou, Xiaochuan. 2009. 'Understanding China's High Savings Rate.' Speech, High-Level Conference on the International Financial Crisis and Asia, People's Bank of China and International Monetary Fund, Dalian, China, 27–28 March.

Zhou, Yixiao, and Ligang Song. 2016. 'Income Inequality in China: Causes and Policy Responses.' *China Economic Journal* 9, no. 2: 186–208. doi.org/10.1080/17538963.2016.1168203.

12

Rural transformation in the developing world and China's strategic role: A cross-country comparative perspective

Moyu Chen, Xianneng Ai, Aizhao Wang and Yu Sheng

Over the past 50 years, China has made significant strides in rural revitalisation, particularly achieving a pivotal phase in rural transformation. According to the Food and Agriculture Organization of the United Nations (FAO 2024), China's agricultural labour productivity has risen significantly, from an average of US$10,000 in 1970 to more than US$13,000 in 2022—a growth rate that surpasses that of other developing regions in Asia, Latin America and Africa. Concurrently, there has been a marked shift in employment from agriculture to non-agricultural sectors; China's off-farm employment increased from 20 per cent in 1970 to approximately 80 per cent in 2022, indicating a key structural change necessary for rural modernisation (FAO 2024). This transformation has improved living standards in rural communities and strengthened the nation's economic growth and food security.

The success of China's rural development is largely due to strategic market-oriented reforms and well-designed institutional, policy and investment measures. Since the 1990s, China has rapidly reformed its agricultural sector, becoming a leader among developing countries with minimal agricultural support relative to GDP. Coupled with domestic reforms, China's gradual opening of its agricultural markets to international trade has further stimulated growth. Meanwhile, dual market and government adjustment mechanisms have become crucial for stabilising rural and structural transformation processes. The government's role in building agricultural infrastructure has been essential, especially in a context in which small-scale farming is prevalent, ensuring that market mechanisms are effectively supported and rural transformation is stabilised, leading to a robust and productive agricultural sector.

Despite its significant strides, China's rural revitalisation is now facing formidable challenges due to environmental and resource constraints. Climate change, as highlighted by the United Nations (UN Statistics Division 2023), impacts more than one-third of the global population, particularly in regions such as Asia, where water stress exceeds 75 per cent. This, coupled with the degradation of natural resources—for example, about 33 per cent of the world's soils are moderately to severely degraded— and the deterioration of more than 60 per cent of irrigated land in critical regions such as North Africa, South Asia and the Middle East, underscores the urgency of these obstacles. Additionally, ecological shifts, such as the loss of nearly 130 million hectares of agricultural land from 2000 to 2019 and the annual contamination of water bodies with nearly 600,000 tonnes of phosphorus from agricultural runoff, compound the issue. These multifaceted environmental pressures demand a strategic reassessment and adaptation of rural development approaches.

In response to these challenges, China has charted a new course for rural transformation and revitalisation, prioritising the enhancement of 'new quality' agricultural productivity. This innovative approach transcends conventional metrics of output and efficiency by embracing the principles of inclusivity and environmental sustainability. By integrating these core values, China is adopting advanced agricultural technologies—encompassing biotechnology, digital technologies and artificial intelligence (AI)—to increase agricultural productivity while considering ecological preservation. This approach seeks not only to augment current productivity but also to mitigate natural resource degradation and environmental pollution, thereby laying a foundation for sustainable agricultural practices that can endure for future generations. More importantly, especially for the rural sector predominantly comprising small-scale farmers, it ensures that the benefits of agricultural advancement are broadly shared across society.

This chapter delves into the evolution of rural revitalisation in China, examining its strategic role within the broader context of the developing world. It offers a comparative cross-country perspective by using data from authoritative sources such as the FAO's database (FAOSTAT),[1] the US Department of Agriculture's International Agricultural Productivity database (IAP)[2] and the National Bureau of Statistics of China (NBS).[3] By examining these trajectories, we seek to extract valuable lessons and insights that can benefit other developing countries in their quest to transform and revitalise their rural sectors. Ultimately, the goal is to provide a comprehensive overview that not only highlights China's journey but also offers a roadmap for sustainable and inclusive rural development applicable to diverse global contexts.

The chapter is structured as follows. The first section outlines the motivation behind China's rural revitalisation and its importance in the context of developing countries. The second section examines the main drivers of China's rural transformation. The third

---

1   FAOSTAT: www.fao.org/faostat/en/#home.
2   ERS: www.ers.usda.gov/data-products/international-agricultural-productivity/.
3   NBS: data.stats.gov.cn/.

section discusses the new challenges faced by China in its rural revitalisation efforts. Section four delves into new directions for China's rural transformation, focusing on inclusiveness, fairness and sustainability, while the conclusion synthesises the key findings and their implications for other developing countries.

Rural transformation pathways in China and their significance for developing countries

Rural development and revitalisation often hinge on rural transformation, as revealed by an examination of various countries' trajectories. The International Fund for Agricultural Development (IFAD) defines rural transformation as a process intricately linked to the ongoing enhancement of agricultural labour productivity, a shift in agricultural production from low-value food crops to more commercially viable high-value agricultural products and a notable increase in non-agricultural employment and entrepreneurial opportunities (IFAD 2016). This process is also characterised by structural transformation, with the national economy transitioning from an agriculture-centric to an industry and service–oriented model. These transformations are interdependent and collectively pivotal for the advancement and modernisation of rural regions.

Economic studies initially primarily focused on examining the rural transformation triggered by the Industrial Revolution. This transformation was pivotal in shaping the trajectory of agricultural modernisation and its broader implications for national economies (Johnston 1970; Schultz 1964). Over the past two centuries, there has been a notable shift in global agricultural practices from extensive to intensive methods. This shift has been driven by the growing demand for food due to rising populations and urbanisation, occurring within the constraints of limited resources (Boserup 1976). The progression from extensive to intensive agricultural practices has been a critical response to the challenges posed by population growth and urbanisation, demonstrating the adaptability and resilience of the agricultural sector in the face of evolving global demands. Stable agricultural development is vital for ensuring global food security and is especially crucial for countries that rely heavily on agriculture (Block 1994; Dethier and Effenberger 2012). According to the World Bank (2023), the total value of global agricultural output has tripled since the mid-twentieth century. This trend is particularly pronounced among developing countries, with agricultural output in lower middle-income Asia-Pacific nations increasing more than sixfold (Alston and Pardey 2014; Block 1994).

Global rural development faces core challenges, including food security, rural income generation and the pursuit of sustainable agricultural practices, which have recently drawn considerable attention. In response to these challenges, the UN Sustainable Development Goals promote the development and implementation of targeted solutions by developing countries, customised to their unique sociocultural and economic contexts (UNDESA 2015). These solutions aim to curb rural decline, alleviate poverty and improve residents' quality of life. The *World Development Report* highlights the critical role of rural transformation as a primary growth driver, particularly for economies with a strong agricultural base (Byerlee et al. 2008). This transformation is evident in the sizeable changes occurring in rural areas across Asia, Africa and Latin America, where economic and social shifts underscore the importance of strategic rural development in driving broader socioeconomic progress (IFAD 2016).

Theoretically, rural and structural transformation unfold through several developmental phases, each with distinct characteristics. IFAD (2016) divides this process into five evolutionary stages: starting with an agriculture-dominated phase, it advances through early industrialisation, service sector expansion, urbanisation and modernisation and, finally, matures into a knowledge and innovation economy. Each stage is a pivotal moment in the journey, reflecting profound shifts in the economic and social landscape of rural areas.

Nevertheless, it is evident that rural and structural transformations do not follow a one-size-fits-all pattern. As a notable example, China has exemplified successful rural transformation through an accelerated developmental trajectory. Over the past four decades, China has experienced rapid growth in per capita agricultural income and has released a significant agricultural labour force through increased agricultural productivity. This has led to a swift increase in the proportion of non-agricultural employment, achieving structural transformation alongside rural transformation. Nevertheless, not all nations have successfully navigated through the stages of rural transformation. For instance, in sub-Saharan Africa, postcolonial GDP growth has not significantly alleviated poverty and agricultural productivity growth remains subdued (Barrett et al. 2017; Wuyts and Kilama 2015).

To encapsulate the income and marketisation of the agricultural sector, we characterise rural transformation in developing countries by agricultural labour productivity and the share of high-value commodities. Figure 12.1 shows that, over the past 50 years, there has been substantial progress in agricultural labour productivity among developing countries. In Asia, agricultural labour productivity has shown an overall upward trend, with an increase from an average of US$5,000 in 1970 to more than US$9,000 in 2022 (constant 2015 prices), indicating ongoing advancements in agricultural transformation and output. Latin American developing countries have seen a growth pattern similar to that of Asian developing countries, with agricultural labour productivity rising from US$5,000 in 1970 to more than US$15,000 in 2022. Conversely, African developing countries confront challenges of both low agricultural labour productivity and slow growth rates. Despite improvements, Africa's per capita agricultural GDP remained below US$5,000 in 2022, reflecting a slower pace of development than other regions.

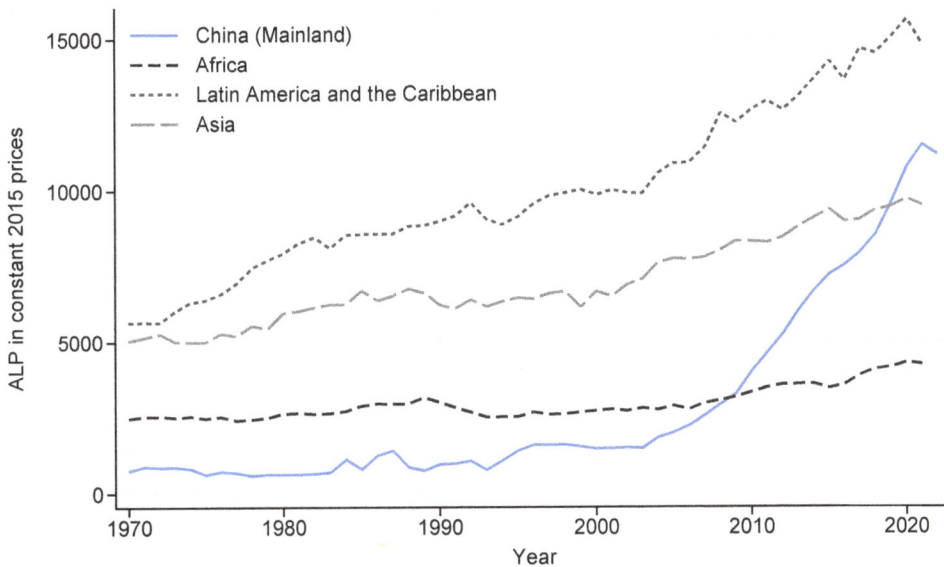

Figure 12.1: Trend of agricultural labour productivity in China and other developing regions, 1970–2022

Sources: Authors' own estimations based on the ERS (www.ers.usda.gov/data-products/international-agricultural-productivity/); and NBS database (data.stats.gov.cn/english/).

Compared with other developing countries, China has exhibited accelerated growth in agricultural labour productivity. From 1970 to 2004, agricultural labour productivity in China fluctuated around US$1,000 to US$2,000. Since 2005, the growth rate has accelerated significantly with the advancement of market reforms and increased trade openness in China, rising from US$2,000 to more than US$10,000. Between 1997 and 2020, China's growth rate surpassed that of other Asian and Latin American countries. However, after 2020, there was a slowdown.

Between 1970 and 2020, the proportion of high-value agricultural products among developing countries grew, with significant regional variations (Figure 12.2). In Asia, the share of high-value agricultural products increased from about 55 per cent in 1970 to nearly 70 per cent by 2020, indicating phased progress in agricultural production reform. In contrast, Latin America had a higher initial share, of more than 60 per cent, in 1970, growing slowly to nearly 70 per cent by 2020. Meanwhile, African countries mainly produced low-value food crops and, although the share of high-value products increased, it did not exceed 50 per cent overall.

As the same time, China's growth rate of high-value agricultural products significantly outpaced other developing countries. Initially, China's agricultural production focused on low-value food crops. However, government and market efforts shifted production towards higher-value crops. By 1990, China's share of high-value agricultural output reached 40 per cent, surpassing the average level of African developing countries. By 2000, this proportion increased to more than 60 per cent, nearing the average level of Asia and Latin America. Over the next two decades, China's share of high-value agricultural products fluctuated slightly but showed a general upward trend.

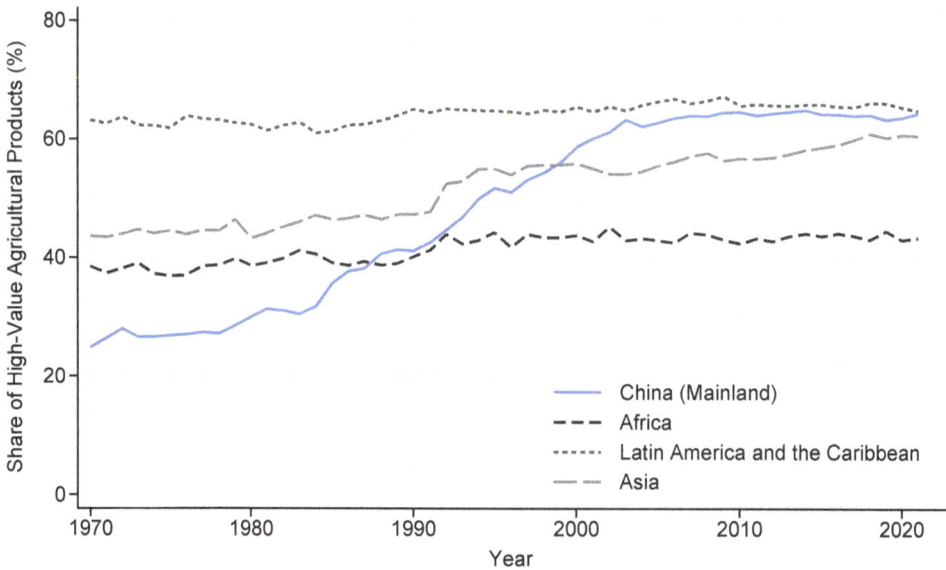

Figure 12.2: Trend of the share of high-value agricultural products in China and other developing regions, 1970–2022

Sources: Authors' own estimations based on FAOSTAT (www.fao.org/faostat/en/#home); and NBS database (data.stats.gov.cn/english/).

With rural transformation, developing countries are also experiencing rapid structural changes, highlighted by the growing importance of non-agricultural GDP and off-farm employment. As shown in Figure 12.3, the proportion of non-agricultural GDP in developing countries has a significant upward trend, reflecting urbanisation and industrialisation advancements. Latin America had the highest proportion, starting at 90 per cent in 1970 and stabilising around 95 per cent between 2010 and 2020. In Asia, the proportion grew rapidly, from about 65 per cent in 1970 to 90 per cent by 2020, surpassing the world average by 2000. This trend shows Asia's progress in rural poverty reduction and structural transformation. In contrast, Africa is still in the early stages of structural transformation, facing obstacles in non-agricultural industry development. In 1970, African countries had just over 70 per cent non-agricultural GDP, which slowly increased to 80 per cent by 2020.

China has achieved significant success in developing its non-agricultural sector, with growth rates far exceeding those of other developing countries. In 1970, China's non-agricultural GDP share was just over 30 per cent—lower than other developing countries. However, over the past five decades, this share grew rapidly, reaching 80 per cent by 2000, surpassing Africa's average. By 2010, it exceeded the average level in Asia, approaching Latin American levels. This trend reflects China's substantial progress in non-agricultural industry development and structural transformation.

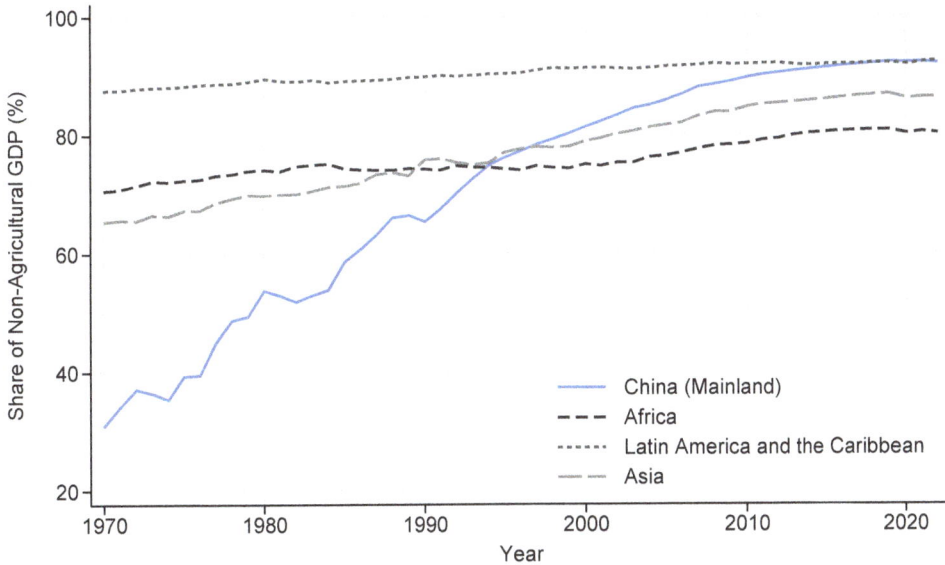

Figure 12.3: Trend in the share of non-agricultural GDP in China and other developing regions, 1970–2022

Sources: Authors' estimations based on FAOSTAT (www.fao.org/faostat/en/#home); and NBS database (data.stats.gov.cn/english/).

In terms of off-farm employment, developing countries in different regions have shown positive growth trends, reflecting the ongoing release of rural labour from farm employment and deepening global rural structural transformation. In Latin America, the share of off-farm employment was as high as 40 per cent in 1970 and grew to about 60 per cent by 2022 (as shown in Figure 12.4). In Asia, the share increased continuously from about 35 per cent in 1970 to nearly 50 per cent in 2022. Africa also experienced a pronounced change in employment structure, but the share of off-farm employment is still lower than in other regions. Starting from a lower base at 20 per cent, the share of off-farm employment in African developing countries just exceeded 40 per cent by 2022, reflecting the fact that African countries are still in an important stage of economic structural adjustment.

The growth rate of China's share of off-farm employment has been even faster. Despite starting from a lower level, by the end of the sample period, China's proportion exceeded the average level in other developing countries. In sync with rural transformation, China's share of off-farm employment increased from about 20 per cent in 1970 to about 70 per cent in 2022, representing nearly a 2.5-fold increase. By the early twenty-first century, China's share had reached 50 per cent, surpassing the average in other developing countries. This continuously adjusting employment structure reflects the rapid progress of urbanisation and structural transformation alongside rural transformation in China.

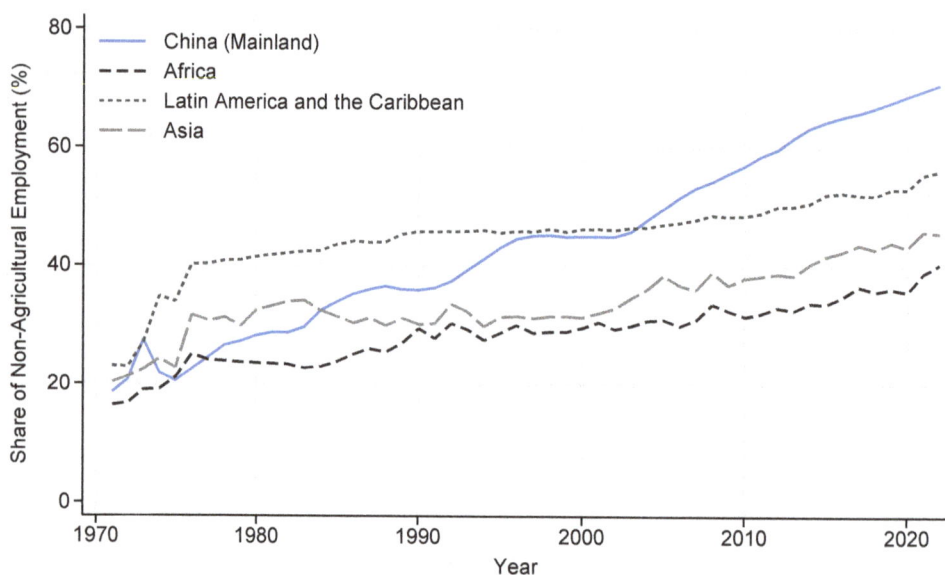

Figure 12.4: Trend in the share of non-agricultural employment in China and other developing regions, 1970–2022

Sources: Authors' estimations based on FAOSTAT (www.fao.org/faostat/en/#home); ERS (www.ers. usda.gov/data-products/international-agricultural-productivity/); and NBS database (data.stats.gov. cn/english/).

The drivers of China's rural transformation

Rural transformation and structural adjustment represent a gradual evolution towards marketisation, where the agricultural sector has shifted from government leadership to market mechanisms. From the end of the last century to the present, the total amount of agricultural support in most developing countries has shown an initial increase followed by a decrease, with its proportion in GDP generally declining, reflecting the decreasing relative importance of agricultural support in the national economy.

Table 12.1 shows a global comparison of government agricultural support in developing countries. In Latin America, the total value of agricultural support was US$8.9 billion between 2000 and 2009, constituting 0.47 per cent of GDP. This decreased to US$6.9 billion between 2010 and 2019, with the proportion of GDP slightly increasing to 0.54 per cent, and further reduced to US$4.4 billion between 2020 and 2023, with the GDP share dropping to 0.41 per cent. In Asia, the total amount of agricultural support was US$14.9 billion between 2000 and 2009, representing 0.75 per cent of GDP. This increased to US$17.1 billion between 2010 and 2019, with the GDP share slightly decreasing to 0.68 per cent, and significantly dropped, to US$13.9 billion, between 2020 and 2023, with the GDP share reduced to 0.59 per cent.

Table 12.1: Agricultural support in China and other developing countries

Region	Agricultural support (US$ million, constant 2015 price)	% of GDP
Panel A: 2000–09		
China	8,205.05	0.16
Africa	258.00	1.12
Latin America and the Caribbean	892.07	0.47
Asia	1,491.94	0.75
Panel B: 2010–19		
China	13,787.12	0.13
Africa	274.17	0.86
Latin America and the Caribbean	694.34	0.54
Asia	1,713.30	0.68
Panel C: 2020–23		
China	8,273.99	0.05
Africa	297.02	0.76
Latin America and the Caribbean	436.67	0.41
Asia	1,389.94	0.59

Note: The level of agricultural support and its GDP ratio in the table pertain solely to mainland China, which is excluded from the 'Asia' sample.

Sources: Authors' estimations based on FAOSTAT (www.fao.org/faostat/en/#home); World Bank (data.worldbank.org/); and NBS database (data.stats.gov.cn/english/).

In contrast, Africa's agricultural sector continues to increase its reliance on public policy, with sustained growth in the total amount of agricultural support and a proportion of GDP significantly higher than in other developing regions. The total value of government agricultural support in Africa was US$2.6 billion between 2000 and 2009, accounting for 1.12 per cent of GDP. It slightly increased, to US$2.7 billion, between 2010 and 2019, with the GDP share reducing to 0.86 per cent, and further increased, to US$3 billion, between 2020 and 2023, constituting 0.76 per cent of GDP.

Since the 1990s, China's agricultural sector has undergone rapid marketisation reforms, making it one of the developing countries with the lowest agricultural support as a proportion of GDP. In China, the level of total agricultural support was about US$8.21 billion between 2000 and 2009, accounting for 0.16 per cent of GDP. From 2010 to 2019, the total increased to US$13.79 billion, while its proportion of GDP declined to 0.13 per cent. Between 2020 and 2023, the total decreased to US$8.27 billion, with its share of GDP reduced, to 0.05 per cent.

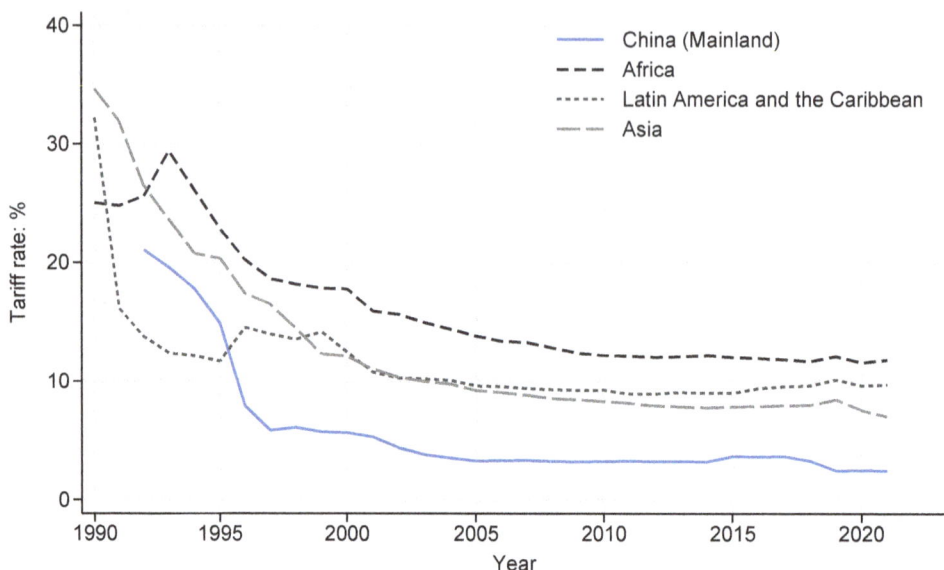

Figure 12.5: Trend of tariff rate changes in China and other developing regions, 1990–2022

Sources: Authors' compilation based on the World Bank (data.worldbank.org/); WTO Stats (stats. wto.org/); and NBS database (data.stats.gov.cn/english/).

In addition to domestic market reforms, China has gradually opened its market for agricultural products to the world. The acceleration of globalisation in the 1990s and the establishment of the World Trade Organization (WTO), along with the impetus of the Doha Round of negotiations in the twenty-first century, have led to a gradual reduction in tariffs and an expansion of market access for agricultural products in developing countries. As shown in Figure 12.5, there was a rapid decline in tariffs in developing countries in the 1990s and a generally downward trend has been maintained in the twenty-first century. Starting in 1996, China reduced its tariffs below the average level of developing countries in Asia, Africa and Latin America, and has since maintained a long-term continuous decline.

Consequently, the reduction in tariffs has promoted international trade and market access for agricultural products from China, enabling these products to enter the international market at a lower cost, thereby increasing export opportunities and competitiveness. At the same time, the domestic market may face more intense competition due to the entry of foreign agricultural products, prompting domestic agricultural producers to improve their efficiency and product quality. Although there may be short-term pressure on some agricultural sectors that depend on protection, in the long run, reducing tariffs is beneficial for enhancing the competitiveness and sustainable development capacity of the entire agricultural sector. This ultimately supports the transition to a more mature stage of rural and structural transformation (Gale 2013; Orden et al. 2003).

However, given the vulnerability of the agricultural and rural sectors in developing countries, relying solely on market adjustment mechanisms for rural and structural transformation is challenging. For example, to ensure food security and meet the basic nutritional needs of a growing population, these regions initially formed production models centred on staple foods such as rice, wheat and corn. This food production–dominated model limited the development of non-agricultural employment, resulting in a concentration of rural labour in agriculture, a single rural development model and generally low incomes dependent mainly on agricultural output (Huang 2020; IFAD 2016). Therefore, governments in various developing countries continuously formulate and adjust institutions, policies (such as technology policy and market reforms) and investments to complement market mechanisms and facilitate smooth rural transformation.

In China, dual market and government adjustment mechanisms have become crucial for stabilising rural and structural transformation processes. The agricultural production model dominated by small farmers makes spontaneous market adjustment difficult for constructing agricultural and rural infrastructure. Consequently, the government has long undertaken the construction of agricultural production infrastructure. Public investment in infrastructure and R&D, due to their significant positive external effects on agricultural production, has been a central focus of rural public investment by the central and local governments in China since the period of reform and opening.

The long-term investment by China's public sector in irrigation infrastructure is a typical and effective strategy for improving the efficiency of the agricultural sector and promoting rural transformation. Since the 1970s, China's investment in irrigation infrastructure has expanded significantly, enhancing agricultural production and achieving stable and increased output. From 1970 to 2022, the proportion of irrigated land in China's total arable land increased from less than 20 per cent to more than 50 per cent—far exceeding the average level of developing countries in Asia, Latin America and Africa (as shown in Figure 12.6). By the end of 2022, China's effective irrigated land area was 70.4 million hectares (Wang et al. 2017). The continuous improvement of irrigation infrastructure has become crucial for ensuring food security, with irrigated agriculture supporting more than 70 per cent of China's grain crops, more than 80 per cent of its cotton and more than 90 per cent of its vegetable production (Wang et al. 2017).

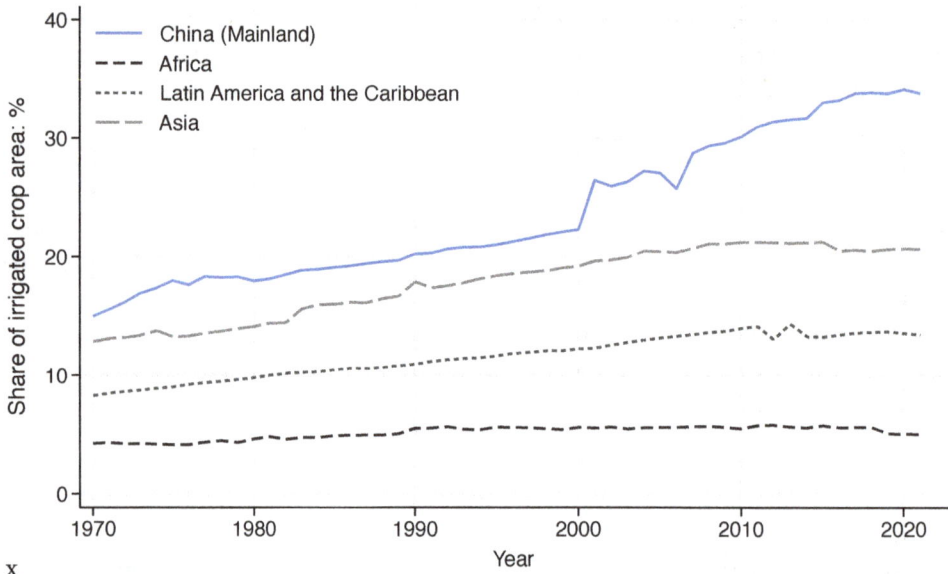

Figure 12.6: Trend of changes in the scale of irrigation infrastructure in China and other developing regions, 1970–2022

Note: The level of irrigation is measured by the proportion of irrigated land in the total area of farmland.

Sources: Authors' estimations based on the ERS (www.ers.usda.gov/data-products/international-agricultural-productivity/); and NBS database (data.stats.gov.cn/english/).

In addition to domestic market reforms, China has gradually opened its market for agricultural products to the world. The acceleration of globalisation in the 1990s and the establishment of the World Trade Organization (WTO), along with the impetus of the Doha Round of negotiations in the twenty-first century, have led to a gradual reduction in tariffs and an expansion of market access for agricultural products in developing countries. As shown in Figure 12.5, there was a rapid decline in tariffs in developing countries in the 1990s and a generally downward trend has been maintained in the twenty-first century. Starting in 1996, China reduced its tariffs below the average level of developing countries in Asia, Africa and Latin America, and has since maintained a long-term continuous decline.

However, the positive impacts of irrigation infrastructure on rural transformation have diminished over time (Sheng et al. 2024). Specifically, from 1980 to 1985, a 1 per cent increase in irrigation infrastructure boosted agricultural productivity by 0.3 per cent; from 2012 to 2018, this impact declined to 0.2 per cent. This trend aligns with the historical development of irrigation infrastructure. In the early 1980s, irrigation construction significantly optimised rural resource allocation and aided adjustment of agricultural structure, enhancing productivity. In the late 1990s, with a solid foundation of irrigation infrastructure, government policies and institutional innovations further accelerated productivity growth. Entering the twenty-first century, the focus on green, efficient and sustainable agricultural development has introduced new requirements for transforming irrigation investment, presenting fresh challenges for productivity growth. Therefore, economic benefits should not be the sole criterion for evaluating rural revitalisation. In promoting dual circulation and sustainable development, the establishment of inclusive and fair sustainable rural development measurement indicators is urgently needed.

New directions in China's rural transformation: Inclusiveness, fairness and sustainability

Climate change, natural resource degradation and ecological alterations have become significant factors constraining rural development due to their adverse impacts on the agricultural sector. Rapid global urbanisation leads to a continuous decrease in the amount of agricultural land and intensifies demand for water resources for agriculture. Long-term reliance on input-intensive strategies, such as the excessive use of chemical fertilisers and pesticides, has caused severe environmental pollution in agricultural production. This creates a 'double constraint' of tightening resource constraints and overloading the environmental carrying capacity, hindering sustainable agricultural development.

The challenges posed by climate change and resource degradation are even more severe for developing countries with large populations and limited cropland. The UN Statistics Division (2023) noted that, in 2020, more than one-third of the world's population lived in countries experiencing water stress, with more than 75 per cent of those in Central and South Asia facing high water stress, and North Africa exceeding 100 per cent. Approximately 33 per cent of soil globally shows moderate to severe degradation and, in North Africa, South and West Asia and the Middle East, more than 60 per cent of irrigated land has been degraded. The impacts of water and soil resource shortages and pollution from pesticides and chemical fertilisers are expanding; globally, the amount of agricultural land decreased by nearly 130 million hectares between 2000 and 2019, with up to 1.5 million hectares of cropland rendered unproductive annually due to soil salinisation (FAO 2021). Furthermore, nearly 600,000 tonnes of phosphorus enters water bodies annually from agricultural production, and a survey of 11 European countries found that nearly 60 per cent of agricultural topsoil contains multiple persistent pesticide residues (FAO 2018, 2021). Climate change also significantly impacts agricultural development, with the Food Security Information Network (FSIN and GNAFC 2024) reporting that nearly one-third of 59 countries (representing about 72 million people) face food crises related to climate—up from 12 countries (and 56.8 million people) in 2022.

Despite some countries adopting measures to address environmental externalities, imbalanced regional development has led to different rural transformation paths. Countries such as India and China have improved agricultural productivity through large-scale irrigation projects, mitigating climate change uncertainties, promoting the commercialisation and diversification of agriculture and accelerating rural transformation. However, in Africa, especially sub-Saharan Africa, irrigation infrastructure is insufficient, with most agriculture relying on rain, leading to low and unstable productivity and hindering rural transformation towards higher value-added agriculture and non-agricultural industries. In these regions, GDP per capita growth has not reduced poverty and agricultural productivity growth remains low (Dercon and Gollin 2014; Wuyts and Kilama 2015). Although the share of agriculture in GDP and employment has declined, the non-agricultural labour force has shifted more towards non-trade service industries rather than manufacturing (IFAD 2016; Rodrik 2015).

Traditionally, rural transformation has been deficient in considering inclusiveness, fairness and sustainability. First, regarding inclusiveness, agricultural labour productivity focuses on overall output efficiency. Policies and strategies often support large agricultural operators who can boost productivity, neglecting the productivity and income of small farmers and vulnerable groups. The emphasis on high-value agricultural products highlights economic benefits but overlooks whether small farmers have the resources to produce these products, potentially excluding them from high-profit production. Second, the focus on the proportion of non-agricultural GDP emphasises agricultural economic growth without reflecting on the fairness of income distribution. This can hide significant income disparities within rural areas, concentrating resources and opportunities among large farmers and affluent groups. Last, traditional agricultural total factor productivity (TFP) emphasises input–output efficiency but ignores potential environmental degradation and resource depletion during production, which is detrimental to the long-term sustainability of agricultural development.

To spearhead a new era of rural revitalisation, China is leading the way among developing countries by focusing on enhancing 'new quality' agricultural productivity—a concept that emphasises inclusion, fairness and sustainability. Inclusiveness is addressed by considering a broader range of input factors, such as natural resources and weather. In terms of fairness, this approach aims to improve resource utilisation efficiency, reduce waste and environmental damage and increase farmers' income through efficient production, promoting a fair distribution of resources and benefits. This is particularly important for developing countries because it encourages small farmers to participate in efficient, low-consumption agricultural production through technological and managerial innovations, thus improving their market competitiveness and achieving more inclusive development. Finally, sustainability is achieved by prioritising the efficient use of resources and environmental protection, which includes reducing chemical inputs and enhancing water efficiency—all aimed at supporting long-term agricultural resilience.

Enhancing 'new quality' agricultural productivity and fostering comprehensive rural revitalisation can take two primary pathways: capital deepening and sustainable environmental development. Capital deepening in agricultural production is the process of increasing the amount of capital, such as machinery, equipment and infrastructure, used in the production process relative to the amount of labour. Historically, developed countries such as the United States and Australia have effectively utilised capital-intensive technologies to spur agricultural productivity growth. The abundant natural resources in these nations, such as land, have diminished the potential for marginal improvements through biological and chemical technologies—a scenario that many land-scarce developing countries cannot replicate.

China has forged a unique path to capital deepening: the emergence of innovative management and business models, such as customised services, is transforming agricultural production methods and propelling a significant leap in productivity in developing countries. This is accomplished by integrating technological advancements into the use of capital goods (Sheng et al. 2022). For example, the development of customised service systems enables the substitution of capital for labour in areas that are densely populated and have limited land. This not only alleviates labour shortages at a reduced cost but also establishes conditions that are favourable for the adoption of sophisticated technologies. In contrast to labour-intensive technological advancements, this new capital-intensive approach, supported by customised services, is gaining popularity as a sustainable engine for agricultural productivity growth.

Concurrently, capital deepening facilitates the spread of agricultural technologies across regions with diverse agricultural ecosystems and climates, thereby reducing disparities in agricultural production efficiency. The natural geographic variations have historically limited the transferability of agricultural technologies between different ecological and climatic zones, leading to international differences in agricultural productivity (Gollin et al. 2014a, 2014b). Nevertheless, investments in R&D, particularly in fields such as biotechnology, AI and information technology, have enabled the broader application of agricultural technologies across regions (Sheng et al. 2015). Genetic modifications, for example, have made certain crops and livestock varieties originally suited for humid conditions more drought-resistant, allowing them to thrive in arid regions and thereby enhancing agricultural output and efficiency.

Figure 12.7 depicts the relationship between agricultural productivity—measured by TFP—and the degree of capital deepening in China relative to major developed countries. Generally, these developed countries and regions have achieved a higher level of capital accumulation. The European Union and Australia, in particular, are currently in a phase in which they are increasing capital investment to bolster productivity. Their agricultural productivity continues to grow at a rate that exceeds that of China, indicating a trend of convergence. However, in the United States, there is a clear negative relationship between agricultural productivity and capital deepening. This suggests that the United States has been the first to complete the phase of promoting productivity growth through increased capital investment, beyond which capital investment has started to decline.

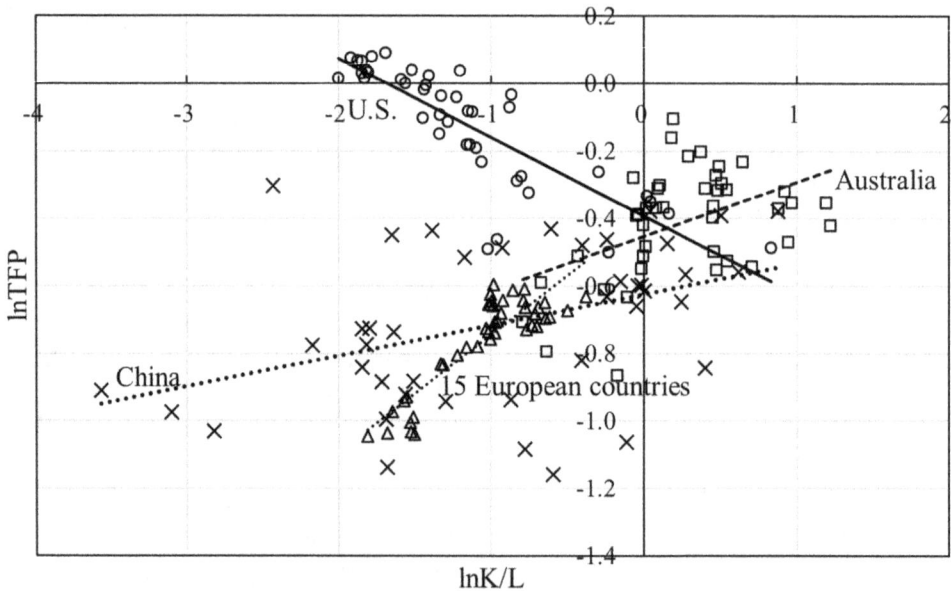

Figure 12.7: The pathways of agricultural capital intensification across countries (regions), 1978–2019

Source: Authors' estimations based on the International Comparison of Agricultural Productivity database (icapproject.com/).

China's agricultural capital accumulation is positively correlated with its agricultural productivity and its capital deepening is still on an upward trajectory, potentially progressing towards an optimal level in comparison with major developed countries and regions. Thus, there remains significant policy space for enhancing agricultural productivity in the future, particularly in the process of improving 'new quality' productivity. However, it should be noted that due to the declining efficiency of existing public capital investment, the current mode of such investment may no longer support the continuous growth of agricultural productivity (as evidenced by the downward trend in the internal rates of return on China's infrastructure investment in the previous section). Therefore, the direction of China's agricultural investment will require substantial reform.

An environmentally sustainable development path is another crucial route to achieving 'new quality' in agricultural productivity. This approach emphasises maintaining or enhancing agricultural output while minimising the social and environmental costs associated with agricultural production. Meeting the growing global demand for food while achieving sustainable agricultural development presents a notable challenge. Throughout the long-term development of agriculture, producers have either increased the input of new land into production or enhanced yields to meet the ever-increasing demand by adding labour, machinery, energy, fertilisers and other inputs. Despite the remarkable achievements in global agricultural development, current production methods, including the massive increase in reactive nitrogen use and changes in land systems, may undermine land productivity, thereby affecting the long-term sustainability of agriculture. Data indicate that if the current trend of global land degradation continues, 1.5 billion hectares of degraded land will need to be restored by 2030 to achieve the zero-growth target for land degradation set in the UN Sustainable Development Goals (UN 2024).

Environmental conditions are crucial determinants of agricultural production, but the heterogeneity of these across countries leads to diverse paths in the pursuit of sustainable agricultural development. Figure 12.8 summarises the general principles of sustainable agricultural development by comparing the relationship between agricultural land productivity and agricultural labour productivity in China and major developed countries and regions. As shown, there is a positive correlation between agricultural land productivity and labour productivity in all countries. However, a weaker positive relationship indicates higher land production efficiency, implying a greater role for sustainable environmental utilisation in agricultural development. Because US agricultural production is capital-intensive and relies more on land to address agricultural development issues, its degree of resource utilisation contributes less to the sustainability of agriculture. In contrast, European countries and Australia, although initially following a similar development path to the United States, have recently begun to shift towards more effective resource utilisation methods.

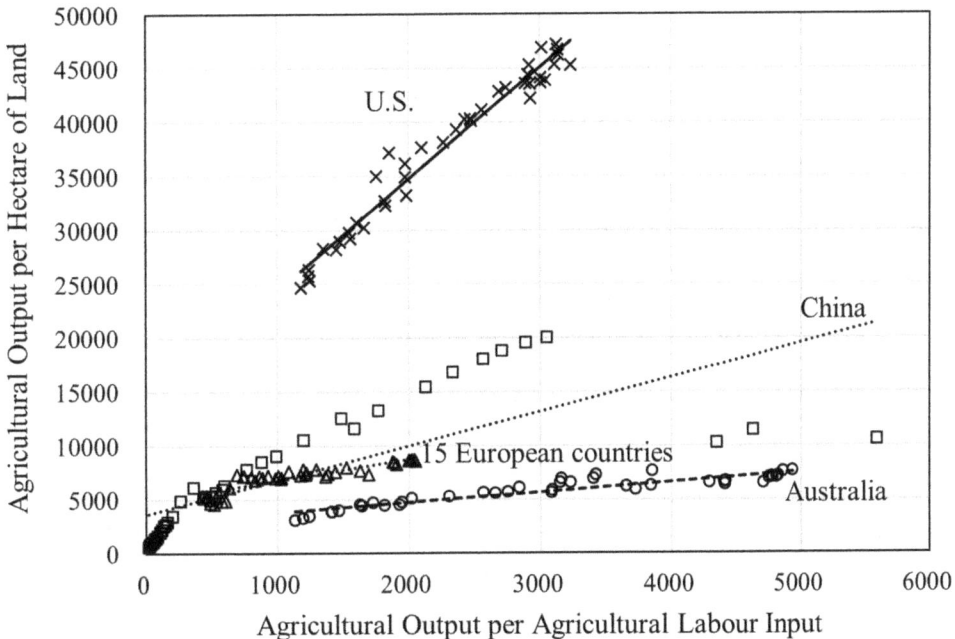

Figure 12.8: Pathways to agricultural productivity across countries (regions), 1978–2019

Source: Authors' estimations based on the International Comparison of Agricultural Productivity database (icapproject.com/).

Compared with major developed countries, China has increased agricultural productivity by relying on land and resource consumption. However, the effect of relying on land resources to enhance agricultural productivity is diminishing, meaning that further increases in land input cannot bring about more growth in productivity. Developed countries such as those in Europe and Australia have already effectively improved the sustainability of agricultural production through policies such as green agriculture. Only the United States still maintains a high dependency on resource consumption, with agricultural income growth coming at the expense of the environment. If the current path continues, the sustainability of China's agricultural production will gradually decrease to below that of the United States.

Since the early twenty-first century, China has been making efforts to improve the efficiency of land and other natural resource utilisation. In 2005, China initiated environmental policies such as emission and fertiliser reduction, leading to a significant improvement in the country's land production efficiency. There is a need to further enhance environmental efficiency by drawing on the sustainable agricultural development models of developed countries and regions such as Europe and Australia.

Conclusion

Over the past half-century, China has made remarkable achievements in rural revitalisation, transforming its agricultural landscape through strategic reforms and investments. The country has witnessed a significant increase in agricultural productivity, which has been instrumental in lifting millions of people out of poverty and enhancing the quality of life in rural areas. China's approach to rural development has been characterised by large-scale infrastructure projects, technological advancements in farming and policies that encourage diversification and modernisation of the agricultural sector. This progress has not only improved the livelihoods of rural populations but also contributed to the nation's overall economic growth and food security.

Currently, China has shifted in a new direction for rural transformation and revitalisation, focusing on enhancing the 'new quality' of agricultural productivity. This concept goes beyond traditional measures of output and efficiency, incorporating principles of inclusiveness and environmental sustainability. By emphasising these values, China aims to ensure that the benefits of agricultural growth are equitably distributed and that the practices adopted do not compromise the environment for future generations. This holistic approach to development seeks to create a balanced and lasting impact on rural communities, promoting economic prosperity in harmony with nature.

China's remarkable progress in rural revitalisation over the past 50 years serves as a beacon for developing countries navigating similar paths. Yet, the road ahead is not without obstacles. The imperative to address environmental and resource degradation, financial constraints and the uneven distribution of skills and resources calls for a comprehensive and nuanced policy approach. To this end, a series of strategic reforms must be pursued with vigour and foresight.

First, it is imperative to reform the science and technology policy system to ensure that it effectively supports the advancement of agricultural practices. This can be achieved by significantly increasing public investment in R&D, with a particular focus on biotechnology, digital technology and agricultural machinery technology. Such a reform will be pivotal in driving innovation that not only enhances productivity but also fortifies the agricultural sector against the impacts of climate change. By prioritising these areas, the potential for sustainable resource use can be maximised, leading to a more resilient and efficient agricultural sector that is better equipped to meet external shocks and can continue to thrive in the face of environmental and economic pressures.

Second, there is a clear need for the government to redefine its role in facilitating rural transformation. Market-oriented reforms should be at the forefront of this strategy, aimed at accelerating the transition to a more efficient and high-quality agricultural sector. It is essential that these reforms are designed to promote a sector that is not only productive but also environmentally sustainable and socially inclusive. To achieve this, the government must also focus on optimising the auxiliary role of institutional, policy and investment measures. These measures should be crafted to support and complement market mechanisms, ensuring that they effectively contribute to the development of the agricultural sector. The goal is to create an environment that encourages sustainable development, where the 'new quality' of agricultural productivity is not just a static objective but an ongoing process that adapts and innovates in response to the ever-evolving challenges of rural development.

Last, and closely intertwined with the previous point, is the necessity to recognise the unique paths to transformation of China's diverse rural regions. The promotion of new agricultural technologies must be sensitive to the varied conditions across the country's vast expanse. By understanding and acknowledging the specific characteristics and stages of development of each region, from pastoral lands to fisheries and the forestry sector, policies can be crafted to resonate with the local context. This tailored approach is essential for the effective empowerment of small farmers, enabling them to harness technological solutions that are not only appropriate but also responsive to their immediate needs. Such an approach allows for the improvement of the productivity of land and water resources, which in turn increases farmers' income and bolsters overall food security. It is through this inclusive and region-specific policy implementation that the interventions can avoid the pitfalls of a one-size-fits-all strategy, ensuring that rural transformation is both sustainable and equitable across China's varied rural landscapes.

Acknowlegements

This research was supported by the Postdoctoral Fellowship Program of CPSF (No. GZB20240034). The views expressed in this chapter are solely the authors' and do not necessarily reflect the views of the supporting agencies and authors' affiliations.

References

Alston, Julian M., and Philip G. Pardey. 2014. 'Agriculture in the Global Economy.' *Journal of Economic Perspectives* 28, no. 1: 121–46. doi.org/10.1257/jep.28.1.121.

Barrett, Christopher B., Luc Christiaensen, Megan Sheahan, and Abebe Shimeles. 2017. 'On the Structural Transformation of Rural Africa.' *Journal of African Economies* 26, [S1]: i11–i35. doi.org/10.1093/jae/ejx009.

Block, Steven A. 1994. 'A New View of Agricultural Productivity in Sub-Saharan Africa.' *American Journal of Agricultural Economics* 76, no. 3: 619–24. doi.org/10.2307/1243676.

Boserup, Ester. 1976. 'Environment, Population, and Technology in Primitive Societies.' *Population and Development Review* 2, no. 1: 21–36. doi.org/10.2307/1971529.

Byerlee, Derek R., Alain De Janvry, Irina Klytchnikova, Elisabeth Marie L. Sadoulet, and Robert Townsend. 2008. *World Development Report 2008: Agriculture for Development.* Washington, DC: World Bank. documents.worldbank.org/en/publication/documents-reports/document detail/587251468175472382/world-development-report-2008-agriculture-for-development.

Dercon, Stefan, and Douglas Gollin. 2014. 'Agriculture in African Development: Theories and Strategies.' *Annual Review of Resource Economics* 6: 471–92. doi.org/10.1146/annurev-resource-100913-012706.

Dethier, Jean-Jacques, and Alexandra Effenberger. 2012. 'Agriculture and Development: A Brief Review of the Literature.' *Economic Systems* 36, no. 2: 175–205. doi.org/10.1016/j.ecosys.2011.09.003.

Food and Agriculture Organization of the United Nations (FAO). 2018. 'Increased Soil Contamination Puts Food Safety and Food Security at Risk.' News, 5 December. Rome: FAO. www.fao.org/newsroom/detail/Increased-soil-contamination-puts-food-safety-and-food-security-at-risk/en#:~:text=According%20to%20FAO%2C%20about%2033,deteriorating %20at%20an%20alarming%20rate.&text=5%20December%202018%2C%20Rome%20 %2D%20Urgent,today%20marking%20World%20Soil%20Day.

Food and Agriculture Organization of the United Nations (FAO). 2021. *The State of the World's Land and Water Resources for Food and Agriculture 2021—Systems at Breaking Point.* Rome: FAO. doi.org/10.4060/cb9910en.

Food and Agriculture Organization of the United Nations (FAO). 2024. *The State of Food and Agriculture 2024: Value-Driven Transformation of Agrifood Systems.* Rome: FAO. doi.org/10.4060/cd2616en.

Food Security Information Network (FSIN), and Global Network Against Food Crises (GNAFC). 2024. *2024 Global Report on Food Crises: Joint Analysis for Better Decisions.* Rome: Food Security Information Network. www.fsinplatform.org/report/global-report-food-crises-2024/.

Gale, Fred. 2013. *Growth and Evolution in China's Agricultural Support Policies.* Economic Research Report No. 153, August. Washington, DC: United States Department of Agriculture. www.ers.usda.gov/publications/pub-details?pubid=45118.

Gatti, Nicolas, Kathy Baylis, and Benjamin Crost. 2021. 'Can Irrigation Infrastructure Mitigate the Effect of Rainfall Shocks on Conflict? Evidence from Indonesia.' *American Journal of Agricultural Economics* 103, no. 1: 211–31. doi.org/10.1002/ajae.12092.

Gollin, Douglas, David Lagakos, and Michael E. Waugh. 2014a. 'Agricultural Productivity Differences across Countries.' *American Economic Review* 104, no. 5: 165–70. doi.org/10.1257/aer.104.5.165.

Gollin, Douglas, David Lagakos, and Michael E. Waugh. 2014b. 'The Agricultural Productivity Gap.' *The Quarterly Journal of Economics* 129, no. 2: 939–93. doi.org/10.1093/qje/qjt056.

Hao, E., K. Hu, and X. Chen. 2015. 'The Income-Increasing Effect of Rural Infrastructure Stock: An Analysis Based on the Panel Data of 30 Provinces in China.' *Rural Economy*: 54–68.

Huang, J.K. 2020. 'Rural Revitalisation: Rural Transformation, Structural Transformation and Government Functions.' *Issues in Agricultural Economy*: 4–16.

International Fund for Agricultural Development (IFAD). 2016. *Rural Development Report 2016: Fostering Inclusive Rural Transformation.* Rome: International Fund for Agricultural Development. www.ifad.org/en/w/publications/rural-development-report-2016.

Johnston, Bruce F. 1970. 'Agriculture and Structural Transformation in Developing Countries: A Survey of Research.' *Journal of Economic Literature* 8: 369–404.

Li, J., Z. Feng, and Q. Wu. 2019. 'Study on the Cost Saving Effect of Farmland Water Conservancy Facilities on Grain Production.' *Reform*: 102–13.

Luo, Y., X. Luo, and L. Wang. 2020. 'Rural Infrastructure, Labour Productivity Gap between Industry and Agriculture, and Non-Agricultural Employment.' *Journal of Management World* 36: 91–121.

Orden, David, Rashid S. Kaukab, and Eugenio Diaz-Bonilla. 2003. *Liberalizing Agricultural Trade and Developing Countries.* Global Policy Program Issue Brief, November. Washington, DC: Carnegie Endowment for International Peace. carnegieendowment.org/posts/2003/03/liberalizing-agricultural-trade-and-developing-countries-march-13-2003?lang=en.

Rodrik, Dani. 2015. 'Premature Deindustrialization.' *Journal of Economic Growth* 21: 1–33. doi.org/10.1007/s10887-015-9122-3.

Schultz, Theodore W. 1964. *Transforming Traditional Agriculture.* New Haven: Yale University Press.

Sheng, Yu, V. Eldon Ball, Kenneth Erickson, and Carlos San Juan Mesonada. 2022. 'Cross-Country Agricultural TFP Convergence and Capital Deepening: Evidence for Induced Innovation from 17 OECD Countries.' *Journal of Productivity Analysis* 58: 185–202. doi.org/10.1007/s11123-022-00646-z.

Sheng, Yu, Eldon Ball, and Katerina Nossal. 2015. 'Comparing Agricultural Total Factor Productivity between Australia, Canada, and the United States, 1961–2006.' *International Productivity Monitor*, no. 29: 38–59. www.csls.ca/ipm/29/shengballandnossal.pdf.

Sheng, Y., T.T. Zhou, A.Z. Wang, and H.Y. Deng. 2024. 'Dynamic Influence of Irrigation Infrastructure on Total Factor Productivity of China's Crop Industry.' *Journal of Huazhong Agricultural University (Social Sciences Edition)*: 39–48.

United Nations (UN). 2024. *The Sustainable Development Goals Report 2024*. New York: United Nations. unstats.un.org/sdgs/report/2024/.

United Nations Department of Economic and Social Affairs (UNDESA). 2015. *Global Sustainable Development Report: 2015 Edition*. New York: United Nations Department of Economic and Social Affairs. www.un.org/en/development/desa/publications/global-sustainable-development-report-2015-edition.html.

United Nations Statistics Division. 2023. '6. Clean Water and Sanitation.' *The Sustainable Development Goals Report 2023: Special Edition*. New York: United Nations Department of Economic and Social Affairs. unstats.un.org/sdgs/report/2023/Goal-06/.

Wang, Yangjie, Jikun Huang, Jinxia Wang, and Christopher Findlay. 2017. 'Mitigating Rice Production Risks from Drought Through Improving Irrigation Infrastructure and Management in China.' *The Australian Journal of Agricultural and Resource Economics* 62, no. 1: 161–76. doi.org/10.1111/1467-8489.12241.

World Bank. 2023. *World Development Indicators*. Washington, DC: World Bank. databank. worldbank.org/source/world-development-indicators.

Wu, Q., G. Li, X. Zhou, and Z. Feng. 2015. 'Infrastructure, Agricultural Location and Crop Structure Adjustment: An Empirical Study Based on Provincial Panel Data from 1995 to 2013.' *Journal of Agrotechnical Economics*: 25–32.

Wu, Q., X. Zhou, and Z. Feng. 2014. 'Does Infrastructure Reduce Agricultural Production Costs? Based on Quantile Regression.' *Journal of Huazhong Agricultural University (Social Sciences Edition)*: 53–59.

Wuyts, Marc, and Blandina Kilama. 2015. 'Planning for Agricultural Change and Economic Transformation in Tanzania?' *Journal of Agrarian Change* 16, no. 2: 318–41. doi.org/10.1111/joac.12111.

13

Can China's high growth continue? Considering global value chain reconstruction

Ran Wang, Jinjun Xue and Yang Zhou

The history of China's economic development

Over the past four decades, China's economic development has undergone a remarkable evolution—from an agrarian and planned economy to socialist market economy and then to the world's second-largest economic powerhouse. Yet, beneath this headline success lies a more complex trajectory marked by multiple structural transitions. This chapter examines two key inflection points: the transition from high-speed growth to high-quality but medium-to-low-speed growth, and the gradual shift from an export-oriented growth model to an economic structure oriented more to the domestic market.

The end of China's high growth

For more than three decades, China sustained extraordinary economic growth, averaging double-digit annual increases in GDP. This period of rapid expansion was fuelled by industrialisation, urbanisation, export-led development and investment-driven stimulus. However, after 2010, a structural deceleration began to emerge.

According to IMF data, China's annual GDP growth has declined from a peak of more than 14 per cent in the early 1990s to below 5 per cent in recent years (see Figure 13.1). The sharp drop in 2020 due to the Covid-19 pandemic was followed by a short-lived rebound, but growth has since moderated. IMF projections suggest this trend will continue, with China entering a phase of medium to low growth in the coming years. This slowdown reflects multiple headwinds: demographic ageing, diminishing marginal returns on infrastructure investment, rising labour costs, environmental constraints and geopolitical pressures—most notably, the US–China trade war, which has accelerated the restructuring of global value chains (GVCs) and reduced external demand for Chinese exports.

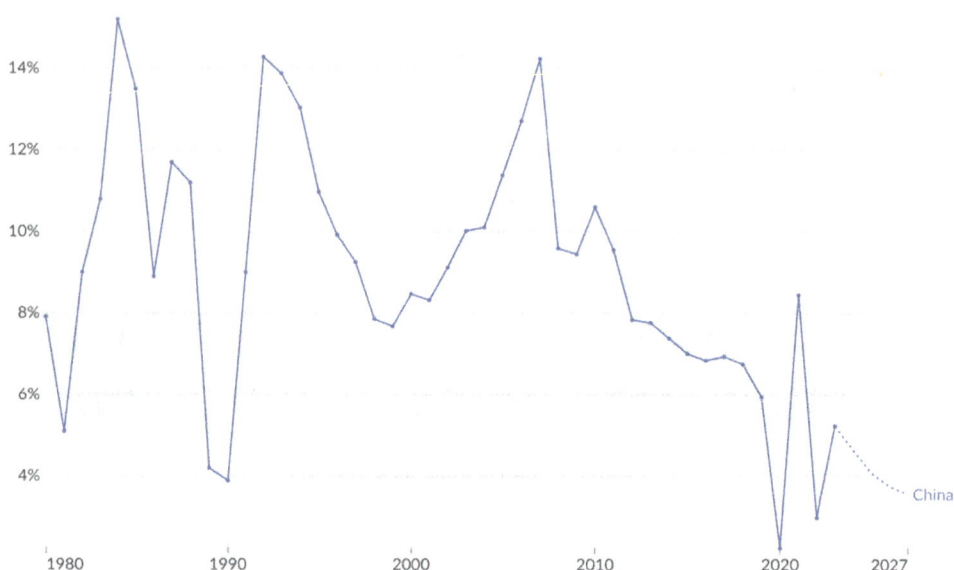

Figure 13.1: Annual GDP growth in China, 1980–2027 (%)
Source: IMF data (data.imf.org/en).

This structural deceleration is underpinned by several factors. First, China's demographic dividend is fading, with a shrinking working-age population and rising dependency ratios. Second, the marginal returns on capital investment have diminished, especially in infrastructure and real estate. Third, the external environment has become more volatile and uncertain, with rising trade frictions, supply chain restructuring and geopolitical tensions. As a result, China's economy is transitioning from being driven by quantity and scale to being led by quality and efficiency.

Despite the slowdown, this shift does not necessarily imply weakness. Rather, it reflects a maturing economy moving towards sustainable development. The government has increasingly emphasised innovation, green technology and domestic consumption as new growth engines.

Transition from export-led growth to a domestic demand–led economy

One of the most significant hallmarks of China's earlier growth phase was its export-led strategy. From the 1990s to the mid-2000s, international trade was the primary engine propelling economic expansion. China's accession to the World Trade Organization in 2001 catalysed its integration into global value chains, leading to a sharp increase in trade activity. As shown in Figure 13.2, the trade-to-GDP ratio surged and peaked around 64 per cent in 2006—an indicator of China's deep dependence on external demand.

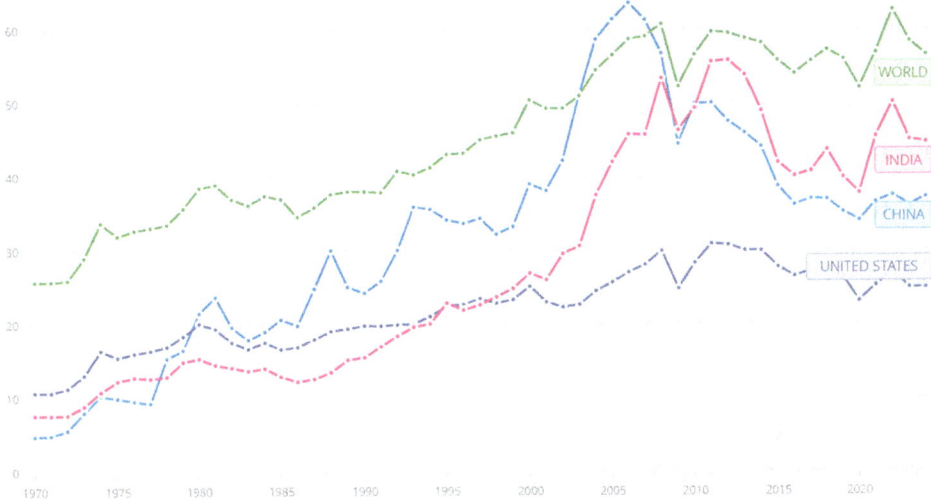

Figure 13.2: Trade as a share of GDP, 1970–2024 (%)

Source: World Bank data (data.worldbank.org/).

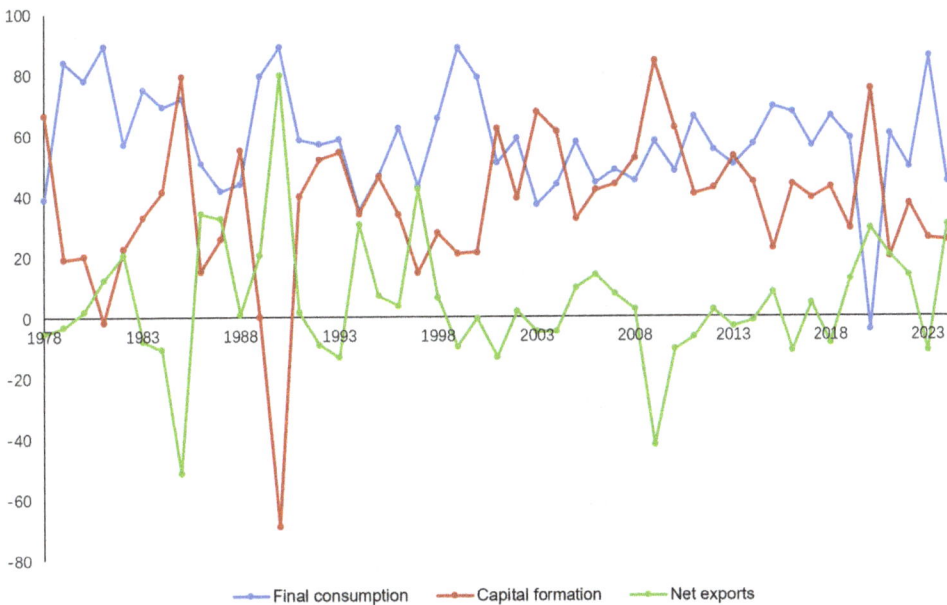

Figure 13.3: Decomposition of GDP growth contribution rate, 1978–2024 (%)

Source: National Bureau of Statistics of China (data.stats.gov.cn/english/).

However, after the 2008 GFC, the limits of the export-led model became increasingly apparent. External demand volatility combined with rising global protectionism exposed the risks of overreliance on foreign markets. Since then, the trade-to-GDP ratio has steadily declined, reaching approximately 37 per cent in 2024. This marked shift reflects both structural constraints and deliberate policy adjustments aimed at rebalancing China's growth model.

The pivot from export to domestic orientation is also reflected in the changing composition of GDP growth. According to Figure 13.3, the contribution of net exports has been declining and has become increasingly volatile over time. In contrast, final consumption has become a more stable and consistent contributor, while gross capital formation continues to play a significant—though more fluctuating—role.

This structural transformation is further supported by policy initiatives such as the 'dual circulation' strategy, which emphasises domestic demand as the primary growth driver, while maintaining openness to global trade and investment. Rather than a retreat from globalisation, the strategy represents a recalibration of economic priorities: leveraging the scale of China's domestic market to build resilience in the face of global uncertainty.

The remarkable rise of China's role in global production networks

China's economic transition is also reflected in its changing position within global production networks. From the late 1970s to the early 2000s, China served primarily as a low-cost manufacturing hub—the so-called world's factory—providing labour-intensive assembly services for multinational firms. Visual network analysis for 1995 and 2000 shows China as a peripheral node in the global production network (see Figure 13.4).

After its accession to the WTO in 2001, China rapidly climbed the value chain. With a more complete industrial base, coordinated development between SOEs and private firms and increased foreign direct investment (FDI), China became an integral part of global supply chains. By 2011, its network centrality had significantly increased, reflecting deeper integration (Figure 13.5).

Since 2012, China has evolved into a regional manufacturing centre, surpassing Japan in Asia's production network. It now maintains strong linkages not only with advanced economies such as the United States and Germany, but also with emerging markets (see Figure 13.6). This shift from quantity to quality is a key backdrop for understanding China's current focus on domestic upgrading and resilience.

These shifts in the pace and structure of growth reflect the broader changes discussed above: rising costs, weaker external demand and policy adjustments. Together, they mark a turning point in China's development path. The evolution of China's role in global production networks—from assembler to regional hub—further illustrates this transition, highlighting the deep link between domestic economic changes and GVC reconstruction.

Year 1995

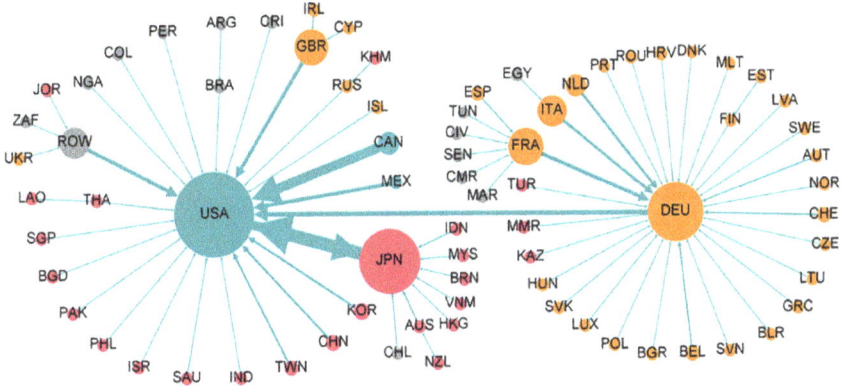

Year 2000

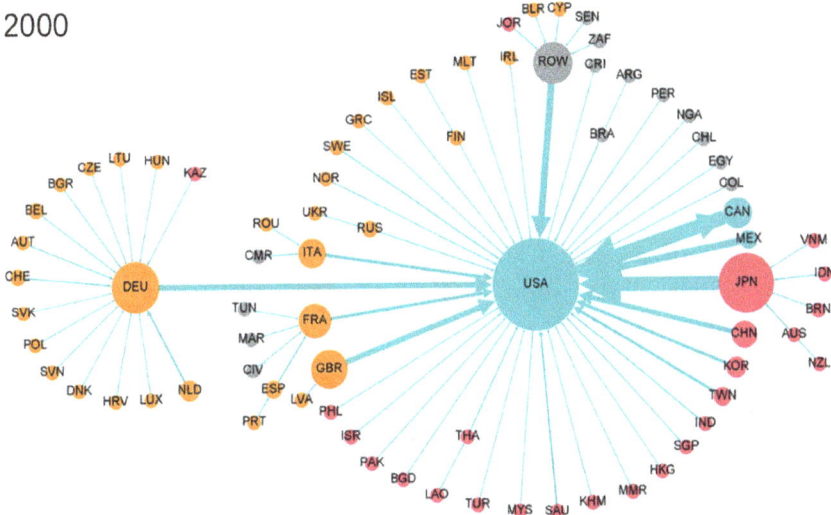

Figure 13.4: Global production network, 1995 and 2000 (all goods and services)

Source: Calculated based on OECD *Input–Output Tables* (www.oecd.org/en/data/datasets/input-output-tables.html).

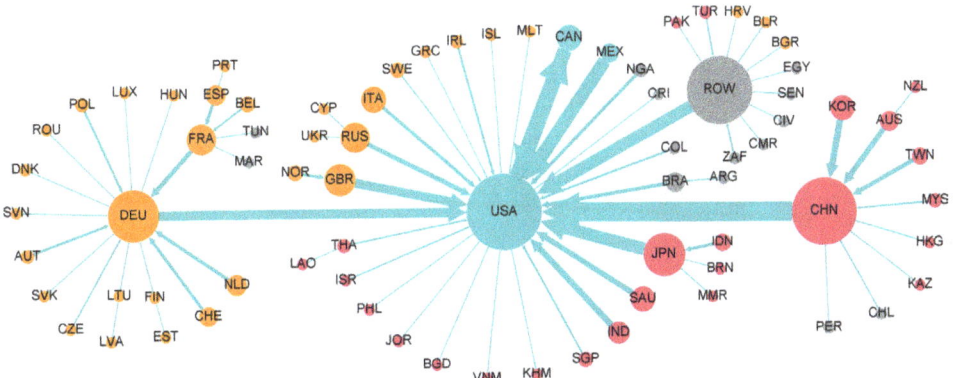

Figure 13.5: Global production network, 2011 (all goods and services)

Source: Calculated based on OECD *Input–Output Tables* (www.oecd.org/en/data/datasets/input-output-tables.html).

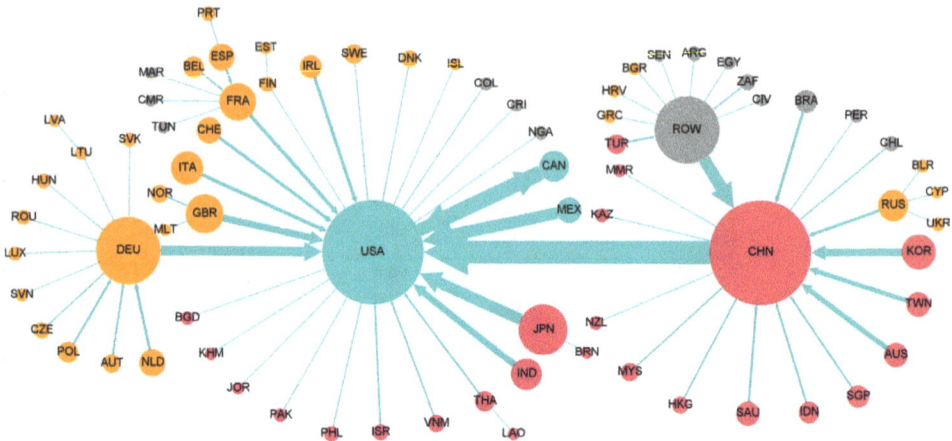

Figure 13.6: Global production network, 2017 (all goods and services)

Source: Calculated based on OECD *Input–Output Tables* (www.oecd.org/en/data/datasets/input-output-tables.html).

The impacts of GVC restructuring on China

Over the past four decades, China has rapidly industrialised to become a global manufacturing hub and the 'world's factory', greatly enhancing the efficiency of global production. However, with rising geopolitical tensions and escalating trade frictions, major economies such as the United States and Japan are promoting 'de-Sinicisation' and diversification of supply chains, resulting in reduced reliance on Chinese intermediate goods and a decline in foreign investment inflows to China. Against this backdrop, China is facing numerous challenges brought about by the restructuring of GVCs.

China, the world's factory, becomes the target of decoupling and de-risking

After more than 40 years of rapid economic growth, China has established the world's most comprehensive and integrated supply chain system, spanning virtually all industries and sectors and earning its reputation as the 'world's factory'. This unparalleled industrial ecosystem has brought significant benefits to the global economy by lowering costs and improving efficiency for countries around the world. However, it has also resulted in a high degree of dependence on Chinese intermediate goods among major manufacturing nations.

		Factory North America			Factory Europe								Factory Asia									
		usa	can	mex	deu	gbr	fra	ita	esp	tur	nld	che	chn	jpn	kor	ind	twn	aus	idn	bra	rus	sau
Factory North America	usa												4.5									0.6
	can	3.0											3.7									
	mex	4.4											8.5		0.6							
Factory Europe	deu	0.8											3.2									
	gbr						-0.5						2.7									
	fra	1.1			1.5	0.6			0.6				2.6									
	ita				1.1		0.6		0.8	0.5			3.4								1.8	
	esp	0.8			0.9	0.8	1.2						3.8									
	tur												4.0		0.8	0.6					-1.1	
	nld	2.4			2.4	0.7	1.0						2.7	0.9							1.9	
	che				-4.1		-1.9	-1.6		0.5	-0.7		4.2	-0.6								
Factory Asia	chn																-0.7					
	jpn	0.7											4.2				0.6			0.8	-0.8	
	kor												6.3	-5.5			-0.7	-0.9			-0.5	
	ind	0.5											4.8				-3.9					
	twn	-1.4											4.8	-6.1	-2.3			-0.9				
	aus												4.0									
	idn												4.9	-0.5								
	bra	2.3										1.5	3.9									
	rus				-1.2								3.0									
	sau	-0.7			0.6	0.5	-1.1						2.6	-0.8	-0.9					-0.6		

Figure 13.7: Percentage-point difference between the reliance of one country's manufacturing sector on another country's inputs, 2009–15

Note: The figure shows the percentage-point difference between the reliance of a country's manufacturing sector (in the horizontal row) on the inputs from another country (vertical columns).

Source: Freeman and Baldwin (2020).

As highlighted in Figure 13.7, from 2009 to 2015, every major manufacturing country increased its reliance on Chinese inputs, with positive and often substantial percentage-point gains seen in China's column. This trend, while economically advantageous, has heightened concerns among some countries about supply chain security and strategic autonomy, especially amid rising geopolitical tensions. As a result, there is growing support for 'de-Sinicisation' and 'de-risking' policies, encapsulated in the 'China plus N' strategy, in which countries and multinational firms seek to diversify their supply chains by incorporating additional sourcing destinations outside China.

New dynamic trends in global trade and investment

The United States and Japan have begun to decouple from China (Figure 13.8). China's share of US intermediate goods imports fell from 18.5 per cent in 2018 to 14.1 per cent in 2022, and further declined, to 11.4 per cent, in early 2023, with the most significant decrease occurring during the 2018–19 period when the Trump administration imposed additional tariffs. During the same period, China's share of Japanese intermediate goods imports also declined, from 26.5 per cent to 24 per cent, reflecting Japan's 'de-risking' strategy—such as the implementation of technology export controls and targeted investments in the semiconductor industry after the G7 summit in Hiroshima in 2023.

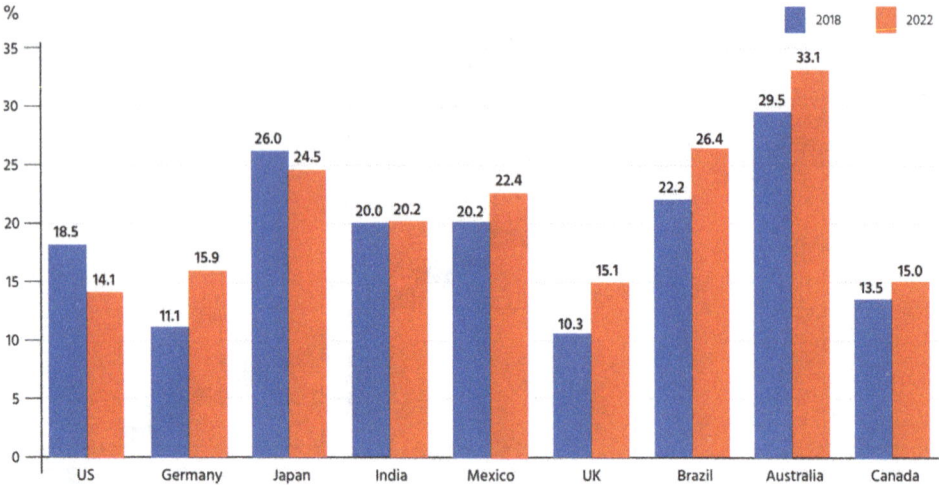

Figure 13.8: China's share in total intermediate goods imports of major economies (%)
Source: Nguyen-Quoc (2023).

In contrast to the United States and Japan, China's importance as a supplier of production inputs has increased markedly in other major economies, including Germany, Brazil, Australia and the United Kingdom, driving up these countries' shares of intermediate goods imports from China. This growth has been primarily driven by expanded trade in inputs related to electronics, machinery and chemical products. At the same time, China has been actively adjusting its export destinations for intermediate goods, with Vietnam and Malaysia seeing the most notable increases in their shares. This shift can be attributed both to rising labour costs in China and the relocation of downstream manufacturing activities in response to US–China trade barriers, as well as multinational enterprises investing in additional production capacity in other countries.

The US–China trade war and subsequent policy adjustments have led to significant shifts in the composition of US import sources. Between 2017 and 2022, China's share of the US import market declined substantially across all goods and strategic industries, with a drop of nearly 6 percentage points—far exceeding that of any other country (see Figure 13.9). In contrast, countries such as Vietnam, Taiwan, Mexico, South Korea and India saw notable increases in their US market shares, with Vietnam and Taiwan standing out as the primary beneficiaries of the supply chain reallocation. This pattern suggests that trade tensions have accelerated the regional restructuring of global production networks, driving supply chain diversification and reinforcing trends towards 'nearshoring' and 'friend-shoring'. The effect is particularly pronounced in strategic industries, reflecting the United States' deliberate efforts to reduce reliance on China in critical sectors and strengthen its sourcing from emerging Asian economies.

All goods

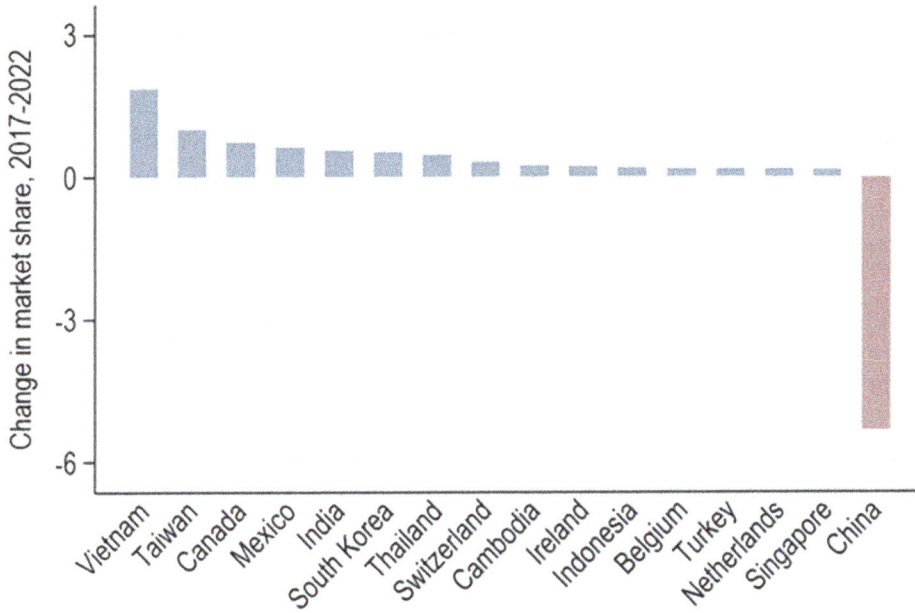

Strategic industries

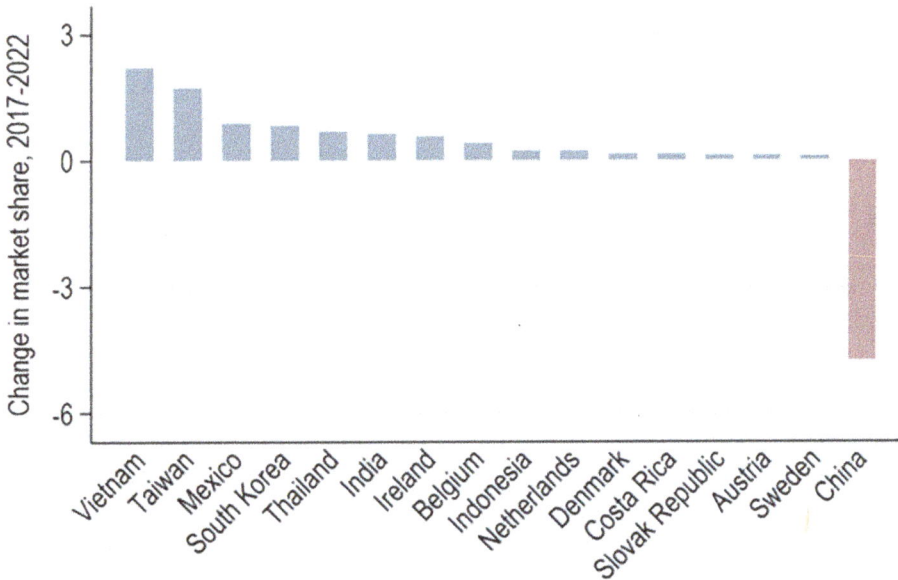

Figure 13.9: US import share in top 10 countries and regions, 2017 and 2022 (%)

Source: Freund et al. (2024).

Since the end of the GFC, rising labour costs in China, restrictions on foreign ownership (such as joint-venture requirements) and concerns about intellectual property protection have contributed to a sustained decline in greenfield FDI inflows into China's manufacturing sector (Figure 13.10). This downward trend accelerated markedly after the onset of the US–China trade war in 2018. Investment from the United States, one of China's major FDI sources, dropped sharply from 2018 and fell to nearly zero after 2020, underscoring the direct impact of geopolitical tensions on bilateral investment flows. Meanwhile, FDI from other key economies such as Japan, South Korea and Taiwan has also contracted significantly. This pattern suggests that the US–China trade conflict has not only affected bilateral investment but also, through supply chain restructuring and policy spillover effects, dampened the willingness of other advanced economies to invest in China. Overall, the escalation of the US–China trade war and ongoing geopolitical tensions have accelerated the diversification of China's FDI sources. However, in the short term, emerging economies have not yet been able to compensate for the decline in investment from developed countries.

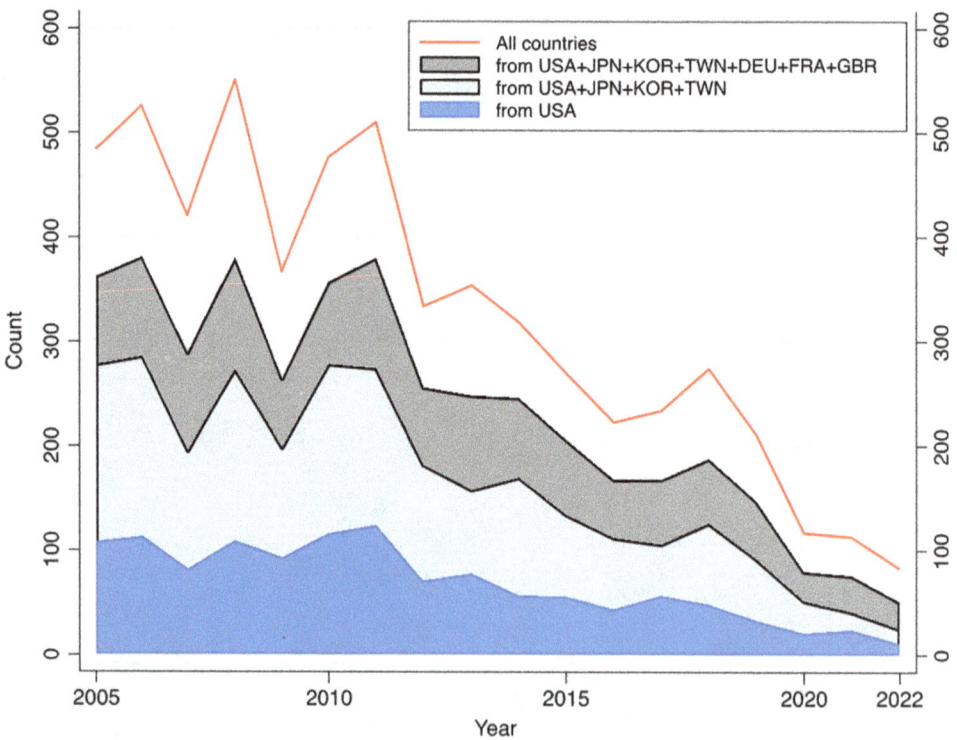

Figure 13.10: Evolution of China's inward greenfield FDI projects in the manufacturing sector, 2005–22

Source: Alfaro and Chor (2023).

Expected pathways for China to sustain economic growth

China's economy has undergone remarkable transformations over the past few decades, evolving from a low-cost manufacturing hub to a global leader in technology and innovation. However, in recent years, China has faced multiple challenges, including trade tensions with the United States, demographic shifts and slowing domestic demand. To sustain economic growth, China must explore new pathways that leverage its strengths in global trade, outward foreign direct investment (OFDI) and emerging high-tech industries.

China's new strategy of trade and investment via 'third countries'

China has increasingly relied on third countries to maintain its export volumes in recent years, particularly in response to US tariffs imposed during the first Trump administration and continued under Joe Biden. According to UN Comtrade data (comtradeplus.un.org/), China's direct exports to the United States declined from 19 per cent of total exports in 2018 to 16 per cent in 2023, while exports to the United States through South-East Asia and Mexico surged. For instance, in 2017, China's domestic value-added exports to the United States through Mexico and Vietnam were worth about US$500 million and US$9 billion, respectively, while in 2022, these values were about US$11 billion and US$18 billion, respectively (see Figure 13.11). Vietnam and Mexico have emerged as key intermediaries, absorbing Chinese components for final assembly before re-export to the United States. This strategy allows Chinese firms to avoid direct tariffs while maintaining access to the American market.

Beyond simply rerouting exports through third countries, China has actively increased FDI in key intermediary markets—such as Vietnam and Mexico—to establish production bases that can directly export to the United States while avoiding tariffs. This strategy helps Chinese firms not only circumvent trade barriers but also strengthen their integration into global supply chains. As illustrated in Figure 13.12, China's FDI flows into Vietnam surged from US$500 million in 2017 to nearly US$1.6 billion in 2020, reflecting a deliberate shift in manufacturing relocation. Since Vietnam's geographic and logistical advantages allow Chinese firms to maintain just-in-time production while reducing costs and its participation in the Comprehensive and Progressive Agreement for Trans-Pacific Partnership and the European Union–Vietnam free-trade agreement provides tariff-free access to Western markets.

(billion USD)

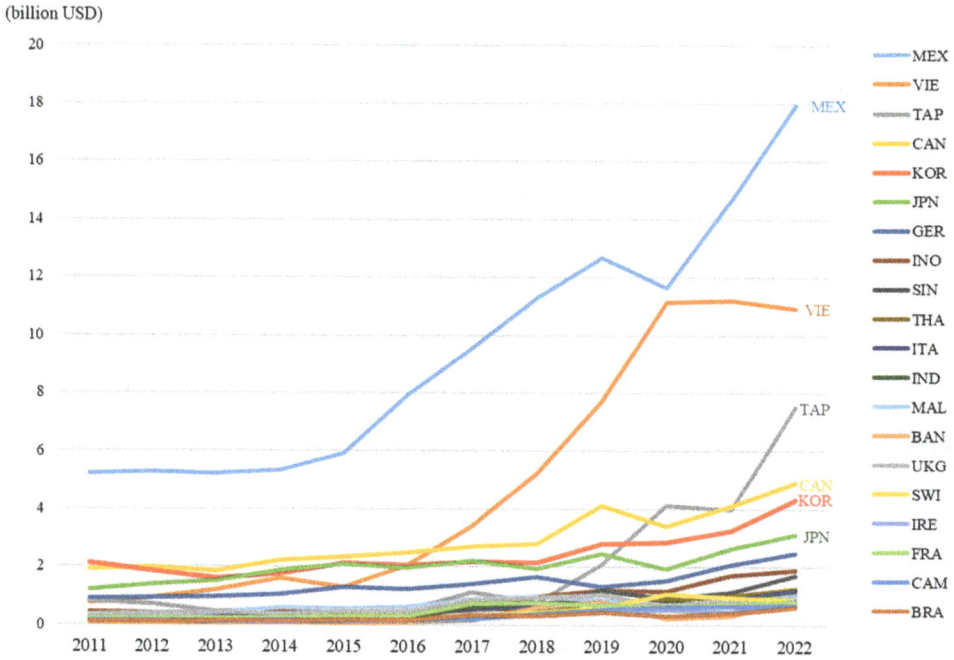

Figure 13.11: China's domestic value added through third countries to the United States, 2011–22

Source: Authors' calculations.

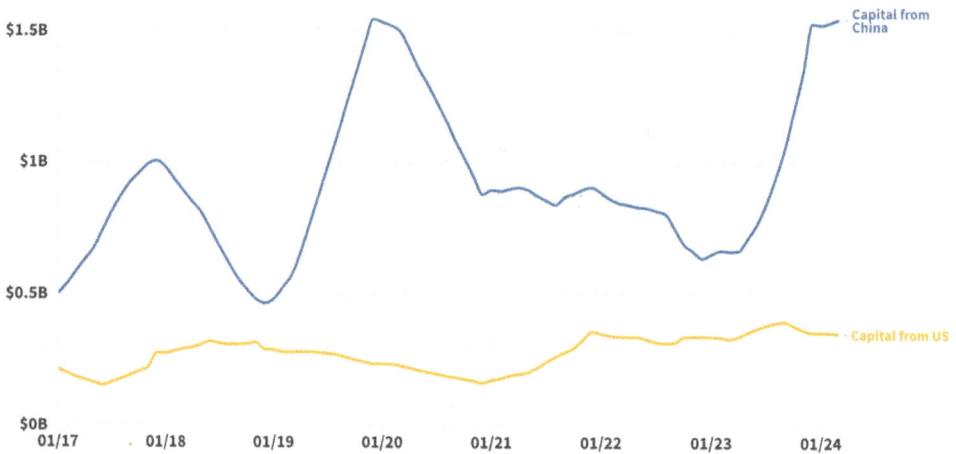

Figure 13.12: Foreign direct investment in Vietnam from China and the United States, 2017–24 (US$ billion)

Sources: Vietnam Ministry of Commerce (vntr.moit.gov.vn/news/post/Statistics); LiveMint (www.livemint.com/); *South China Morning Post* (www.scmp.com/); USI Global (www.usiglobal.com/en); Capco (www.capco.com/); *Sohu* (www.sohu.com/); *Vietnam Investment Review* (vir.com.vn/).

Promoting reverse technological spillovers and industrial upgrading through OFDI

While China is no longer the world's top destination for foreign investment, it has strategically shifted towards OFDI—particularly in developed economies (such as the United States, Germany and the United Kingdom; Figure 13.13)—to acquire advanced technologies and accelerate domestic industrial upgrading. Empirical studies demonstrate that Chinese firms gain reverse technology spillovers through overseas mergers and acquisitions, joint ventures and research and development collaborations, which in turn enhance productivity and innovation back home (Li and Fabus 2019; Zhang et al. 2022). For instance, after acquiring Volvo in 2010, Chinese carmaker Geely gained access to patents for safety systems, hybrid engines, and so on, which were later integrated into its own brands. This not only promotes Geely's technological upgrading but also drives it towards internationalisation. Despite these benefits, China's OFDI faces obstacles. The first is Western scrutiny; the United States and the European Union have tightened FDI screening—through, for example, the Committee on Foreign Investment in the United States and the European Union's FDI regulation—to block sensitive technology transfers. The second are intellectual property protection concerns; some acquired firms resist full technology sharing due to fears of forced transfers. The third obstacle is geopolitical tension, with US export controls (such as semiconductor bans) limiting China's access to cutting-edge technology.

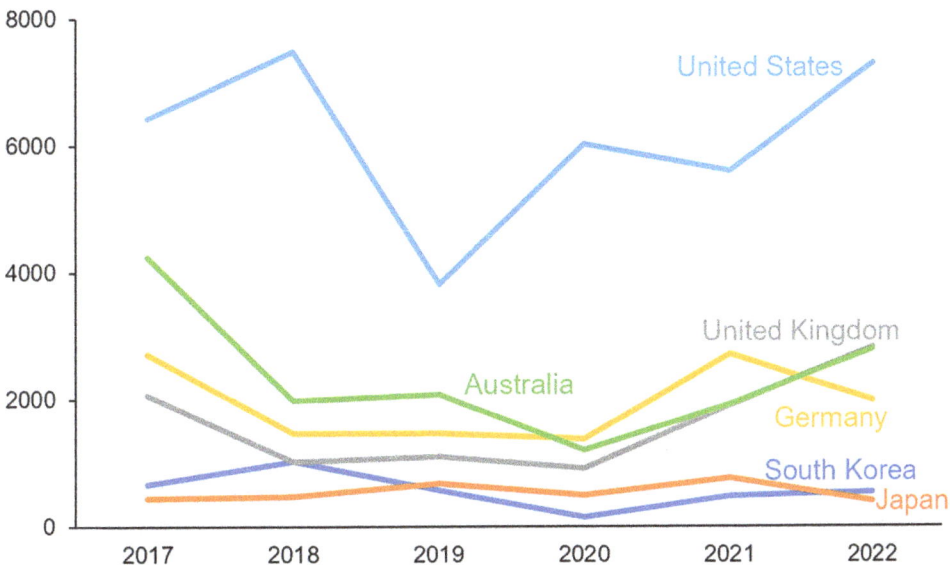

Figure 13.13: Destination countries for China's outward FDI, 2017–22 (US$ million)
Source: National Bureau of Statistics of China (data.stats.gov.cn/english/).

The 'new three' bring new momentum

China's economy is undergoing a structural shift, with the 'new three' industries—electric vehicles (EVs), lithium-ion batteries and photovoltaic (PV) products (or solar cells)—emerging as the country's most dynamic export drivers. Meanwhile, the contribution to exports of the 'old three' industries—clothing, home appliances and furniture—continues to weaken (Figure 13.14). In the first 10 months of 2024, the export growth of the new three contributed 18 per cent to total export growth (Figure 13.15). These industries not only reinforce China's dominance in global green technology but also provide a crucial counterbalance to slowing growth in traditional industries such as real estate and low-end manufacturing.

China's rapid expansion in the 'new three' industries has not only reshaped its own economic landscape but also catalysed growth in developing economies along the global supply chain. By integrating upstream resource extraction, midstream manufacturing and downstream applications, China's industrial strategy has enabled developing countries to participate in high-value segments of global production networks, fostering technological spillovers and economic diversification. For example, China's demand for raw materials such as lithium, cobalt and polysilicon has driven investments in resource-rich developing nations. Chinese companies such as Ganfeng Lithium and Tianqi Lithium have established mining operations in Africa and South America, creating jobs and infrastructure while transferring sustainable extraction technologies. This vertical integration ensures stable supply chains for battery production and solar panel manufacturing, while also helping countries such as Chile and Democratic Republic of Congo to shift from raw material exports to intermediate processing in the future.

The new three's share in exports has been rising, while the old three's moderates

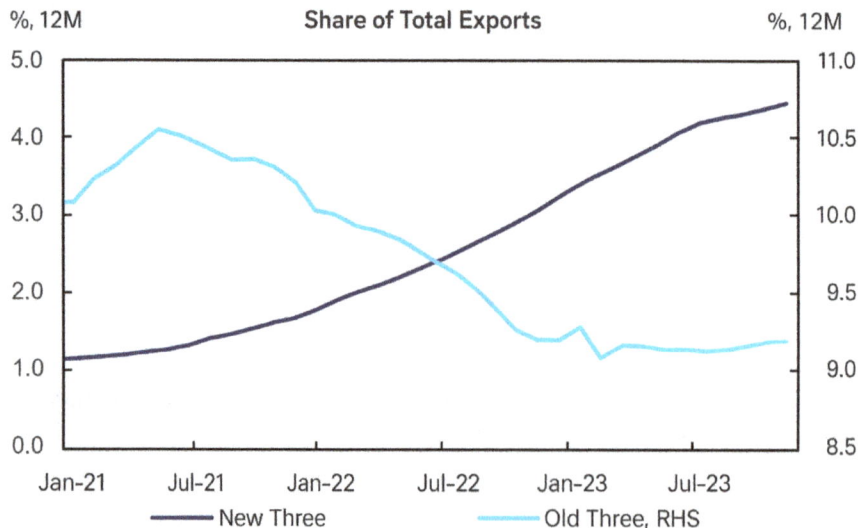

Figure 13.14: Shares of the 'new three' and 'old three' industries in total Chinese exports
Source: Citi Global Insights (2024).

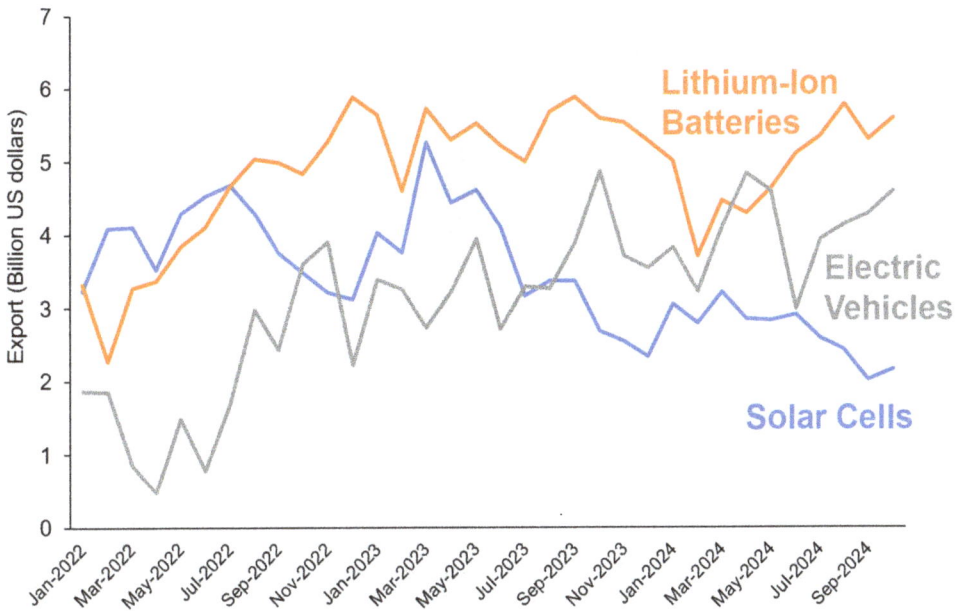

Figure 13.15: Chinese exports of the 'new three' industries

Source: China Customs Statistics database (english.customs.gov.cn/Statistics/Statistics?ColumnId=1).

At the same time, the development of China's 'new three' also faces various challenges from both domestic and international sources. On the one hand, due to excessive competition, the new-energy industry has experienced frequent price wars, leading to the risk of profit compression and overcapacity. On the other hand, geopolitical risks and trade barriers continue to rise, and some European and North and South American countries restrict the entry of Chinese products under the pretext of 'national security', which hinders the development of national markets. In addition, some key mineral resources (such as lithium) and non-core technologies (such as high-purity electrolytes and high-end separators) have a high degree of external dependence and face the risk of supply disruption.

Conclusion and discussion

Global value chains are undergoing significant restructuring, characterised by trends such as 'de-Sinicisation', 'friend-shoring' and 'nearshoring'. At the same time, the dependence of the United States and Japan on Chinese intermediate goods imports has declined notably, and the scale of FDI inflows into China has decreased. Nevertheless, China's export share in so-called friendly markets—including Vietnam, Mexico and the European Union—has been rising. For example, since 2018, China's exports to Vietnam and Mexico have increased by 72 per cent and 155 per cent, respectively, and, from 2017 to 2023, China's share of EU imports grew faster than that of any other country.

In response to the evolving GVC landscape, China can pursue multiple strategies to sustain its economic growth, including maintaining exports through third-country trade and overseas production, expanding OFDI to promote industrial upgrading and technology spillovers and vigorously developing the 'new three' export industries: EVs, lithium-ion batteries and PV products. However, China currently faces numerous challenges, such as US trade restrictions on the 'new three', the European Union's anti-dumping investigation into Chinese EVs, as well as external dependence on key resources and advanced technologies. In summary, China may be able to maintain a certain degree of growth, but achieving long-term sustainable growth will require effectively addressing the risks and uncertainties posed by GVC restructuring, economic decoupling and related external barriers.

References

Alfaro, Laura, and Davin Chor. 2023. *Global Supply Chains: The Looming 'Great Reallocation'.* NBER Working Paper No. 31661. Cambridge: National Bureau of Economic Research. doi.org/10.3386/w31661.

Citi Global Insights. 2024. 'China Economics: Out with the Old Three and In with the New Three.' *Citi Global Insights*, 8 January. New York: Citi Institute.

Freeman, Rebecca, and Richard Baldwin. 2020. 'Trade Conflict in the Age of Covid-19.' *VoxEU*, 22 May. London: Centre for Economic Policy Research. cepr.org/voxeu/columns/trade-conflict-age-covid-19.

Freund, Caroline, Aaditya Mattoo, Alen Mulabdic, and Michele Ruta. 2024. 'Is US Trade Policy Reshaping Global Supply Chains?' *Journal of International Economics* 152: 104011. doi.org/10.1016/j.jinteco.2024.104011.

Li, Shuyan, and Michal Fabus. 2019. 'The Impact of OFDI Reverse Technology Spillover on China's Technological Progress: Analysis of Provincial Panel Data.' *Journal of International Studies* 12, no. 4: 325–36. doi.org/10.14254/2071-8330.2019/12-4/21.

Nguyen-Quoc, Thang. 2023. 'How Asia's Supply Chains Are Changing.' *Techonomics Talks*, 16 April. [Online]. Oxford: Oxford Economics. www.oxfordeconomics.com/resource/how-asias-supply-chains-are-changing/.

Zhang, Wenyue, Jianan Li, and Chuanwang Sun. 2022. 'The Impact of OFDI Reverse Technology Spillovers on China's Energy Intensity: Analysis of Provincial Panel Data.' *Energy Economics* 116: 106400. doi.org/10.1016/j.eneco.2022.106400.

Index

Page numbers in **bold** indicate figures or tables.